The Bioarchaeology of Metabolic Bone Disease

The Bioarchaeology of Metabolic Bone Disease

Megan Brickley and Rachel Ives

AMSTERDAM • BOSTON • HEIDELBERG • LONDON
NEW YORK • OXFORD • PARIS • SAN DIEGO
SAN FRANCISCO • SINGAPORE • SYDNEY • TOKYO
Academic Press is an imprint of Elsevier

Academic Press is an imprint of Elsevier
Linacre House, Jordan Hill, Oxford OX2 8DP, UK
84 Theobald's Road, London WC1X 8RR, UK
525 B Street, Suite 1900, San Diego, CA 92101-4495, USA
30 Corporate Drive, Suite 400, Burlington, MA 01803, USA

First edition 2008

British Library Cataloguing in Publication Data
A catalogue record for this book is available from the British Library

Library of Congress Cataloging-in-Publication Data
A catalog record for this book is available from the Library of Congress

ISBN: 978-0-12-370486-3

For information on all Academic Press publications
visit our web site at books.elsevier.com

Typeset by Charon Tec Ltd (A Macmillan Company), Chennai, India
www.charontec.com

Printed and bound in Hungary
08 09 10   10 9 8 7 6 5 4 3 2 1

Contents

nowledgements

The authors would like to thank the following people for their kind permission to reproduce figures and providing photographs for inclusion in this book: Lynne Bell, Simon Fraser University for providing Figures 8.5 and 8.6. Alan Boyde, Queen Mary, University of London for Figure 5.20. Jo Buckberry, Anthea Boylston, Paola Ponce, Alan Ogden, Iraia Arabaolaza and staff at BARC, Archaeological Sciences, University of Bradford for giving us Figures 7.4, 8.4 and 8.8. Anwen Caffel, York Osteoarchaeology, for supplying us with Figures 4.3b and 4.6. Judith Littleton, the University of Auckland for providing us with Figures 9.1–9.4. Antonietta Cerroni, University of Toronto, for providing Figure 6.8. David Bowsher, Museum of London, for permission to publish Figure 8.7. Staff at Birmingham Archaeology, particularly Alex Jones for permission to use Figure 2.7. Jackie McKinley of Wessex Archaeology for providing us with Figure 2.6b. Cathy Patrick of CgMs Consulting on behalf of the Birmingham Alliance for permission to use Figure 2.5. Bill White and staff at the Centre for Human Bioarchaeology, Museum of London, for granting us access to the collections held by the Museum, providing constant help and support during our investigations and granting permission to include over 30 images from their collections. Thanks to Natasha Powers, Museum of London Archaeology Service Head of Osteology, for access to the material from St. Marylebone, Westminster, and permission to include Figures 5.10a–5.10c and 5.10f. Bob Paine and Barrett Brenton for supplying us with Figure 9.5. Michael Parfitt for permission to reproduce his conceptual ideas as Figures 3.4 and 3.5. Simon Mays for his ready agreement for us to use a wide range of images that were made during a joint project on vitamin D deficiency (images from St. Martin's, Birmingham), and permission to reproduce images from his study of the human skeletal remains from Wharram Percy, England, Figure 5.13. Alan Saville, National Museums of Scotland, who provided Figure 2.6a. The World Health Organization for allowing us to use Figure 5.8. Michael Weinstein, Departments of Paediatrics and Radiology, Hospital for Sick Children, Toronto for supplying us with Figure 4.8. Kazutaka Yamada, Obihiro University of Agriculture and Veterinary Medicine for supplying us with Figure 8.1. Thanks are owed to the Food and Agriculture Organization of the United Nations, for supplying us with a copy of Figure 4.1.

We would like to thank staff at both the Natural History Museum in Basel, Switzerland and the Federal Museum for Pathological Anatomy, Vienna, Austria for their help and support during our visit, in particular Gerhard Hotz and Beatrix Patzak. We would also like to thank both these institutions for allowing us to photograph material in their collections.

Many colleagues also generously provided us with data and information, sometimes in advance of publication, which we would otherwise have had difficulty obtaining. Thanks are owed to Chryssa Bourbou, Della Collins Cook, Per Holck, Maria Teresa Ferreira, Judith Littleton, George Maat, Simon Mays, Robert Mensforth, Piers Mitchell, Jeffrey O'Riordan, Robert Paine, Rebecca Redfern, Charlotte Roberts, Jerry Rose, Doris Schamall, Michael Schultz, Peter Sheldrick, Martin Smith and Richard Thomas. Jelena Bekvalac, Tania Kausmally, Lynne Cowal and Richard Mikulski, Centre for Human Bioarchaeology, Museum of London, for provision of data from the WORD osteological research database.

We would like to thank Don Ortner for his continued help and support throughout this project, providing us with access to his research in advance of publication and for the Foreword to this book. Special thanks are owed to Graham Norrie, Harry Buglass and Martin Smith, University of Birmingham who provided considerable technical help and advice regarding many of the illustrations published in this volume.

We are particularly grateful to Jane Buikstra, University of Arizona, Anne Grauer, Chicago Loyola University, Tania Kausmally and Jelena Bekvalac, Centre for Human Bioarchaeology, Museum of London, for reading drafts of several chapters during the preparation of this book. We thank them for all of their valuable comments and note that any omissions remain the sole responsibility of the authors.

The authors thank staff at the University of Birmingham library, the British Library and Ealing Library.

Megan Brickley would like to thank the Natural Environment Research Council (NERC) and the British Academy, who both helped the research presented in this book through various grants over the years.

Megan Brickley and Rachel Ives

Foreword

In this book Brickley and Ives bring together both clinical and palaeopathological research on the effects of metabolic diseases that affect the skeleton and, in doing so, provide an important reference work that has the potential to enhance bioarchaeological research on the importance of these diseases in antiquity. Although the major emphasis is on the effect of these disorders in human remains, the authors also emphasise that metabolic diseases can affect non-human skeletons; zooarchaeologists will benefit from the contents of the book. The authors correctly emphasise the role of metabolic disease as an important dimension of the challenges that faced past human populations. This is particularly true during the last 10,000 years when monumental changes took place in human society as our ancestors gradually but incompletely migrated from a hunting–gathering economy to an economy based on agriculture. This change affected the relationship between human individuals and groups but also between human populations and the environment in which they lived.

There is general agreement among biological anthropologists that diet changed profoundly with the advent of agriculture. With that change the specter of malnutrition became a much more serious potential problem. However, the evidence for this conclusion remains ephemeral, resting uncomfortably on a very thin base of data. The quest for additional and better evidence for malnutrition should represent a major emphasis for the biological anthropologist. This makes the subject of this book a timely step in the right direction in our understanding of this important category of bioarchaeological human disease and in our ability to gather new data on the presence and prevalence of metabolic diseases in past human groups.

Metabolic disorders include an assemblage of diseases whose pathogenesis is varied to say the least. The relationships between disorders included in this category require a tolerant attitude regarding the scientific rigors of classification. These disorders include vitamin-related diseases such as those that result from deficiencies in vitamins C and D, disorders related to changing hormonal balances such as post-menopausal osteoporosis and disorders caused by too much or too little of crucial dietary components including trace elements such as fluorine. The authors provide a brief exploration of the effect of severe protein-calorie malnutrition on the skeleton. Also included is a review of some diseases that may have a metabolic component but the cause of which remains unclear. Paget's disease, for example, is often classified with the metabolic diseases and is reviewed in this book. The disorder is associated with greatly accelerated osteoclastic and osteoblastic activity. However, the cause of the disease is unknown and its relationship with other metabolic diseases remains unclear.

The disparity of pathological conditions included in the category of metabolic disease illustrates some of the problems in creating and using a classificatory system for skeletal disorders. Classification is a crucial exercise in all the biological sciences and skeletal pathology is no exception. Nevertheless it remains very important to be aware of the limitations of any classificatory system and equally important to not permit the system to oversimplify our understanding of the diversity of skeletal disorders both in their skeletal manifestations and pathogenesis.

The problems in classification are well illustrated by a disease not included in this book. Iron deficiency anaemia is one of the most common dietary disorders today. In most medical reference works, it is classified with the other haemopoietic diseases largely because it has, in common with some, defective haemoglobin. However, the basic cause is a deficiency in dietary iron intake and the disorder appropriately could be classified with the metabolic diseases such as scurvy and rickets because it is a manifestation of malnutrition. Indeed one of the challenging issues in human skeletal palaeopathology is to distinguish between skeletal manifestations of anaemia, scurvy and rickets.

Research linking various stable isotopes to variation in diet represents a significant development in reconstructing some aspects of diet in bioarchaeological populations. Since this methodology began to be applied to bioarchaeological problems about 30 years ago with the publication of a paper on early evidence of maize cultivation in North America (Vogel and van der Merwe, 1977), we have learned a great deal about food resources in earlier human societies. However, this evidence does not address adequately the need for data regarding both the type and severity of malnutrition in various past human groups. There have been attempts to use variation in amino acid residues to identify some types of dietary deficiency such as iron deficiency anaemia (Ortner, 2003). However, what has been lacking is a greater knowledge of the bioarchaeology of metabolic diseases and particularly those whose cause is malnutrition. The major factor in this gap in our knowledge has been that the skeletal manifestations of some of the metabolic disorders have not been carefully defined and described. For sub-adult scurvy, rickets and osteomalacia, this gap in our knowledge has been filled partially although more research is needed, particularly on early stage skeletal manifestations of these disorders.

Another disorder reviewed in this book is osteoporosis. As with many abnormal skeletal manifestations, both osteoporosis and osteopenia are essentially symptoms that can be caused by several pathological conditions; determining the most likely cause is a challenging exercise and is not always possible.

The skeletal manifestations in osteoporosis and osteopenia are relatively clear. Trabecular bone is the most metabolically active skeletal tissue and usually is affected first and most severely when abnormal bone loss occurs. Additional bone changes in osteoporosis and osteopenia include reduced thickness of cortical bone and diminished bone density, i.e. subnormal mass per unit volume of bone. All of these manifestations result in reduced biomechanical function of bone with an increased risk of fracture. Increasing availability of CT radiography for research in palaeopathology offers the potential of significant research on abnormal bone loss in human bioarchaeological skeletal samples.

A problem in need of research and resolution is distinguishing between the various potential causes of osteoporosis and osteopenia. Clearly the hormonal changes linked to aging are the most common cause in the Western world today as life expectancy has risen dramatically and with that rise, the risk factors for age-related loss of bone have increased as well. In research on the palaeopathology of age-related osteoporosis, the major criteria for this diagnosis is the estimated age of a bioarchaeological burial. However, as Brickley and Ives emphasise, other disorders including starvation are also potential causes of bone loss and at least some of these can occur at any age in either sex

Fluorosis is a metabolic disease caused by excessive amounts of fluorine primarily in human water supplies. Contamination of drinking water by fluorine occurs in many areas of the world, often as the result of leaching into well water. Although it is not a common disorder, it can produce skeletal and dental changes that are specific to this disorder and diagnosis of skeletal abnormalities caused by this disease is relatively certain and uncomplicated. Generally fluorine concentrations in excess of about 3 ppm in the water supply are toxic, but the severe skeletal manifestations apparent in some cases of fluorosis are probably the result of higher concentrations. In some individuals, mineralisation of connective tissue other than bone occurs, resulting in bony growths on many surfaces of bone including the spinal canal. The latter manifestation is associated with neurologic deficiency that can be severe in some patients. In severe cases abnormal bone development may result in complete fusion of the spine. The very spiky abnormal bone distinguishes it from ankylosing spondylitis. Fluorosis can also affect developing teeth resulting in a brown colour of, or brown spots in, the enamel. Excessive intake of fluorine can have an adverse effect on compact bone histology; osteon remodelling is affected with zones of poor mineralisation of osteoid associated with toxic levels of fluorine at the time a layer of bone is being mineralised in a forming osteon.

One of the surprising observations that emerges in the study of bioarchaeological skeletal samples from areas where fluorosis is endemic, is that many of the burials show no evidence of excessive fluorine intake. Clearly if well water is the source of the fluorine, all people using a well are drinking the contaminated water presumably in about the same amounts but not all burials show skeletal manifestations of abnormal fluorine intake. Why this variation exists is an important problem that deserves attention both in clinical situations and in bioarchaeological skeletal samples. However, variation in bone involvement occurs in many disorders that have the potential to affect the skeleton.

Skeletal evidence of metabolic diseases has the potential to provide important insight into dietary adequacy and health in earlier human populations. Making inferences about the prevalence of metabolic disorders in a past human population on the basis of the prevalence of these diseases in bioarchaeological skeletal samples will always be a challenging exercise. What is needed is a better sense of how severe a metabolic disease has to be before the skeleton is affected and, related to this, what is the relationship between the prevalence of skeletal metabolic disease and metabolic disease prevalence in the living population represented by the archaeological skeletal sample? Several lines of evidence are relevant, including gross and microscopic skeletal changes caused by metabolic diseases, isotopic evidence that has the potential of clarifying dietary variation that might be related to metabolic disease and bone protein research because of the effect of some types of dietary deficiency on protein synthesis. We are unlikely ever to have the certainty that would be ideal, but approaching the problem of metabolic disease prevalence in archaeological human populations by evaluating for multiple types of evidence does provide a strategy that at least would permit generalisations about the relative importance of these disorders in a given bioarchaeological population.

Brickley and Ives have provided a significant review of what is known about metabolic diseases as they affect the human skeleton. This review provides an important tool as biological anthropologists attempt to reconstruct in greater detail the dietary problems that human groups encountered in past human societies.

References

Ortner DJ. 2003. Identification of pathological conditions in human skeletal remains. Amsterdam: Academic Press.

Vogel JC, and van der Merwe NJ. 1977. Isotopic evidence for early maize cultivation in New York State. American Antiquity 42:238–242.

Donald J. Ortner

Chapter 1

Introduction

The impairment of health is of great importance in the modern world. Determining the extent to which disease also affected past populations is a fundamental goal of the study of bioarchaeology. The direct physical evidence derived from the examination of human remains provides a vital means of understanding illness and its impact on the lives of those who lived previously. In writing this book, we aim to develop the knowledge of past human health through a detailed consideration of factors that can mediate the expression of a particular group of diseases, the metabolic bone diseases.

This chapter introduces the specific nature of these complex and intriguing conditions, and highlights the insight they can provide into past life and health. We define the terminology crucial to the understanding of these diseases and outline the format that this book takes in illustrating the challenges faced in the reconstruction of health from many past contexts.

The scale of disease expression is vast and investigations of health in the past have expanded enormously over recent years (e.g. Cohen, 1989; Goodman et al., 1988; Goodman, 1993; Larsen, 1997; Ortner, 2003; Roberts and Cox, 2003; Buikstra and Beck, 2006). Specific study of the metabolic bone diseases provides a fascinating means of determining how factors inherent within a given lifestyle, including diet/nutrition, cultural practices, socio-economic status and environmental surrounding, can impact on health both at an individual and population level.

A significant focus of this research aims to highlight how an integrated appreciation of the metabolic bone diseases within many modern cultures can broaden the interpretative frameworks that are considered within bioarchaeology. Variation in disease expression may become manifest by additional factors involved in the onset and development of these conditions, including age, sex, ancestry and geographic habitat. Heritability can contribute to the development of various metabolic bone diseases, but interactions with many components of the environment may exacerbate or prevent disease expression. Recent research into the characteristics of these diseases in many modern populations, in addition to increasing investigation in zoological research, has significantly expanded the recognition and understanding of these conditions.

Many of the diseases examined in this book may be well known, such as osteoporosis; others are likely to be less so, including fluorosis. The spectrum of diseases that we discuss, demonstrate how the bioarchaeological evidence of human health can contribute to the overall study of anthropology. The distinctive goal of anthropology is a better understanding of humanity, whether it is the nature of social or cultural organisation and change, determining the role of human development in economic growth, or the adaptation to environmental situations and nutritional resources. Anthropology can be defined in a number of ways, but is used throughout this book to mean the study of all aspects of humankind, both past and present. Consideration of health challenges faced by non-human primates and other zoological species can also further our understanding of the diverse range of causative variables within a specific environment. Such observations are likely to make an important contribution to the interpretation of the many pathological conditions that can affect humans. The potential for consideration of metabolic bone disease manifestations in zoology will be highlighted through various avenues in this book.

METABOLIC BONE DISEASE: A DEFINITION

The term 'metabolic bone disease' has been used since 1948, when Albright and Reifenstein introduced it to describe conditions that affected the processes of bone formation and remodelling involving the whole skeleton (Albright and Reifenstein, 1948). As with many medical terms, subsequent alterations in the inherent meaning ascribed to this term have become apparent. Recently, there has been considerable interest in a group of cardiovascular diseases that pose significant health problems throughout the world. This group of closely related conditions has been termed the 'metabolic syndrome', and has become the focus of intense study (e.g. Byrne and Wild, 2006). Many of the conditions investigated in the present book could illuminate aspects of health and lifestyle that may impact on conditions within metabolic syndrome. However, our research does not detail obesity, high triglyceride levels, high blood pressure and low HDL (high-density lipoproteins)-cholesterol; conditions inherent within the metabolic syndrome (World Health Organization (WHO), 1999a).

In this book, the term metabolic bone disease is used to indicate conditions in which the processes of bone modelling and remodelling are specifically disrupted. These fundamental processes determining the structural organisation of the skeleton are discussed in Chapter 3. However, pathological changes may not systematically affect the whole skeleton. Particular conditions, such as Paget's disease of bone (Chapter 8), tend to exhibit fairly localised skeletal changes rather than demonstrate involvement of the whole skeleton.

Various investigations have included anaemia as a manifestation of the metabolic bone diseases (e.g. Roberts and Manchester, 2005). This inclusion largely derives from the role of dietary iron deficiency in contributing to anaemia. The extensive involvement of bone marrow tissue in order to regenerate blood cell supply results in secondary effects on bone cells and structure. The specific diagnosis of iron deficiency anaemia in past populations is complex and likely over-estimated, particularly as the wide range of potential causative factors, which includes intestinal parasitic infections and/or excessive blood loss, are frequently overlooked. The diagnosis of the condition can be challenging and cannot be simply attributed to cortical bone porosity in the skull, most notably the orbits (cribra orbitalia). As such, anaemia has not been considered as a specific example of the metabolic bone diseases in this book. However, the secondary effects of this condition can result in the development of various other metabolic conditions, including vitamin C deficiency and osteopenia, and it will be discussed where relevant to these conditions (Chapters 4 and 7, respectively).

FORMAT OF THE BOOK

The discussions throughout this book aim to demonstrate the important role that the analysis of metabolic bone disease has in developing a comprehensive understanding of health, particularly as an indicator of life within past populations. The study of paleopathology is vital in advancing the knowledge of health in the past. However, it is critical that such work should not be viewed as an end in itself, but rather as a step in a wider process enabling increasingly accurate interpretations of health in the past to be ascertained. As has been emphasised above, consideration of manifestations of metabolic bone disease and health within various modern contexts will develop the interpretative framework to which paleopathology substantially contributes. Chapter 2 introduces the study of metabolic disease in bioarchaeology and highlights the benefits and challenges of the evidence and methods that are available to enable investigation of these diseases in the past. Consideration is also given to using clinical data in bioarchaeological research and the potential benefits and pitfalls of this type of information. This chapter further outlines the wider relationship that bioarchaeological research can have in investigations within the many sub-disciplines of anthropology.

Accurate interpretations of the metabolic bone diseases are enhanced by recognition of the normal and pathological processes evident at the bone cell level. Chapter 3 presents an introduction to bone biology with a review of the mechanisms inherent in bone growth (modelling) and maintenance (remodelling), as well as the functions of individual cells and the systemic and local factors that can influence cell behaviour. Recent years have seen a number of important advances in the understanding of bone structures and processes. Continued awareness of these advances made in clinical and scientific research has important consequences for those working within bioarchaeology. As such, this chapter contains references to recent investigations at the bone cell level to create a basis for more detailed consideration of this subject.

The subsequent chapters of this book each discuss a specific metabolic bone disease, considering the nature of the disease process and etiology. Within each chapter an emphasis is placed on the modern understanding of the disease, together with extensive discussion of prominent anthropological considerations and debates. The existing archaeological evidence for each disease is reviewed, together with recommendations for directions of future developments necessary for improving current knowledge. Each chapter concludes with the macroscopic, radiological and histological features that are required for the paleopathological diagnosis of these conditions.

The chapters contain a series of Box Features designed to illustrate how the analysis of the metabolic bone diseases can provide a range of insights into issues considered by many anthropologists. Variation in health can have widespread impacts in nutritional, cultural, social and biological/physical anthropology. The Box Features aim to explore a range of key concepts which could be investigated by considering disease interactions, modern versus past perspectives or animal versus human disease expression, in order to enhance the study of anthropology.

Discussion of vitamin C deficiency (scurvy) in Chapter 4 demonstrates the potential for this disease to become manifest across very diverse contexts. Recent advances in paleopathological diagnosis have demonstrated that scurvy cannot be considered merely as a disease of mariners, or as a simplistic consequence of warfare. The etiologies of many of the metabolic bone diseases are complex, enabling them to exist in a wide range of situations, as well as in non-human primates. Chapter 5 discusses the nature of vitamin D deficiency and its manifestation in children and adults. This chapter illustrates how various metabolic bone diseases can provide substantial data on the range of factors, including living conditions, pollution levels, cultural practices and nutritional quality, that can have a detrimental impact on health.

Osteopenia, or abnormal bone loss, has recently attracted considerable interest in modern medicine. Chapter 6 investigates the age-related onset of osteoporosis, a condition that has a very prominent place in relation to women, in both clinical and mass media. This chapter tackles the wider issues related to the onset of this condition including its expression in males and considers to what extent variation in the causative factors may alter the timing and prevalence of this disease in different societies. Whilst humans appear to have always suffered from bone loss as they age, bioarchaeological evidence from around the world indicates the inherent variations in rates of bone loss between different population groups, with implications for the onset and identification of pathological outcomes. Chapter 7 considers the wider range of circumstances that can result in osteopenia as a secondary pathological consequence. Loss of reliable food resources and the complex outcomes of dietary interactions are considered, together with the effects of limited treatments previously available for a number of conditions, as significant factors likely to have contributed to secondary osteopenia in many past populations.

Chapter 8 explores the nature of Paget's disease of bone. This is a particularly important condition to consider, as despite investigations spanning over 100 years, there is still no known cause for this disease. Whilst presently limiting potential bioarchaeological interpretations, it does provide an exciting opportunity for accumulating paleopathological evidence to substantially enhance what is currently known of the causative mechanisms. A number of lesser known metabolic bone diseases are evaluated in Chapter 9, such as fluorosis and pellagra. With further investigation the understanding and identification of these conditions in past societies can make central developments to the study of health.

The final chapter of the book, Chapter 10, draws together an overview of the principal components of the findings and concepts within this research. This chapter provides a reference for those who wish to discover the fields in which future research on these subjects might most fruitfully be directed. Ultimately, the range of information brought together in this book aims to demonstrate that the metabolic bone diseases have an increasingly important role to play in bioarchaeology, through enabling many anthropologists to gain a fuller understanding of health and its complex interaction in past human life.

Chapter 2

The Study of Metabolic Bone Disease in Bioarchaeology

The study of metabolic bone disease has an important role to play in improving the understanding of a variety of aspects of life in both past and present human societies. There are many branches of anthropology in which information regarding metabolic disease is assuming an increasingly significant role. For example, data on disease prevalence will broaden understanding of subsistence strategies, living and environmental conditions, social and cultural practices, as well as the impact and effects of the aging process. In this book we have utilised a range of evidence that contributes to the investigation of metabolic bone disease, and have provided specific examples that demonstrate how improved understanding of these conditions benefits the study of bioarchaeology. The primary source of evidence that we have used to investigate past health derives from paleopathology, the study of archaeological human remains for evidence of disease. This chapter outlines the potential and limitations of utilising the various forms of evidence available to those investigating the metabolic bone diseases in bioarchaeology.

APPROACHES TO THE STUDY OF METABOLIC BONE DISEASE

Metabolic bone diseases are a particularly valuable source of information for those who have adopted a 'life course' approach to the study of individuals in the past (e.g. Harlow and Laurence, 2002). With a life course approach, the transition through life is recognised as being culturally construed and the life course itself may vary depending on status and sex, together with the surrounding constraints of the community or society. Childhood in particular is both a biological phenomenon and a social construct, with huge differences observed between societies (Panter-Brick, 1998). Within the biological interpretation of the life course, culturally and behaviourally mediated influences on life and health will be inherently interlinked with the surrounding environment and dietary practices. However, both factors may further vary depending on the position of the individual in the community. Such interactions may be reflected in the expression of metabolic bone diseases within a society; for example, possible changes in clothing types as individuals aged or varying attitudes towards exposure of skin and the relation to vitamin D deficiency (Chapter 5).

The onset of the menopause, which has clear links to the development of conditions such as age-related bone loss and osteoporosis may have marked a new stage in the life course of women in past communities. There needs to be greater awareness of the effects of aging on disease processes, and the extent to which males and females might be differently affected. Factors that may impact on the development of metabolic bone diseases are plotted in Figure 2.1.

There are a number of variables that require consideration in relation to bioarchaeological material. For example, are the older adults represented from archaeological sites a more robust group than those who were buried as juveniles? To what extent can pathology experienced in childhood affect the long-term growth and development of individuals? It has been suggested that malnutrition experienced during growth and development may leave affected individuals more prone to developing age-related bone loss and osteoporosis in later life (Grech et al., 1985). These factors are considered further throughout this book (see Box Feature 3.1 and Chapter 6).

When considered in the context of information about a society, the presence of metabolic bone diseases may yield insight into behavioural treatment of specific groups within a community, such as juveniles. Child abuse and neglect in modern communities are problems that have received increased recognition in recent years. Whilst physical abuse may be evident on the skeleton, additional evidence of malnutrition would demonstrate prolonged neglect of dietary provision. Through wider publication of information on cases, anthropologists from a range of fields are now more aware of such issues. Unfortunately what would now be termed 'child abuse and neglect' are not new developments and increasing evidence suggests that such factors may have been present in many

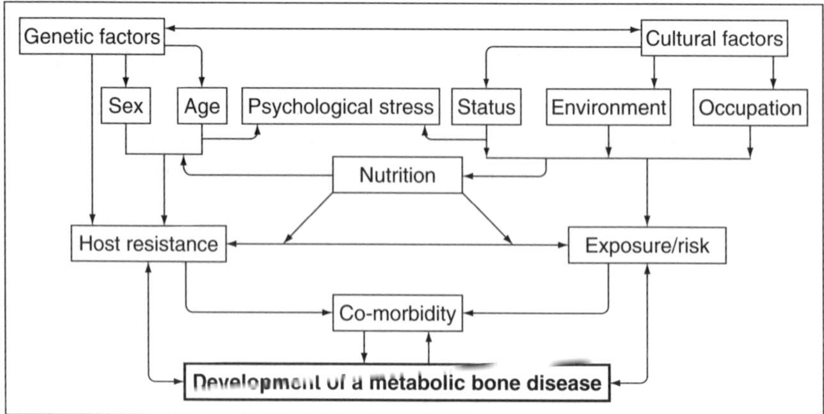

FIGURE 2.1 Schematic relationship of some of the key factors that may impact on the development of metabolic bone diseases. The position of the various factors included in this diagram does not denote importance.

past human communities (e.g. Blondiaux et al., 2002). Abuse is not specific to children and may occur at any stage of the life course. The potential for recognition of abuse and/or deliberate neglect is discussed in Box Feature 5.2.

Information on vitamin D deficiency has been used to infer information on the treatment of children in past societies (e.g. Mays et al., 2006a). For example, swaddling children could have had an impact on the development of vitamin D deficiency. Recent research has indicated that childhood malnutrition and ill-health may affect longer-term survival and adult health (Cameron and Demerath, 2002). Some of the conditions covered will impact on human growth and biological development, for example the ability to attain peak bone mass as an adult (discussed in Chapter 6).

As today, certain groups within past communities may have been more at risk of malnutrition and/or disease co-morbidities, such as women (particularly during childbearing) and the elderly. A greater understanding of the presence of metabolic bone diseases would help interpret how health varied over different time periods or across different regions. The types of research questions being generated by nutritional anthropologists relating to diet will play a fundamental part in a greater understanding of health patterns in both past and present communities. For example, changes in human subsistence practices at a micro- and macro-scale, will have impacted on many aspects of life in past communities. Utilising a life course approach will enable the impact of various conditions on particular demographic groups and their consequences to be considered. For example, vitamin deficiencies in parous females can result in congenital forms of the disease in children and this is discussed in detail in Chapters 4 and 5. Breastfeeding and weaning practices may have had an influence on the development of conditions such as scurvy and rickets in infants (as discussed in Box Feature 4.1). In adults, factors such as vitamin D deficiency will have an impact on a woman's ability to successfully give birth.

As discussed in Chapters 4 and 9, the study of metabolic bone diseases can also contribute to understanding the health consequences of natural disasters that result in various types of pollution and destruction of crops and stored foods. Food shortages have a number of consequences including vitamin deficiencies, and the health problems associated with displaced peoples are discussed in Box Feature 9.1. However, as the information referred to here demonstrates, issues of illnesses caused by food shortages may be complicated by co-morbidities caused by overcrowding and poor hygiene. Factors such as these are likely to exacerbate the risk of infection, and may result in death before metabolic conditions become visible in the skeleton (see discussion in Chapters 4 and 9).

The various branches of anthropology draw upon a wide range of evidence. This diversity brings with it very real opportunities for work to be undertaken that will allow a better understanding of the contribution of metabolic bone diseases to bioarchaeological investigations. In the section below, some of the key biases that affect the types of data used in the study of metabolic bone disease, particularly in past societies, are identified.

CHALLENGES IN THE INVESTIGATION OF METABOLIC BONE DISEASE

Museum Collections

There are a number of important museum collections around the world in which recent human bone is held. Museum collections fall into two broad categories; those in which examples of pathological conditions are curated, some of them from documented individuals, for example the Galler collection, Switzerland (Rühli et al., 2003) and museum collections comprising of documented individuals that were collected from autopsy room cadavers. Many of the latter collections were established to allow research and teaching on subjects relating to human biology to be undertaken, for example the Terry collection, held by the Department of Anthropology of the National Museum of Natural History of the Smithsonian Institution, Washington, US. Anatomical collections also occasionally contain individuals who had metabolic bone diseases (Figure 2.2). Both types of museum collection form excellent research collections, and many of the individuals housed within them lived prior to the adoption of medical treatments that are now available. For those interested in disease processes in past populations, these collections provide a valuable resource. However,

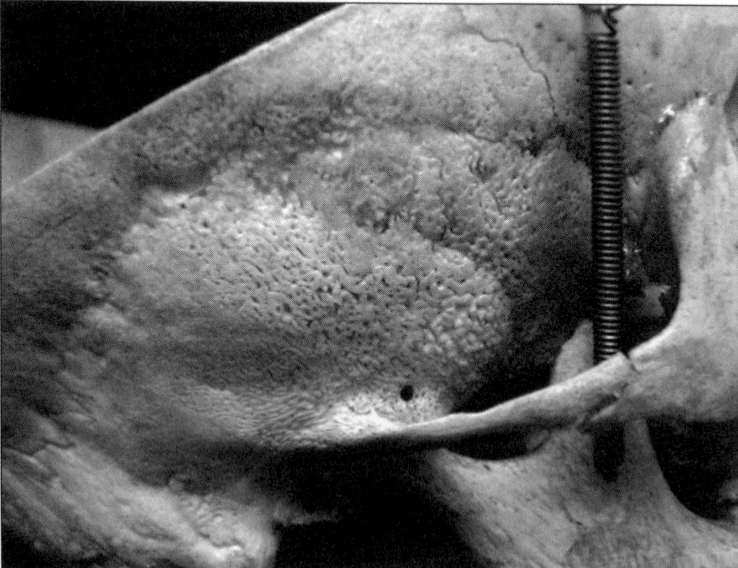

FIGURE 2.2 Cranium of a two-year old child held in the Federal Museum of Pathological Anatomy, Vienna. This cranium is recorded as being from a child with a 'rachitis cranium', but porosity on the sphenoid and other changes across the cranium indicate that this child may also have been deficient in vitamin C. The co-occurrence of rickets and scurvy is discussed in more detail in Chapter 4. Photograph by the authors courtesy of the Federal Museum of Pathological Anatomy, Vienna.

although there is interesting material housed in museum collections, there are a number of limitations that should be kept in mind when using information from such collections.

Although the aims and purpose of the museum collections held around the world varies enormously, individuals selected for inclusion frequently had a 'classic' or very interesting case of a medical condition. For those who study human skeletal remains from an archaeological context, it would be fair to say that there would be little chance of human remains displaying the level and types of pathological features exhibited by human remains in these collections, surviving in a buried environment.

At the time of collection many treatments that are currently available were not present, but diagnostic tests now used by clinicians were also absent. Not all the diagnoses undertaken at the time of collection would be accepted today. Many individuals may also have had multiple conditions, but the person who collected the specimen only diagnosed one. In the experience of the authors, quite a number of individuals held within these collections almost certainly had more than one pathological condition. However, the catalogue of the collection specimens will often only have one of the conditions recorded. In such cases the effects of the co-existence of a number of pathological conditions need to be considered.

Many of the metabolic bone diseases are chronic rather than acute. Individuals in past societies would often have lived with a condition for some time prior to death, frequently developing severe bone changes. Given the long period an individual could have lived with a condition, co-morbidities may well have developed. For example, older individuals who may well have had age-related bone loss could also have been at risk of developing osteomalacia (see Chapter 5). The exact expression of pathological conditions in the skeleton, particularly where diseases co-exist, are still a very poorly understood area and are discussed in the various chapters of this book.

The socio-economic status of those who were collected also needs to be considered when examining human remains in museum collections. As discussed later in this book, low socio-economic status (which often characterises individuals in such collections) would have had important consequences for wider health issues. Individuals included in such collections may have suffered more than individuals of a higher status from inadequate housing conditions and poor levels of nutrition.

Archaeological Human Bone

Archaeological human bones are an invaluable resource for scientific study, and are found across much of the world (e.g. see discussion by Armelagos et al., 1992; Aufderheide and Rodríguez-Martín, 1998; Ortner, 2003). Indeed the study of human skeletal remains is the primary source of evidence relating to metabolic bone diseases in pre-historic communities. Even when texts are available, written records have many problems. Many texts relate predominantly to small elite sections of society and detail issues involved with government and control of power, as discussed in the case of written records relating to cases of scurvy considered in Chapter 4.

One of the aspects of archaeological human skeletal remains that make them particularly valuable is the fact they are frequently associated with a particular cultural or social group. Within the cultures and societies around the world from which human remains have been excavated, there is often differential burial and association of artefacts that can contribute to interpretations of the socio-economic status of the individual (as discussed by Robb et al., 2001; Pechenkina and Delgado, 2006; and illustrated in Figure 2.3). Over the last few years a number of historical cemeteries have been excavated with input from biological or physical anthropologists; for example, Christchurch Spitalfields, London (Molleson and Cox, 1993), St. Thomas' Anglican Church, Belleville, Ontario (Saunders et al., 2002) and more recently St. Martin's, Birmingham, England (Brickley and Buteux, 2006). Not only have these collections helped to determine the accuracy of techniques used to establish the demography of

FIGURE 2.3 A Roman skeleton (approximately AD 0–500) with associated grave goods excavated in advance of building work in 1925, Cologne, Germany. Fremersdorf-Köln (1927:263).

individuals from past societies, they have also contributed to our understanding of the link between socio-economic status and health. However, it should be remembered that there are still large gaps in our knowledge linked to archaeological human bone. There are extensive geographical regions in which no structured excavations utilising recent developments in fieldwork techniques have been undertaken.

One problem with studying the metabolic bone diseases in archaeological skeletons is that many of the conditions covered in this volume cause considerable weakening of the skeleton (discussed in Chapters 5 and 8). It is therefore likely that not all the evidence for these conditions in past communities will survive in a state that will make recovery and interpretation of the conditions easy. As was demonstrated at St. Martin's, Birmingham, England, the decision to concentrate the available resources on the better preserved and more complete individuals (Brickley and Buteux, 2006) introduced the potential that quite a few individuals with pathological conditions were excluded from the recording undertaken for the site report. Analysis undertaken as part of a research project on vitamin D deficiency at St. Martin's demonstrated that some individuals who were very poorly preserved, and so not recorded for the report, did have metabolic bone diseases (Brickley et al., 2007).

Various factors that affect human bone when it is buried in the ground will result in less than ideal preservation, and may result in the loss of certain skeletal elements through taphonomic processes. A number of papers have covered aspects of bone preservation and survival from archaeological sites (e.g. Waldron, 1987; Mays, 1992; Stojanowski et al., 2002). Lyman (1994) reviewed many aspects of vertebrate taphonomy, but this is still a poorly understood area. At a very basic level it appears that smaller bones survive less well and bones with higher trabecular bone content may also be prone to loss. However, as discussed by Mays (1992), some skeletal elements may be lost during recovery of skeletons from archaeological sites.

At a microscopic level there may also be changes in archaeological human bone that make the study of disease processes difficult. Diagenetic alterations, such as those shown in Figure 2.4 can severely disrupt the histological appearance of bone, and as are demonstrated in a number of chapters in this book, histological investigations can be invaluable for assisting with the diagnosis of pathological conditions. Diagenetic changes include disruption of bone microstructure by fungi and bacterial activity, but visual examination of bone will not allow the detection of such changes. Bell and Piper (2000) provide an introduction to the use of paleohistological studies within bioarchaeology. It is also possible that mineral replacement may take place, which will render the results of investigations that utilise non-invasive measures of 'bone mineral' inaccurate (Farquharson and Brickley, 1997).

In addition to the various taphonomic factors that may affect the preservation of bones from archaeological sites, there are a number of factors linked to human actions that may impact on the survival of skeletons. Where societies use designated places to bury individuals, and these sites are intensively used,

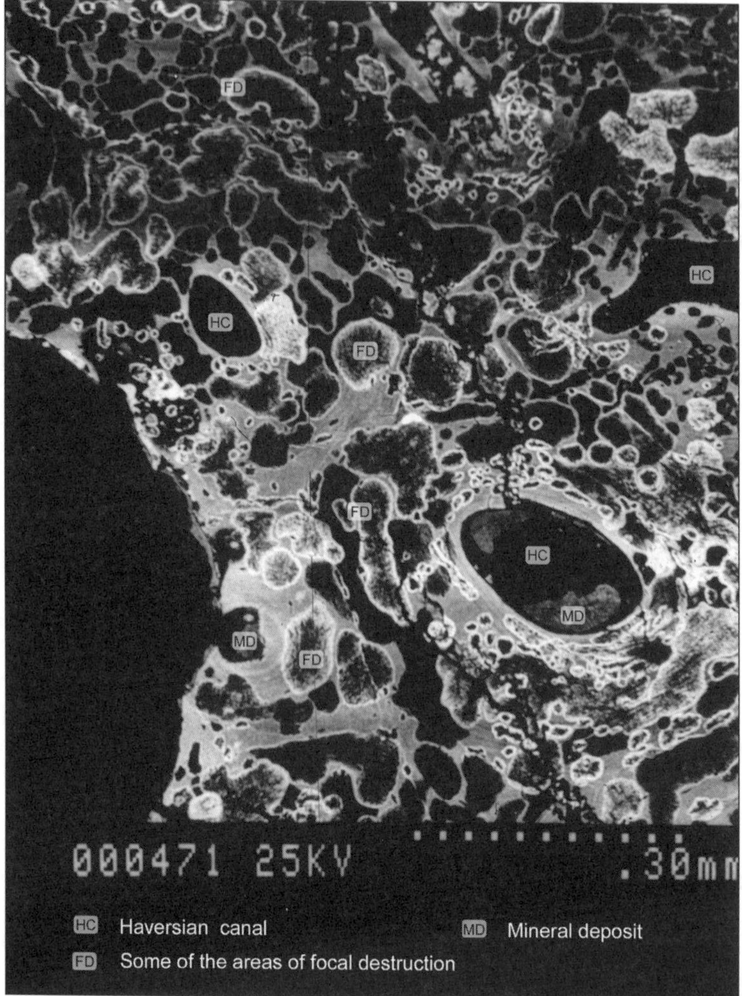

FIGURE 2.4 Back-scattered scanning electron microscopy (BSE-SEM) image of diagenetically altered archaeological bone (femoral shaft). Extensive areas of focal destruction have obliterated most structural features in this piece of archaeological human bone.

earlier burials may be disturbed or truncated (Figure 2.5). Having a complete skeleton is often very important for a possible diagnosis to be suggested, as the pattern of pathological changes across the skeleton needs to be observed. Incomplete or partial skeletons will be a hindrance to paleopathological investigations, and will also reduce the amount and quality of biological information available. Large assemblages of skeletons from clearly designated burial areas are also less likely from small mobile band societies. In other groups funerary processes that involve excarnation and disarticulation or cremation of bodies may be undertaken. McKinley has demonstrated that it is worth studying

FIGURE 2.5 Truncated skeletons from the historic St. Martin's cemetery, Birmingham, England. Use of the cemetery was intensive and as can be seen in this image, many earlier graves were disturbed by later burials. Reproduced with permission of CgMs Limited on behalf of the Birmingham Alliance.

cremated human bone from archaeological contexts (e.g. McKinley, 2000). However, the amount of information on pathological conditions and factors such as the age and sex of the individuals that can be derived from burned or disarticulated bone will be more limited than that which can be obtained from complete skeletons (see images of such material in Figure 2.6).

The demographic composition of a skeletal collection is important to consider in the analysis of the bioarchaeology of the metabolic bone diseases. Waldron (1994) covers some of the key issues involved when dealing with the study of pathological conditions in archaeological human remains. There are many factors that affect the 'sample' that is available for analysis from different cultures. For example, there may be different burial treatments according to age or sex, or specific burial treatment for individuals of different socio-economic status (see Figure 2.7). Such factors should always be considered before drawing any conclusions from data obtained from archaeological human remains. Cohen (1989) and Goodman (1993) provide fuller discussion of the problems and potential of investigating archaeological skeletal material.

Although the authors would encourage people to use the information provided by archaeological skeletons and would argue that such data are of interest to those in a wide range of fields of anthropology, the collections available to us are at best very partial. Such collections are never fully representative of the living community from which they came. The ideas introduced by Wood et al.

FIGURE 2.6 (a) Disarticulated human bone at the third clearance level in the south chamber of Hazleton North, a Neolithic long Cairn in England. Photograph by Alan Saville (Saville, 1990:82). (b) Romano-British lidded urned cremation burial from Dorset, England. Lack of soil infiltration means the bone has retained the fragment-size as at the time of original deposition. Courtesy of Jackie McKinley, Wessex Archaeology UK.

(1992) relating to the osteological paradox also need to be considered. It may be possible that weaker individuals in past communities may have died prior to displaying clear pathological changes in their skeleton. In the study of the human bone from St Martin's, Brickley and Buteux (2006) noted several instances where the osteological paradox appeared to play a role in disease expression. Issues relating to selective mortality are discussed further by Larsen (1997).

FIGURE 2.7 Cultural practices have a significant impact on the manner of burial, as illustrated by this burial of a decapitated infant from the Romano-British site of Godmanchester, Cambridgeshire, England. As discussed by Boylston et al. (2000) decapitation was a common practice during this period. Courtesy of Birmingham Archaeology, UK (Jones, 2003:219).

Paleopathological Diagnoses

There are still many problems with accurately diagnosing disease in archaeological bone, but recently far more information has been assembled on the range and type of expressions of disease in the skeleton. However, much of the information that would be available to clinicians when making a diagnosis is missing in archaeological bone. A review of recent clinical literature demonstrates that there is still considerable debate surrounding making the correct diagnosis, and misdiagnosis does happen. For example, Grewar (1965) reported on numerous instances of misdiagnosis in cases of scurvy. Akikusa et al. (2003) also report that lack of awareness of scurvy among clinicians often leads to a delay in diagnosis. There are in fact a number of conditions that can produce similar changes in the skeleton, and distinguishing between these can be very difficult. In some cases it may be possible to suggest which of the pathological conditions that could have produced a change may have been responsible, using a variety of cultural historical and other contextual information.

With the definition of the metabolic bone diseases used in this book (discussed in Chapter 1), there should be evidence of pathological changes at a gross, radiological or histological level. However, given the fairly limited repertoire of changes that can be produced in bone (discussed in detail in Chapter 3), suggesting a definite diagnosis may not be possible even if skeletal changes are present and histological and radiological analyses have been undertaken. Although improvements have been made there is still a lack of uniformity in the way in which a diagnosis is suggested for many conditions, and many older reports fail to provide consistent information, even at the level of the recording criteria used. These issues are discussed in detail by Roberts and Cox (2003).

Demographic Issues

Collections of archaeological skeletons consist of dead individuals, and the likelihood of those of various ages to die in past communities will have been very different to that observable in many modern populations. In addition, social and cultural practices may result in certain parts of the dead population being treated in a way that will prevent their recovery by archaeologists. Hoppa and Vaupel (2002) discuss a number of the issues relating to the demographics of archaeological collections.

In particular, our lack of ability to pick out and distinguish accurately between the many groups who must reside in the category of 'old adult' (50+ years), which researchers are advised to use (Buikstra and Ubelaker, 1994; Brickley and McKinley, 2004) will limit research that can be undertaken. Recent work has moved towards addressing this problem (e.g. Buckberry and Chamberlain, 2002; Falys et al., 2006), but lack of precise aging techniques for older individuals and absence of other biological information, such as the age of menopause, are serious problems. Far more information could be derived about a number of important conditions (such as age-related bone loss, discussed in Chapter 6) if these data were available (see Box Feature 6.3). This is not the appropriate place to enter the extensive debate surrounding paleodemography, but issues relating to demography do have quite profound implications for investigations of the bioarchaeology of metabolic bone diseases. Although less has been written about sex determination from archaeological human bone, the potential for difficulties for the accurate determination of sex in archaeological skeletons should also be considered. It has been suggested that a number of metabolic bone diseases have very different distributions in males and females. In some conditions, such as age-related bone loss, possible sex differences have been investigated. However, in others, such as osteomalacia, possible sex differences are still not well understood.

MODERN MEDICAL DATA

All anthropological investigations of metabolic bone diseases will clearly rely very heavily on modern medical data on these conditions. The value of these data can be seen throughout this book, but there are a number of important considerations that need to be considered when using this information. Care needs to be taken as the use of terminology frequently differs between anthropologists and medical professionals. For example, the term 'osteomalacia' (discussed in Chapter 5) is used differently by both groups. Terminology also changes through time, as has already been discussed in relation to the term 'metabolic disease' (Chapter 1).

Anthropologists frequently view medical diagnosis with a degree of certainty that they would never ascribe to their own work. However, as discussed in Chapter 4, even clinically there can be problems with making a direct diagnosis, and the existence of co-morbidities may not always be picked up on. Modern medical data also tend to focus on conditions affecting individuals in the developed countries of the world. As discussed in Chapter 7, the effects of starvation and malnutrition on the development of osteopenia remain almost completely un-investigated, because generally these are conditions that affect individuals in the poorest countries of the world. The amount of research undertaken is also dictated to a certain extent by funding available. As discussed in Chapter 6, osteoporosis in women has been extensively studied in comparison to the condition in men where there is no one-drug treatment that can be sold. Whilst medical data offer an invaluable resource to anthropologists, these and the many other limitations of these data need to be considered.

GENETICS AND ANTHROPOLOGY

The extent to which the timing or onset of genetic abnormalities in programming may result in disease, or may predispose or increase the risk of disease in later life, are important factors to consider in relation to metabolic bone diseases. The range of issues that genetics can help researchers address, and the implications of genetics for an understanding of human health and disease are discussed in detail by Jobling et al. (2004). Conditions such as age-related bone loss, discussed in Chapter 6, almost certainly have some genetic components (Ralston, 2005; Huang and Wai Chee Kung, 2006) and the amount of melanin in the skin, and hence vitamin D synthesis is also genetically determined. Factors such as the heritability of transmission of conditions, the cause of some hereditary diseases (some of which are discussed in Chapter 9) and the mechanism through which such conditions have evolved, are becoming better understood as knowledge of genetics advances. Recent years have seen a number of important developments and these along with advances in the study of ancient DNA can provide valuable information (e.g. Montiel et al., 2001; Zink et al., 2002; Gilbert et al., 2005).

CULTURAL AND SOCIAL ANTHROPOLOGY

Given the limitations of some of the types of information discussed above, a number of researchers have drawn on ethnographic data as a primary form of evidence, particularly in cultural and social anthropology (e.g. studies by Dufour et al., 1998). Increasingly, biological anthropologists have used such data as a supporting form of evidence where other sources have failed to provide adequate information (Dettwyler, 2005). The study of diseases in modern populations provides a dynamic and interesting addition to our knowledge about various conditions, and the study of health and diseases in both the past and present have considerable mutual benefit in terms of increasing our knowledge and understanding regarding human health. Many of the issues such as modes of subsistence and adaptation, family structure, peasants and cities, covered in texts such as Keesing and Strathern (1998) would be enhanced through consideration of health. Approaches taken by cultural anthropologists allow aspects such as gender roles in subsistence, access to dietary resources, clothing and costume, as well as housing types and working practices to be explored.

NUTRITIONAL AND MEDICAL ANTHROPOLOGY

Another important linked field is nutritional anthropology; the multidisciplinary approach to the study of human diets and related issues. As will be apparent throughout this book, diet is a key factor in human health issues and so considerations of diets, both past and present, are vital in any assessment of health (see discussions in Chapters 4, 5, 7 and 9). Nutritional anthropology has seen many important developments over the past 15 years (e.g. Ulijaszek and Strickland, 1993). In order to gain the fullest possible understanding of past and present health, researchers should consider these factors. The area of cultural anthropology referred to as medical anthropology; the study of the relationship between aspects of cultures and health and disease can also provide very valuable information for investigations of metabolic bone disease (e.g. studies by Benyshek and Watson, 2006).

PRIMATOLOGY

Many of the metabolic bone diseases covered by this book also affect some of the non-human primates, and as a result are of interest to those engaged in primatology. Research in primatology also enables a number of different aspects of metabolic bone disease to be investigated. For example, research on age-related bone loss and osteoporosis in non-human primates has considerable potential to contribute to our understanding of such conditions in modern humans (e.g. Sumner et al., 1989; Cerroni et al., 2000). Where possible, throughout this book, examples of the various disease processes in non-human primates will be discussed as these can provide important insights into the evolution and development of some of the metabolic bone diseases.

CONCLUSIONS

This chapter has highlighted the diverse range of evidence that exists for the study of bioarchaeology and anthropology. As long as researchers are aware of the factors that may challenge or bias investigations of metabolic bone disease, there is considerable potential for future studies of these conditions. As discussed throughout the book and summarised in Chapter 10, it is clear that investigation of the metabolic bone diseases can make a significant contribution to understanding the impact of disease in many anthropological contexts.

Chapter 3

Background to Bone Biology and Mineral Metabolism

A clear understanding of underlying bone biology is vital to enable differentiation between normal and pathological conditions. During life, bone is not a static tissue and whilst the bone cells themselves do not survive after death, evidence of recent cell activity can be recognised in skeletal tissue. This chapter reviews processes within bone biology in order to facilitate the accurate understanding and identification of the manifestations of the metabolic bone diseases.

BONE TISSUE: CORTICAL AND TRABECULAR BONE

Bone is comprised of organic collagen fibres, together with non-collagenous proteins and water, which provide elasticity and resistance to tension, as well as mineral tissue comprising of hydroxyapatite, a calcium–phosphorous mineral. The mineral tissue enables the skeleton to be strong and resistant to compression (see reviews Sommerfeldt and Rubin, 2001; Pearson and Lieberman, 2004:67–68). The skeleton is comprised of two tissue types: cortical bone and trabecular bone (see Figure 3.1). Cortical bone forms the outermost solid layer of the skeleton and forms thick, dense walls of the long bones. The cortical bone of the long bones encloses the medullary cavity, which during life contains the haematopoietic bone marrow (see Table 3.1) (Martin et al., 1998:31; Downey and Siegel, 2006). Trabecular bone is formed as a honeycomb network of bone tissue located at the epiphyseal ends of the long bones and within areas of the axial skeleton, such as the vertebrae. An individual spur of bone within this network is called a trabecula. The spaces within the trabecular network also contain haematopoietic bone marrow. Joint surfaces at the epiphyses are covered with uncalcified articular cartilage (see Schwartz, 1995:4–7; Baron, 1999:3). Bone is surrounded by an osteogenic tissue containing cells that affect the production and removal of bone tissue (see below). The layer of tissue in contact with the external bone surface is called the periosteal membrane or periosteum, while the internal bone surface in contact with the medullary canal is covered with the endosteal membrane (see Table 3.1) (Vaughan, 1975; Baron, 1999; Allen et al., 2004). A comprehensive review of the structure and composition of bone is given by Sevitt (1981). Various descriptive terminology utilised to describe the structural components of bone are outlined further in Schwartz (1995) and White (2000).

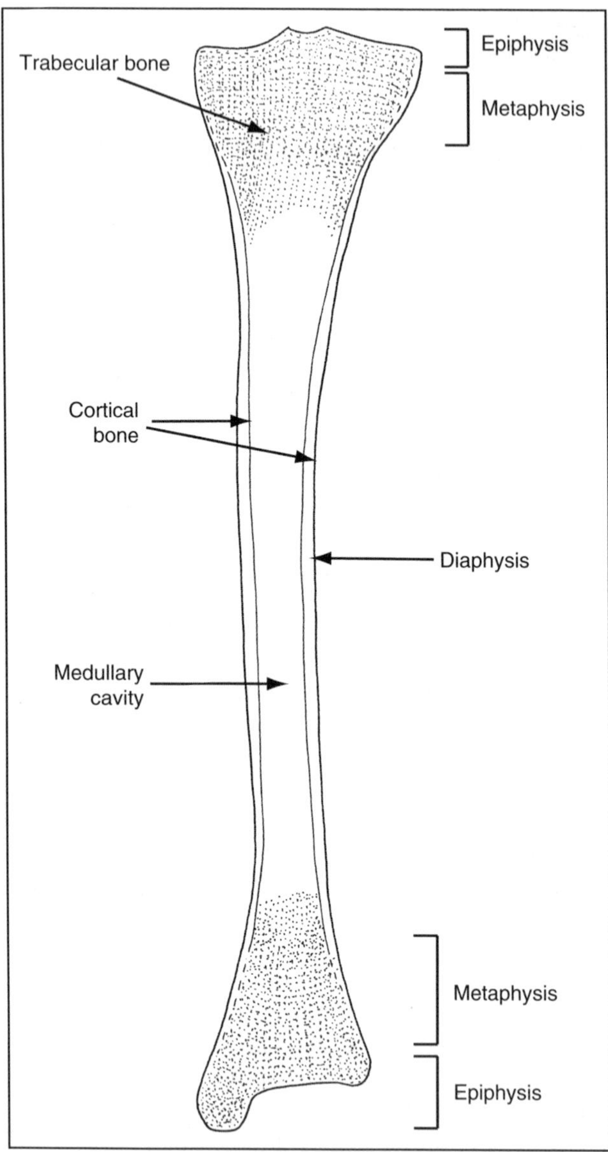

FIGURE 3.1 Schematic section of the major components of an adult human long bone.

DIFFERENT TYPES OF BONE STRUCTURE: WOVEN BONE AND LAMELLAR BONE

During periods of rapid bone formation, particularly for example during growth or in certain pathological conditions such as Paget's disease (see Chapter 8) or in fracture healing, the bone produced is not well organised. Collagen bundles are irregular in formation and are randomly orientated, lacking well-defined arrangement (Martin et al., 1998; Baron, 1999; Sommerfeldt and Rubin, 2001). This type of bone is called woven bone (or alternatively fibre bone; see Ortner, 2003:19). The importance of the formation of woven bone is in providing a means of rapidly responding to changes, whether growth, or through reacting to functional activity or various disease states, and this bone type also plays a role in fracture healing and callus formation (Sommerfeldt and Rubin, 2001). However, the disorganised structure of woven bone is unable to provide long-term, efficient structural support.

In order to provide bone strength, a second type of bone is formed later during childhood and throughout adulthood comprising of mineral crystals and collagen fibres that are highly organised and orientated into layers called lamellae, forming lamellar bone (Martin et al., 1998:34). Through processes of bone turnover described below, lamellar bone eventually replaces woven bone. Lamellae are formed as parallel layers at the sub-periosteal margin of cortical bone and along trabeculae (circumferential lamellae). In cortical bone, lamellae are arranged in successive concentric layers in which the collagen fibres alternate in alignment (see Baron, 1999; Pfeiffer, 2000; Ortner, 2003:19–21). The structural unit containing the concentric lamellae is called an osteon. At the centre of an osteon is a Haversian canal, which contains a blood vessel. Some canals may also contain nerves and lymph vessels (Downey and Siegel, 2006:81). Positioned at relative right angles to the Haversian canals are networks of much smaller blood vessels called Volkmann's canals. These provide additional links between the blood supply throughout cortical bone. In contrast, trabecular bone is not arranged surrounding specific canals but instead accesses blood supplies and nutrients directly from the haemopoietic tissue, which surrounds the bone structure. The boundary of each osteon is defined by a cement line, which denotes the extent of bone removal and replacement undertaken during modelling and remodelling (see below).

When lamellar bone is originally formed, it effectively fills in the spaces apparent in the structure of the rapidly formed woven bone. These lamellar units are called primary osteons (Sevitt, 1981:19; Martin et al., 1998:37; Ortner, 2003:25), and tend to be quite small and are not well defined by a cement line. However, bone tissue needs to be regenerated throughout life and lamellar osteons are continually altered and replaced by new osteons. These osteons are formed by a specific process called remodelling (see below) and are clearly defined by the cement line. These structures are called secondary osteons. Fragments of lamellar bone located between different osteons are called interstitial bone (Sevitt, 1981:19). Secondary osteons may show numerous cement lines, which represent repeated remodelling events, or may show an osteon having formed within an osteon unit (type II osteon). As Ortner (2003:25–26) has discussed, this type of remodelling does not occur uniformly throughout secondary osteons and may well represent a remodelling event triggered by a physiological cause. Anthropological research has particularly investigated the various factors which may affect the shape and size of osteons and Haversian canals during remodelling, including age and potentially mechanical strain (see reviews by Pankovich et al., 1974; Stout and Teitelbaum, 1976a; Cho and Stout, 2003; Robling and Stout, 2003).

BONE CELLS

Disease expression in skeletal tissue can only become manifest in a limited number of ways; through bone formation, bone removal or combination of both processes. The metabolic bone diseases are characterised as those which reflect disruptions in bone formation, maintenance (remodelling) or mineralisation, or comprise conditions which involve a combination of these factors (see Chapters 1 and 2). The specific bone cells that are responsible for these processes are discussed below and summarised together with relevant tissues in Table 3.1.

Osteoclasts remove bone tissue via the process of bone resorption (Sommerfeldt and Rubin, 2001). At a determined site an osteoclast seals a portion of bone surface using its apical membrane, thus defining the extent of bone that is to be removed. The osteoclast's membrane creates an irregular, ruffled or scalloped-shaped border of the resorbed area, which is visible histologically (Parfitt, 1994; Phan et al., 2004). In order to resorb bone, hydrogen ions and lysomal enzymes are secreted and activated by the lowering of the pH level beneath the osteoclast (Baron, 1999; Raisz, 1999). Mineralised bone is effectively degraded or dissolved (Raisz, 1999) and the resulting resorption bay with a ruffled border is called a Howship's lacuna. The number of active resorbing sites may vary between pathological conditions that increase bone resorption and ones which inhibit the initiation of this process.

Osteoblasts synthesise new bone through secreting organic bone matrix (osteoid) and regulating its subsequent mineralisation (discussed below) (Parfitt, 1994; Baron, 1999). Osteoblasts maintain extensive communication with surrounding cells via transmembranous proteins, specific cell receptors and through the actions of cytokines, hormones and growth factors (Linkhart et al., 1996). Such communication facilitates cellular function and responses to metabolic and/or mechanical stimuli (Sommerfeldt and Rubin, 2001). Research also suggests that osteoblasts exert an influence on osteoclasts by using receptors that can control the chemical or hormonal signals which would either initiate or inhibit osteoclastic action (Marks and Odgren, 2002). These inherent qualities of the osteoblast aid the 'coupling' of osteoblasts to osteoclast activity which is vital in remodelling (see below) (Lian et al., 1999; Parfitt, 2000). Once the osteoblast has completed bone formation it can undergo various transformations, either becoming a flattened bone lining cell, undergo programmed cell death (apoptosis), or it can become embedded within the newly formed bone tissue transforming into an osteocyte (see below) (Parfitt, 1994; Baron, 1999; Sommerfeldt and Rubin, 2001).

TABLE 3.1 Summary of Bone Cells and Tissues Involved in Bone Formation and Remodelling

Cells and tissues	Summary of function	Sources
Bone cell origins: haematopoietic tissue	Production of stem cells Production of blood cells in bone marrow in the medullary cavity in adults, in widespread tissue (extra-medullary haemopoiesis) in children Contains osteogenic tissue to generate osteoclast bone cells Can create (differentiate) macrophage cells (host defence)	Vaughan (1975); Martin et al. (1998); Baron (1999); Sommerfeldt & Rubin (2001)
Bone cell origins: mesenchymal tissue	Loosely organised, undifferentiated tissue Production of non-haematopoietic stromal cells for bone, cartilage, tendon and muscle tissue Contains non-specific repair cells, forming connective tissue (e.g. trauma, fracture, inflammation, necrosis and tumours) Cells reside in tissue of bone marrow, spleen and thymus, cartilage, periosteum, endosteum, synovium, muscles and tendons Produces osteoblasts, osteocytes and bone lining cells	Vaughan (1975); Martin et al. (1998); Baron (1999); Sommerfeldt & Rubin, (2001); Pountos & Giannoudis (2005)
Periosteal membrane	Covers external bone surface Outer layer contains nerves, blood vessels and collagen fibres Internal layer adjacent to bone (cambium layer) contains mesenchymal progenitor cells, osteoblasts and micro-vessels Varying thickness across skeletal regions	Baron (1999); Allen et al. (2004)
Endosteal membrane	Covers internal surfaces of bones	Baron (1999)
Osteoblasts	Forms new bone tissue Produces unmineralised organic bone matrix (osteoid) (see text) Regulates matrix mineralisation 'Coupled' with activity of osteoclasts in remodelling (see text) Derive from mesenchymal cells	Peck & Woods (1988:8); Parfitt (1994); Lian et al. (1999)

Continued

TABLE 3.1 *(Continued)*

Cells and tissues	Summary of function	Sources
Osteoclasts	Remove (resorb) existing bone (see text) Specific attributes for resorption – apical membrane, digestive enzymes Derive from haematopoietic cells 'Coupled' to osteoblasts for remodelling (see text)	Peck & Woods (1988:8); Parfitt (1994); Lian et al. (1999)
Bone lining cells	Members of the osteoblast family Do not excrete bone matrix Covers resting (quiescent) bone surfaces Removes bone membranes to enable resorption	Parfitt (1994)
Osteocytes	Transformed from mature osteoblast cells No longer significant role in forming new bone matrix Reside in osteocyte lacunae (Figure 3.2) Maintain extensive communication (osteocytes, bone lining cells, osteoblasts) May recognise bone fatigue, aiding initiation of remodelling	Lian et al. (1999)
Chondrocytes	Differentiated from mesenchymal tissue Secrete extracellular matrix forming hyaline cartilage Basis of endochondral ossification Cartilage matrix is easily degraded by blood vessels needed for bone cell proliferation Sequential development: chondrocyte proliferation and differentiation, synthesis of cartilage matrix (hypertrophic activity), cell death, osteoblast activity, bone formation on cartilage template	Olsen (1999); Scheuer & Black (2000a)

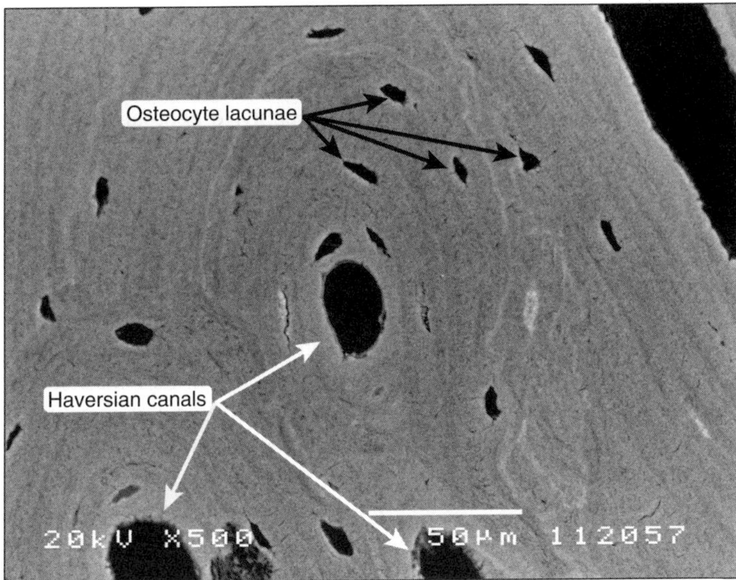

FIGURE 3.2 BSE-SEM image identifying numerous osteocyte lacunae and Haversian canals in secondary osteons. Sample from adult human cortical bone. Archaeological femur, St. Martin's Churchyard, Birmingham.

Osteocytes are former bone-producing osteoblasts that no longer synthesise osteoid and have instead become trapped within the calcified bone matrix. Osteocytes, however, remain important in facilitating extensive cellular communication via long cytoplasmic extensions (canaliculi), which contact other cells (see Curtis et al., 1985; Baron, 1999; Lian et al., 1999; Martin, 2000; Sommerfeldt and Rubin, 2001). These networks are likely to be important in initiating cellular stimuli for bone remodelling (Raisz, 1999; Sommerfeldt and Rubin, 2001). During life, osteocytes reside in lacunae filled with extracellular fluid (Baron, 1999; see also Pearson and Lieberman, 2004) and these lacunae remain visible at the histological level in archaeological bone as shown in Figure 3.2

There are numerous other factors that can influence bone growth and maintenance, which are beyond the scope of this review. These factors include various hormones and proteins, such as macrophages and lymphocytes. These may aid the preparation of bone surfaces for resorption, enable cellular differentiation into specific bone cells, or may carry receptors for factors (e.g. parathyroid hormone (PTH), vitamin D) that can regulate bone remodelling (see below). Proteins such as osteocalcin and osteonectin, expressed by osteoblasts can encourage crystal growth and collagen and hydroxyapatite binding, playing an important role in the mineralisation process. Furthermore, groups of polypeptide growth factors, insulin-like growth factors, transforming growth factor-β (TGF-β), bone morphogenetic proteins as well as growth hormone, have significant effects on the bone cells and play an important role in the maintenance of bone remodelling. For detailed reviews of these additional factors see Linkhart et al. (1996), Lian et al. (1999), Mundy (1999), Ortner (2003:30–31), Ueland (2004) and Downey and Siegel (2006:84). Reviews of the blood supply to bone are introduced by Sevitt (1981) and Milgram (1990).

MODELLING AND REMODELLING: GROWTH AND ADULTHOOD

Mechanisms of Growth

There are several mechanisms of skeletal growth (see reviews Scheuer and Black, 2000a, 2000b; Downey and Siegel, 2006). Intramembranous bone formation requires mesenchymal cells within embryonic connective tissue to proliferate and eventually differentiate into osteoblasts. Osteoblasts form woven bone matrix using connective tissue membranes (Baron, 1999; Ortner, 2003:16). Blood vessels can become incorporated in between the woven bone trabeculae from haematopoietic bone marrow. The woven bone is eventually replaced with more mature and organised lamellar bone (Baron, 1999). Woven bone formation is particularly responsible for the development of the skull and the maxilla (Scheuer and Black, 2000a, 2000b). Endochondral bone formation involves the differentiation of mesenchymal cells into chondroblasts, which create a cartilaginous template forming the shape of the bone. Cartilaginous templates are subsequently replaced by bone. This process proceeds in the long bones and vertebrae (Sommerfeldt and Rubin, 2001; Marks and Odgren, 2002; Downey and Siegel, 2006). A third type of bone growth combines the two above processes, whereby intramembranous ossification occurs initially, but is subsequently replaced by endochondral bone growth, as occurs in the development of the clavicle and mandible (see Scheuer and Black, 2000a, 2000b).

Modelling

Whilst the template of the long bones is formed by cartilage, this does not account for the need of the width of bones to constantly increase during growth. As such, it is necessary to remove bone tissue from some locations and deposit bone in new ones in order to meet changing mechanical demands, particularly in the shaping of the metaphyses (Parfitt, 1994; Baron, 1999; Raisz, 1999; Sommerfeldt and Rubin, 2001; Ortner, 2003). This process of skeletal development during growth is called modelling. Modelling can involve bone being deposited without requiring previous resorption (Baron, 1999). Extensive reviews of bone growth are provided by Martin et al. (1998), Scheuer and Black (2000a) and Marks and Odgren (2002). In modelling, bone is deposited and then removed. This is the opposite to the process of remodelling (see Parfitt, 2003:4).

Remodelling

Once growth has ceased, bone cells continue to function to remove existing bone and create new bone, but in contrast to modelling, the bone tissue selected for removal is targeted as structurally weakened and needing replacement. Old bone is removed and replaced with new bone in order to maintain the integrity and functionality of the skeleton. This process of skeletal maintenance is called remodelling or bone turnover.

Bone remodelling occurs within temporary units called basic multicellular units (BMU). Within each BMU, osteoclasts are created to resorb bone and new osteoblasts then follow to replace bone in the same location. The relationship between osteoclast action followed by osteoblast function in one location is known as coupling, and is an essential concept in bone biology (Parfitt, 1994, 2000; Frost, 2003; Parfitt, 2003). Remodelling occurs on a resting bone surface (quiescent), which is covered with bone lining cells (Table 3.1; see Parfitt, 1994). This process is initiated by an activation stimulus received by mesenchymal bone lining cells, which subsequently prepare a bone surface and remove the covering membrane (Parfitt, 1994; Raisz, 1999; Hauge et al., 2001; Everts et al., 2002; Parfitt, 2003). The exact mechanism of this stimulus is presently unknown but may relate to the recognition of fatigue damage or a mechanical strain stimulus (Martin, 2000; Parfitt, 2003; Pearson and Lieberman, 2004). The bone lining cells then withdraw and a local blood supply enables the differentiation of osteoclasts (Parfitt, 2003:10).

Osteoclasts resorb a pre-determined amount of bone (Parfitt, 2000; Phan et al., 2004). In the subsequent reversal phase bone lining cells enter the resorption lacuna and clean the bone surface, digesting any remaining collagen prior to osteoblastic bone formation (Everts et al., 2002). A thin layer of fibrillar collagen is excreted by the latter cells to create a smooth surface layer, which is the cement line marking the boundary of resorption (Baron, 1999; Parfitt, 2000; Everts et al., 2002; see also Frost, 1963). Figure 3.3 demonstrates scalloped bays in bone surface indicative of resorption together with cement lines in an image of archaeological bone taken with back scattered electron microscopy (BSE-SEM).

Following reversal, osteoblasts assemble and secrete unmineralised organic bone matrix (osteoid) and regulate its mineralisation (see below). Eventually the resorption cavity is re-filled with new bone tissue (Parfitt, 2000). The duration of each remodelling cycle has been estimated to take three to six months to complete (Baron, 1999), but may vary between the bone tissue types and under pathological conditions.

Whilst the mechanisms of bone remodelling are essentially the same in both bone tissue types, the structural BMU in which remodelling is carried out varies slightly between cortical and trabecular bone (Parfitt, 1994, 2003). Cortical bone remodelling is undertaken in an elongated and cylindrical structure BMU (Figure 3.4) (for detailed discussion see Parfitt, 1994). This unit

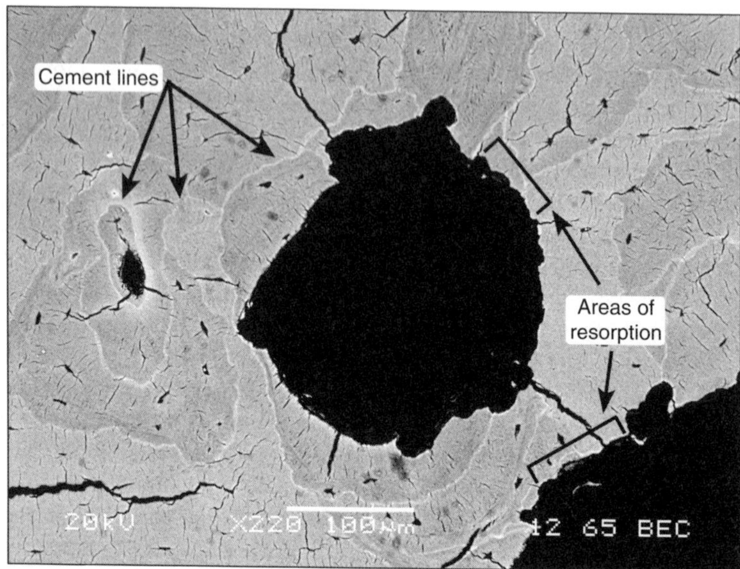

FIGURE 3.3 BSE-SEM image of bone resorption sites with scalloped borders and cement lines. Sample from adult human cortical bone. Archaeological rib, St. Bride's Lower Churchyard, London. SEM by authors, courtesy of the Museum of London.

progresses through bone aligned with the long axis of the bone. Osteoclasts are present in the front 'cutting cone'. Following this is the 'closing zone' where osteoblasts secrete osteoid. In the centre of the BMU is a blood capillary providing pre-cursor cells. The final result of a BMUs' work in cortical bone is a new osteon, with central Haversian canal and thin layer of cement marking the remodelling boundary (osteonal remodelling) (Jowsey et al., 1965).

Trabecular bone remodelling occurs in a unit which progresses over the surface of the bone (see Parfitt, 1994; Hauge et al., 2001; Parfitt, 2003). Whereas in cortical bone a circular tunnel is cut through the bone, in trabecular remodelling a trench comprising effectively a semi-circular shape is excavated while the osteoclasts travel along the surface of a trabecula. The trench is subsequently filled moving in the same manner using the cutting cone and closing zone as in cortical remodelling. The surrounding haematopoietic tissue replaces the need for a specific blood capillary for pre-cursor cells. This process is called hemi-osteonal remodelling (Parfitt, 1994, 2003) (see Figure 3.5).

Bone Mineralisation: The Extracellular Matrix (osteoid)

Osteoblast cells secrete unmineralised organic bone matrix called osteoid. This matrix contains a high amount of collagen (94%) together with a number of proteins, growth factors and cytokines required for mineralised tissue (see reviews in Termine and Robney, 1996; Sommerfeldt and Rubin, 2001; Marks and Odgren, 2002; Downey and Siegel, 2006). Various enzymes are required in order to aid

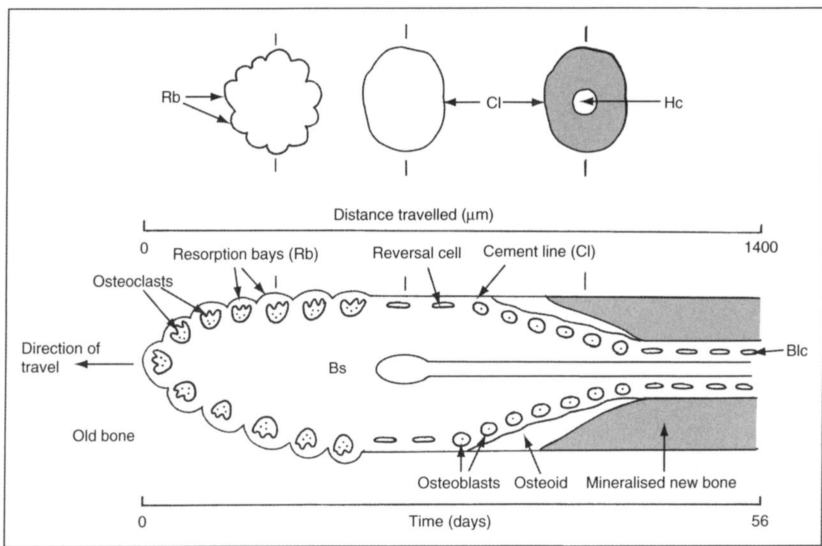

FIGURE 3.4 Schematic drawing of the process of cortical bone (osteonal remodelling). Osteoclasts resorb old bone creating resorption bays (Rb). Reversal cells (Blc) prepare the bone surface and create a smooth cement line (Cl). (Bs) blood supply at central focus of BMU. Osteoblasts secrete bone matrix (osteoid), which is subsequently mineralised. The changes occur as the remodelling unit (BMU) passes through a section of cortical bone over a period of approximately 56 days. Above the unit are cross-sectional features observable in archaeological bone, the resorption bays, the cement line and the infilling of the osteon with newly mineralised bone and Haversian canal (Hc). Courtesy of Michael Parfitt (1994), reprinted with permission from Wiley-Liss, Inc.

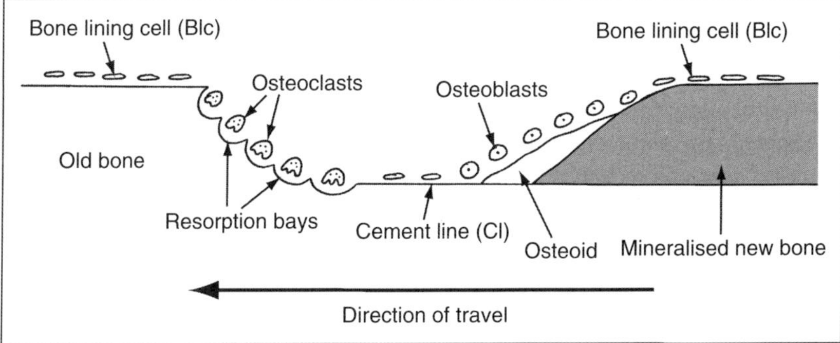

FIGURE 3.5 Schematic drawing of the process of trabecular bone (hemi-osteonal remodelling). Osteoclasts resorb old bone creating resorption bays before the formation of the cement line and secretion of osteoid. Mineralisation of new bone occurs shortly afterwards. The unit passes along the surface of a trabecula rather than remaining stationary. Courtesy of Michael Parfitt (1994), reprinted with permission from Wiley-Liss, Inc.

the proper formation of collagen strands in osteoid and to enable subsequent maturation. Amongst the elements necessary for osteoid production and collagen formation is vitamin C, which is essential for effective enzyme function (Stini, 1990; Sakamoto and Takano, 2002; see Chapter 4). In contrast, mature lamellar bone contains approximately 25% organic bone matrix, 5% water and 70% inorganic bone mineral, hydroxyapatite (Sommerfeldt and Rubin, 2001). The process of bone mineralisation is complex and to date is incompletely understood. Recent reviews are provided by Sommerfeldt and Rubin (2001) and Marks and Odgren (2002).

Within 10–15 days after osteoid secretion, mineralisation of the collagen starts to occur. It is during this period that bone rapidly attains a large amount (70%) of its mineral content. The remaining 30% is acquired subsequently over several months before becoming completely mineralised (Sommerfeldt and Rubin, 2001). In order to facilitate this rapid accumulation of bone mineral, concentrations of mineral ions are needed to create sites for initial calcification. Supplies of calcium and phosphate are available in the extracellular fluid which surrounds bone (see below), but the concentrations of these minerals are not large enough to enable calcification (Vaughan, 1975). Instead, the initiators of mineralisation are small extracellular organelles, which bud from chondrocytes or osteoblasts, and which form a matrix vesicle containing phosphatases, phospholipids and calcium ions. Disintegration of the vesicles introduces these minerals to the organic osteoid, which enables mineral crystallisation to begin. During this process, collagen fibrils and glyocoproteins act to determine the organisation and structure of bone mineral crystals (see review Sommerfeldt and Rubin, 2001). The mineralisation process is particularly affected by adequate amounts of vitamin D, without which osteoid will remain unmineralised (Chapter 5).

Tooth Formation and Mineralisation

Detailed reviews of the biological processes of tooth formation are beyond the scope of this chapter. Both Hillson (1996) and Jones and Boyde (1999) are useful sources for this discussion. Pathological alterations of the dentition associated with the metabolic bond diseases are briefly highlighted in specific chapters on pathology in this book, including vitamin C deficiency (Chapter 4), vitamin D deficiency (Chapter 5) and fluorosis (Chapter 9). However, the relationship between the many factors that can result in defects in tooth formation and mineralisation needs further detailed investigation.

REASONS FOR REMODELLING

It is of functional importance that the skeleton is strong and able to resist mechanical failure (fracture) (see Einhorn, 1992; Martin et al., 1998; Ruff, 2000; Currey, 2002; Pearson and Lieberman, 2004). Mechanical loading has a significant impact on the skeleton and is likely to play an important role in the stimulus of remodelling (Turner, 1998; Martin, 2000; Frost, 2003; Martin, 2003). Bone is subject to fatigue damage, with the attainment of micro-cracks and fractures evident at the structural level (see Schaffler et al., 1995; Burr et al.,

1997; Martin, 2003; Grynpas, 2003). Recognition of the accumulation of such fatigue damage and attempts to repair structurally weakened bone may act to trigger bone remodelling (Martin, 2000; Martin, 2003:127; Qui et al., 2005). In particular, cellular communication between osteocytes, osteoblasts and bone lining cells may stimulate remodelling to enable the bone to respond to such loading and micro-damage (Martin, 2000; Frost, 2003; Martin, 2003:125; Pearson and Lieberman, 2004). The effects of micro-damage on bone strength may have greater implications with aging, and research into this relationship is developing (Schaffler et al., 1995; Burr et al., 1997; Diab et al., 2006).

An important concept long investigated in anthropological research is that mechanical loading can alter a bone's shape, and this is often designated as 'Wolff's Law' or more recently, bone functional adaptation (as reviewed by Ruff, 2000; Pearson and Lieberman, 2004; Ruff et al., 2006). This premise is of particular interest in the study of various metabolic bone diseases in which bone shape can be altered by mechanical loading in response to an underlying pathological condition, such as in vitamin D deficiency (see Chapter 5), or alternatively where bone shape may alter over the life course via mechanical loading with potential for this to alter the susceptibility to fractures as in osteoporosis (see Chapter 6). Remodelling will also occur due to various physiological demands, such as hormonal status (e.g. oestrogen deficiency), pathological conditions, physiological status (e.g. pregnancy) and nutritional state (Box Feature 3.1). It is significant that various factors in the burial environment can affect bone tissue with potential impacts on the recognition of the processes of modelling and remodelling in bioarchaeological samples. These problems are discussed further in Chapter 2 and see also Bell (1990).

Box Feature 3.1. Bone Biology in Context of the Life Course

The processes of modelling and remodelling clearly demonstrate that bone is not a static tissue. The mechanisms influencing bone do not remain the same throughout the life course, and it is the nature and variation of these processes which form the essential base for the understanding of both normal and pathological tissue and which shape subsequent interpretations. Variation in bone morphology and mineral content are clearly evident between childhood and adulthood. There are further sources of alteration in bone tissue between young adulthood, with the attainment of peak bone mass (discussed further in Chapter 6), and the onset of aging, with effects on the actions of bone cells and the efficiency of mineral metabolism.

It is becoming clear that episodes of ill-health and growth disturbances during childhood may have lasting effects into adulthood, including stunted growth, increased morbidity, delayed mental and physical development with potential impacts on intellectual performance and the ability to work (Ulijaszek, 1998:417–418;

Continued

Cameron and Demerath, 2002; Kuh and Hardy, 2002; Dewey, 2005). A cyclical development of malnutrition in pregnant mothers may also affect offspring producing low birth-weight infants with poor health (Krishnamachari and Lyengar, 1975; Cameron, 1996; WHO, 2002:13). There may be complications in the clear identification of such processes within past populations owing to the nature of the demographic composition of the 'non-survivors'. The challenges in interpretations of the presence or absence of pathological changes have been reviewed by Wood et al. (1992; see Chapter 2). However, increasing consideration of these concepts could prove valuable in developing a better understanding of demographic trends of disease. Further analysis of health variation across the life course and between different comparative frameworks may provide an important means to improving interpretations both in modern and past contexts.

BONE BIOLOGY IN FRACTURE HEALING

Fractures occur in bone via two mechanisms. One is traumatic, in which an accidental overload of the normal range of bone loading occurs, whether through a fall or intentional force applied to the bone. The second is a pathological fracture, occurring when normal loads are applied to a bone that is substantially weakened by an underlying pathological condition. These latter fractures are particularly evident in metabolic diseases, including osteoporosis, osteomalacia and Paget's disease. A brief review of normal fracture healing is presented here and summarised in Table 3.2.

Fracture healing involves the combined actions of endochondral and intramembranous bone formation to enable a rapid response to bone recovery (Table 3.2) (see especially McKibbin, 1978; Sevitt, 1981: Chapters 4–7; Martin et al., 1998; Salter, 1999: Chapter 15; Gerstenfeld and Einhorn, 2003). Osseous response to fracture can occur very quickly with some bone formed at approximately four weeks after a fracture, although this may not be easily visible during macroscopic examination. A range of factors will affect the rate of healing, including the type and location of the fracture, and the alignment or degree of overlap of bone fragments. Adequate blood supply at the fracture site is essential for healing as is the supply of calcium and a good nutritional status. The availability of treatment to reduce and re-align fractured bones may be a significant factor affecting healing as is the potential for bacterial infection to complicate open (compound) fractures (see further Chapter 7). These factors are discussed in bioarchaeological contexts in Grauer and Roberts (1996) and Lovell (1997a).

Various deficiency states will impede fracture healing and these are reviewed by Sevitt (1981:52–63). For example, individuals who are vitamin C deficient (scurvy, Chapter 4) are likely to exhibit delayed fracture healing. Cartilage proliferation is not affected but the osteoid required for initial woven bone repair and ossification of the cartilage framework will fail to form owing

TABLE 3.2 Mechanisms of Fracture Healing

Phases of healing and actions	Mechanism	Phase duration
Inflammatory phase		
Immobilise fractured bone Activate cells required for bone repair	Pain and swelling Mesenchymal tissue differentiation Thickening of periosteum aids differentiation mesenchymal and osteoprogenitor cells	3–7 days
Haematoma formation, clotting broken blood vessels	Cytokines and growth factors aid climate for cell differentiation	
Reparative phase		
Formation of periosteal callus (intramembranous ossification)	Osteoblasts differentiated from cells in periosteum form woven trabecular bone to bridge the external fracture gap at periosteum	1 month
Formation of medullary callus (intramembranous ossification)	Osteoblast differentiation from bone marrow. Formation of initial bone union at the medullary cavity	
Increase in vascular proliferation	Facilitates cell differentiation and supply of minerals	
Provisional callus formation	Cell differentiation into chondroblasts. Formation of cartilage framework. Bridging fractured bone ends Template for endochondral ossification Initial callus and woven bone created	
Bone callus formation	Continuing calcification, greater rigidity. Bony callus achieving bone union	
Remodelling phase		
Bone is mechanically strong and functional but is greater in mass than original shape	Remodelling to restore shape and structure Removal of periosteal and medullary calluses. Replacement of woven bone and cartilage by lamellar bone	Continual
Modelling drifts	Attempts at increasing mechanical efficiency by removing excess bone mass, or malalignment	

Sources: Sevitt (1981:40), Martin *et al.* (1998:69–70), Li *et al.* (2002), Gerstenfeld and Einhorn (2003).

to defective collagen synthesis. Clinical studies have demonstrated increased retention of vitamin C in individuals having suffered trauma and increased dietary intakes are also required to aid soft tissue repair (see review in Sevitt, 1981:61). Vitamin D deficiency is a significant factor in inefficient fracture healing. This hormone (see Chapter 5) is vital for the mineralisation of osteoid. In active deficiency states osteoid in the fracture callus is produced normally but is poorly mineralised. The fissure fractures evident in osteomalacia (pseudofractures) may show some evidence of fracture healing, but the callus bridges comprise irregular and poorly mineralised trabeculae (see Sevitt, 1981:61; and Chapter 5). Recent studies have demonstrated that fracture healing in individuals with osteoporosis may be delayed, owing to the effects of oestrogen deficiency and calcium deficiencies (Namkung-Matthal et al., 2001; Chapter 6). Immobilisation as a means of fracture healing may further lead to localised bone loss and osteopenia (see Chapter 7).

MINERAL METABOLISM DURING LIFE

Growing body mass and maintenance during life are dependent on the dietary supply of minerals such as calcium, phosphate and magnesium and the complex management and utility of these minerals. These minerals are absorbed by the intestine in quantities dictated by the proportions available in the diet and influenced by the efficiency of intestinal absorption (see Lemann and Favus, 1999:63). Proportions of the minerals can be stored in the skeleton, in small amounts in the extracellular fluid (see below) and also in small supplies at the margin of the bone lining cells and quiescent bone surfaces (see discussion by Parfitt, 2003). Minerals not utilised by the endocrine system are excreted by kidney function and faecal excretion. Dietary intake needs to be maintained above the level of renal mineral excretion. Detailed discussions of the mechanisms of mineral homeostasis can be found in Wasserman (1997), Civitelli et al. (1998), Broadus (1999) and Heaney (1997a, 1997b, 2002).

Extracellular Mineral Metabolism

During life, bone is covered by blood plasma and extracellular fluid, which serve to transport oxygen and carbon dioxide. Concentrations of various minerals also form part of these fluids and facilitate the transfer of minerals between bone and the extracellular fluid (Heaney, 1997a, 1997b). PTH and vitamin D (1,25$(OH)_2$D) maintain tight control of the mineral concentrations in the fluid. Without such regulation, serum amounts of calcium would fluctuate to the extremes of hypo- and hypercalcaemia with potentially fatal consequences during periods of feeding and fasting (see Heaney, 1997a:487). Both serum calcium and phosphate are necessary to maintain the normal mineral ion product, facilitating skeletal mineralisation, as well as helping to maintain the stability and permeability of soft tissue plasma membranes (Heaney, 1997a).

Calcium is a vital element in the human diet and it forms a principal component of the skeletal system and plays an important role in extensive cellular communication and activation of molecular catalysts (Heaney, 2002). Calcium present in the extracellular fluid influences many physiological and biological functions, ranging from aiding the process of blood clotting in the soft tissues to initiating the mechanisms of muscle contractions (Stini, 1990; Civitelli et al., 1998; Miller et al., 2001; Heaney, 2002; Chapter 7, Table 7.3).

There is a delicate balance between the amount of calcium in dietary intake, calcium utilised by the endocrine system and calcium subsequently lost during urinary and faecal excretion (Heaney, 1997b:487). Dietary calcium intake must remain greater than those amounts lost in order to maintain a healthy supply of skeletal and extracellular calcium. The recommended daily amounts of calcium are discussed in Chapter 7 in relation to the dietary causes of osteopenia. Following intestinal absorption, calcium enters the blood serum and extracellular fluid. However, some of the calcium exits this system and is passed into the kidney, where it is either excreted, or is re-absorbed and following filtration subsequently re-enters into the extracellular fluid for re-use as outlined in Figure 3.6. It is estimated that in order to maintain the extracellular calcium balance, approximately 98% of the serum calcium filtered through the kidney must be re-absorbed (Hoenderop et al., 2000). A deficiency of calcium will contribute to osteoporosis and can also underlie other deficiency disease states.

Dietary intakes of phosphate and magnesium are also essential in human life. Both phosphate and magnesium are constituent parts of most plant and

TABLE 3.3 Summary of Adequate and Minimum Dietary Intakes of Calcium Phosphate and Magnesium and the Amount of Each Mineral Which is Absorbed by the Intestine

Mineral	Normal range of dietary intake (mg/day)	Amount absorbed by intestine (%)	Low levels of intake causing zero intestinal absorption and greater losses via excretion (mg/day)
Calcium	1000	30*	200
Phosphate	775–1860	60–80	310
Magnesium	168–720	35–40	48

Source: Lemann and Favus (1999:64–67).

**Calcium levels are subject to variation and are especially dependent on vitamin D availability.*

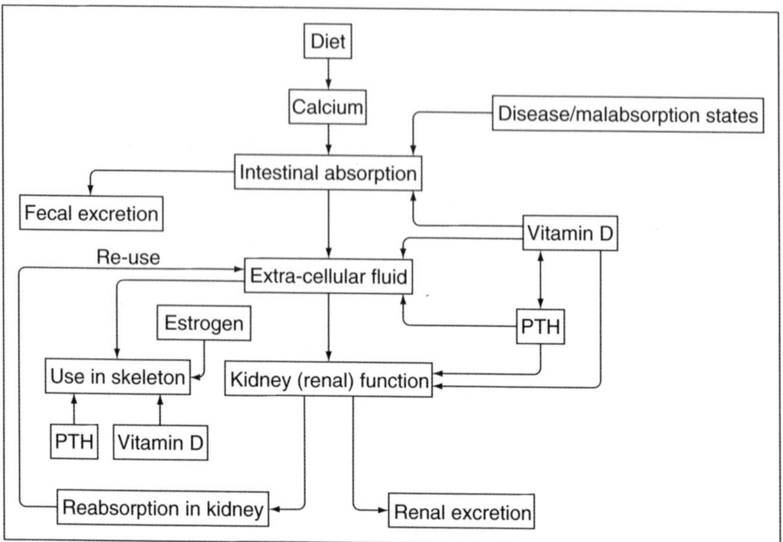

FIGURE 3.6 Flow diagram demonstrating the principal mechanisms of human calcium metabolism.

animal cells and adequate intake and absorption can be usually attained from most diets (see Lemann and Favus, 1999:65–66). Table 3.3 summarises the normal range of dietary intake required for adequate intestinal absorption of the minerals. Phosphate is important in bone remodelling as increased amounts can inhibit bone resorption (see Peck and Woods, 1988:20); therefore action to remove phosphate via kidney excretion can result in resorption activation. This mechanism is under the control of the PTH (see below).

Additional minerals required by the skeleton but in lesser amounts than those above include fluoride, sodium, potassium, strontium and citrate. A review of the role of these variables in human mineral metabolism is provided by Vaughan (1975).

Vitamin D is a pro-hormone which is essential in the skeleton. Vitamin D can be obtained from exposure of the skin to sunlight or from dietary intake. These sources produce inactive pre-vitamin D_3 (cholecalciferol), which requires conversions in the liver and kidneys to become active vitamin D. These processes are reviewed in detail in Chapter 5. It is the active form of vitamin D that can be used by the body in the metabolism of calcium and phosphorous (Mankin, 1974; Francis and Selby, 1997; Hochberg, 2003; Holick, 2003). Active vitamin D affects the intestinal absorption of calcium and phosphate, enabling the transfer of calcium into cellular components (see Heaney, 1997a; Wasserman, 1997; Civitelli et al., 1998:169). If there is a deficiency of vitamin D, or if the intestine does not respond to this hormone, the amount of dietary calcium absorbed by the intestine will decrease, lowering concentrations of calcium and phosphate in the extracellular fluid (Lemann and Favus, 1999). Vitamin D is also essential in facilitating skeletal mineralisation. Without vitamin D, osteoid will accumulate and remain unmineralised. The skeletal effects of this are discussed in Chapter 5. Vitamin D seems to play little role in affecting magnesium absorption, but an active vitamin D deficiency will reduce the intestinal absorption of phosphate (Lemann and Favus, 1999:66).

The hormone oestrogen has an important influence on both the female and male skeletal systems. Oestrogen has recently been implicated in subperiosteal apposition during growth in males and females (Bilezikian, 2006) and during adulthood, oestrogen affects bone maintenance through variously inhibiting bone resorption (Reid, 1999; Pacifici, 2001). Oestrogen can also act on osteoblasts and other factors such as cytokines. As such, rapidly occurring oestrogen withdrawl following the menopause in females and prolonged declines with aging in both sexes have direct impacts on bone remodelling, and makes an important contribution to increased bone loss and fracture risk in both sexes (see discussion in Chapter 6 as well as Civitelli et al., 1998; Reid, 1999; Orwoll and Klein, 2001; Pacifici, 2001; Riggs et al., 2001; Bilezikian, 2006).

PTH is secreted by the parathyroid glands and has a complex role in maintaining calcium homeostasis and bone remodelling. It has a particularly sensitive involvement with calcium and phosphate handling in the kidney as well as in the synthesis of vitamin D, in order to maintain a constant serum calcium level (Silver and Naveh-Many, 1997; Broadus, 1999; Jüppner et al., 1999; Hock, et al., 2002; Houillier et al., 2003). Prolonged PTH action can increase the number and activity of osteoclasts, thus increasing bone resorption, in order to stimulate the release of calcium and phosphorous from bone (Jüppner et al., 1999:84). However, actions by the PTH to increase bone resorption will also increase bone formation via increases in remodelling (see Peck and Woods, 1988:21). The inter-relationship between PTH, vitamin D and calcium metabolism is discussed further in Chapter 5. Pathological conditions affecting PTH are discussed in Chapter 9.

Additional factors affecting mineral metabolism and bone cell function include calcitonin, which is a polypeptide hormone that has a significant inhibitory effect on osteoclasts (see Downey and Siegel, 2006), and can block proliferative responses of the PTH further preventing osteoclast stimulation (Peck and Woods, 1988:22). High levels can prevent bone resorption which also has the effect of preventing the release of calcium and lowering serum calcium levels (see Downey and Siegel, 2006:83). The thyroid hormone is a stimulator of trabecular bone remodelling resulting in increases in both resorption and formation (Downey and Siegel, 2006:84) and vitamin A stimulates bone resorption, potentially through modifying osteoblast activity (Peck and Woods, 1988:27).

CONCLUSIONS

The metabolic bone diseases are inherently linked with cellular defects in the processes of bone resorption, formation and mineralisation. As such, accurate interpretations of skeletal pathological defects are enhanced by a detailed realisation of the processes evident at the underlying bone cell level. Brief reviews of bone growth (modelling) and maintenance (remodelling), and cell function have been discussed in this chapter. An introduction to mineral metabolism highlights the complex inter-relationship of calcium metabolism with other minerals and at the organ and hormonal level as well. Throughout this chapter, references to recent literature are highlighted, which will substantially enhance the understanding of bone biology and its relevance within many disciplines of bioarchaeology.

Chapter 4

Vitamin C Deficiency Scurvy

CAUSES OF VITAMIN C DEFICIENCY

Most mammals can produce their own vitamin C (ascorbic acid). However, humans, other primates, guinea pigs and the fruit-eating bat *Pteropus medius*, need to obtain vitamin C from dietary sources (Stone, 1965). In these animals a range of disease processes (discussed in detail later, and described in Box Feature 4.3) result from a deficiency, and these are referred to as scurvy. It has been suggested that in humans, and the other animals that need to ingest vitamin C, the ability to synthesise ascorbic acid was lost during evolution (Nishikimi and Udenfriend, 1977). Biochemical enzyme systems responsible for the synthesis of vitamin C vary across the animal kingdom; in higher vertebrates, such as mammals, the site of synthesis is the liver, but in lower vertebrates, for example reptiles, synthesis occurs in the kidney. Birds appear transitional, with variation in the site of synthesis. The change in the site of synthesis probably occurred after vertebrates began to evolve temperature regulatory mechanisms, changing from cold-blooded to warm-blooded species (Stone, 1965). Stone (1965) discusses the evolutionary development of the inability to synthesise vitamin C in more detail. As the need for dietary supply of this vitamin is an inborn error of metabolism and universal in humans (Nishikimi and Udenfriend, 1977), the condition has frequently been omitted from the clinical literature on metabolic bone diseases (e.g. Avioli and Krane, 1998) and until recently has been neglected by many anthropologists.

Sources of Vitamin C

Vitamin C is available from a wide range of fresh fruits and vegetables, and small amounts of vitamin C are also available from milk, meat and fish (Fain, 2005; see Table 4.1). Various plants have been suggested as a cure for scurvy from the writings of Pliny the Elder (first century AD) onwards (Hughes, 1990), and knowledge of a range of plants that could be used in this way was probably far more ancient than this. Scurvy grass was frequently referred to in texts from the fifteenth century onwards, and Carpenter suggests that the term was used for a variety of plants including 'spoonwart *Cochlearia curiosa*, or *officinalis*, or cresses such as *Cardamine glacialis* and hirsute' (Carpenter,

TABLE 4.1 Summary of the Vitamin C Content in a Range of Foods

Food source	Raw vitamin C content per 100 ml (liquid)/ 100 mg (solid)	Stored vitamin C content per 100 ml (liquid)/100 mg (solid)	Cooked vitamin C content per 100 ml (liquid)/ 100 mg (solid)
Angelica* (*Angelica sylvestris*)	14–100	nd	0
Apple, crab	8	nd	nd
Apple, desert	6	3.9	0.2 (boiled)
Avocado	8	nd	nd
Banana	9–13	7 (dried)	nd
Bean/grain sprouts*	30–200	na	nd
Bilberry/whortleberry (*Vaccinium myrtillus*)	3	nd	nd
Birch leaves (*Betula* sp.)	460	nd	nd
Black crowberry* (*Empetrum nigrum*)	30	nd	nd
Blackberry (*Rubus* sp.)	6	nd	nd
Bread fruit	29	nd	6.1 (boiled)
Cereals (un-sprouted)	0	0	0
Cloudberry* (*Rubus chamaemorus*)	80	nd	nd
Corn, sweet, yellow (maize)	6.8	0 (cornmeal and tortillas)	6.2 (boiled)
Cranberries (*Vaccinium* sp.)	5–10	nd	nd
Eggs	0	0	0
Fig	2	1.2	1 (canned)
Fish flesh (cod, char)	0.5–2	na	0 (boiled)
Goosegrass* (*Galium aparine*)	78	nd	nd
Honey	0	na	0

Continued

TABLE 4.1 (*Continued*)

Food source	Raw vitamin C content per 100 ml (liquid)/ 100 mg (solid)	Stored vitamin C content per 100 ml (liquid)/100 mg (solid)	Cooked vitamin C content per 100 ml (liquid)/ 100 mg (solid)
Liver, beef	1.3	na	0.7 (fried)
Liver, lamb	4	na	4 (braised)
Liver, seal	18–35	na	14–30 (boiled)
Meat	0–1.8	na	0–0.5 (boiled)
Milk, cow (mature)	2 (raw)	1.4 (evaporated)	1.5 (pasteurised)
Milk, goat/sheep	2–3	0 (cheese)	nd
Milk, guinea pig	12	nd	nd
Milk, human and camel (mature)	3–6	–	nd
Mustard*	70	nd	nd
Nettle juice* (*Urtica* sp.)	160	nd	nd
Nuts, hazel	5	na	nd
Onions	7–30	4.5 (2 months)	0.4 (boiled)
Orange	53	nd	nd
Parsley	150–170	122 (dried)	nd
Parsnip	17	nd	10–17 (boiled)
Peppers, sweet green	80–200	nd	41 (boiled)
Peppers, hot, green	243	2–6	34 (canned)
Potatoes	20–30	8 (8–9 months)	7.4 (peeled/boiled)
Raspberries*	30	nd	8.7 (canned)
Rose hips*	1000–7000	nd	nd
Rowanberries* (*Sorbus* sp.)	80	nd	nd

Continued

TABLE 4.1 *(Continued)*

Food source	Raw vitamin C content per 100 ml (liquid)/ 100 mg (solid)	Stored vitamin C content per 100 ml (liquid)/100 mg (solid)	Cooked vitamin C content per 100 ml (liquid)/ 100 mg (solid)
Sauerkraut	150	10–15 (1 month)	nd
Scurvy grass* (*Cochleare officinalis*)	200	nd	nd
Seal flesh	0.5–3	nd	0.5–2.5
Spruce pine needles*	65–200	<0.5 (fermented infusion)	nd
Sweet potatoes	20–30	nd	13 (boiled)
Tomatoes, red	19–24	39 (sun dried)	23 (canned)
Turnip, greens	60	nd	30 (boiled)
Turnip, root	21	nd	10–17 (boiled)
Watercress* (*Nasturtium officinale*)	60	nd	nd

Sources: Grewar (1965), Norris (1983), Holck (1984), Crawford (1988), USDA National Nutrient Database for Standard Reference (Release 19) and WHO (1999b). *Notes*: na, not applicable; nd, no data available. Values provided are for unfortified foods. Exact vitamin C content of the fruit/vegetable may vary according to variety, growing conditions/methods and ripeness. Once harvested treatment and storage of food will impact on vitamin C content. There are various methods for drying foods and the technique used will determine vitamin C content. Latin names are provided where colloquial names for plants may vary.

*Items used in traditional scurvy remedies.

1986:14). Writings relating to the voyages of Vasco de Gama also make it clear that the use of oranges for treating or preventing scurvy was known from an early date. It is likely that in the past a wide range of plant 'cures' for scurvy was known, particularly in regions where there may have been a seasonal shortage of readily available foods containing vitamin C. Holck (1984) discusses the vitamin C content of some of the plants that were used to treat scurvy in Scandinavia in the past. Norris (1983) has provided detailed information on the vitamin C content of a range of foods in both fresh and cooked states, and in addition Crawford (1988) considers the impact of storage on the availability of vitamin C in a range of foods.

As can be seen from Table 4.1, a number of animal products contain high amounts of vitamin C likely sufficient to prevent the development of scurvy, and these sources may have prevented scurvy developing in a number of past communities. However, processing of food resources can affect the availability of vitamin C. For example, pasteurisation will reduce the vitamin C content of cow's milk (Grewar, 1965). Prior to the domestication of animals, which would provide an alternative source of milk allowing artificial feeding, scurvy would be extremely rare in young infants (Box Feature 4.1). The availability of heat-treated milk and propriety food (such as the early manufactured 'infant foods') caused nutritional problems, including cases of scurvy, towards the end of the nineteenth century (Ratanachu-Ek et al., 2003:S735). In this instance, it was the children of wealthier families in London in the 1870s that suffered adverse consequences from eating such foods (Pimentel, 2003). The use of proprietary foods was also a problem for children in the United States (Rajakumar, 2001). Due to the high vitamin C content of human milk, scurvy is unlikely to develop in societies that have not adopted artificial feeding. Vitamin C is sensitive to heat, oxygen and ultraviolet radiation (Fain, 2005). Frequent consumption of fresh foods is important in preventing scurvy, as vitamin C is water-soluble and cannot be stored by the body. Symptoms of scurvy can become apparent between 29 and 90 days after vitamin C is removed from the diet (Pimentel, 2003).

Box Feature 4.1. Scurvy and Weaning

As human infants are unable to synthesise vitamin C, human milk produced by healthy women receiving adequate nutrition contains good quantities of this nutrient. Grewar (1965) documents levels of vitamin C that occur in a range of milk types (including rat and whale); human milk contains much higher levels than those found in fresh cow's milk (see Table 4.1). Early investigations of scurvy demonstrated that infantile cases rarely occurred if infants were breastfed (Barlow, 1894; Follis et al., 1950; Shorbe, 1953). A more recent study by Severs et al. (1961) found an increase in cases in infants as breastfeeding practices declined in Canada. However, the figures provided in Table 4.1 assume that the mother is not vitamin C deficient herself. Deficiency in the mother will reduce the vitamin C content of breast milk (Fain, 2005).

Thomas Barlow (1894) placed the age at which the first symptoms of scurvy became apparent in juveniles at eight months, and clinical investigations by Follis et al. (1950) suggest a similar age. These researchers also identified a very early case of scurvy that they refer to as a case of 'congenital scurvy'. Although early development of scurvy is rare, a number of cases have been reported. Jackson and Park (1935), Hirsch et al. (1976) and Bhat and Srinivasan (1989) noted cases occurring during the first month of life in children whose mothers were severely malnourished.

Whilst cases of scurvy in very young infants clearly exist in the literature, it is likely that such occurrences will remain relatively rare. However, there are circumstances where nutritional disturbances or early weaning are likely to have occurred in past communities. When examining pathological changes in young infants consideration should be given to the possible effects of malnourishment of mothers. An understanding of the occurrence of scurvy has a significant potential to contribute to debates on weaning and the diets of children.

Despite losses during food processing, the potato was an important source of vitamin C for the post-Medieval working classes of Ireland and England (Crawford, 1988; Cheadle, 1878). In Scandinavia the potato was referred to as the 'Nordic orange' until relatively recently (Holck personal communication). When successive potato crops were destroyed by blight during 1845–8, the resulting famines caused widespread outbreaks of scurvy. Natural disasters that impact on local food supplies (see Figures 4.1 and 4.2), such as crop disease,

FIGURE 4.1 A herd of goats amongst a swarm of desert locusts near Kaedi, Mauritania. Livestock are in competition with the insects for available grazing land. Courtesy of Food and Agriculture Organization (FAO) of the United Nations: FAO/G.

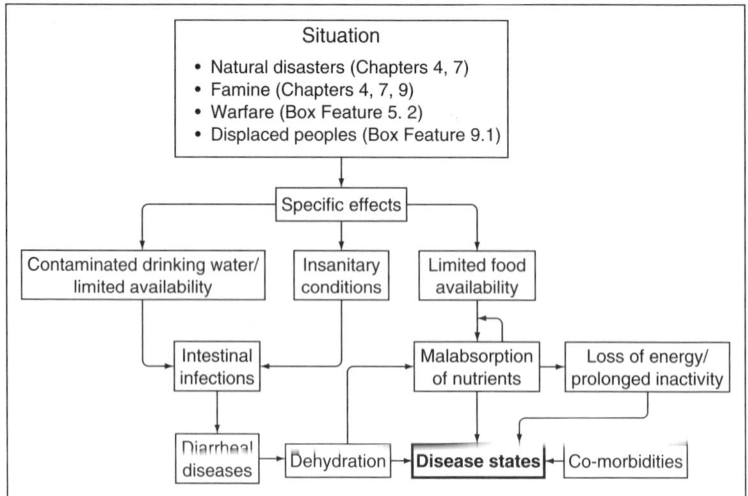

FIGURE 4.2 The manner in which situations may contribute to the development of metabolic bone diseases and the way in which the various factors interact.

or damage through drought, earthquakes or pests are very likely to result in an array of malnourishment states and subsequent dietary diseases including scurvy. The scale of the problems that can occur following a natural disaster can be seen following the earthquake that struck the north of Pakistan and India in 2005 (Brennan and Waldman, 2006). As discussed later, deficiencies of micronutrients such as vitamin C, folate and iron, frequently co-occur and the possibility of more than one condition being present in any individual should always be considered. Although at the time of writing no cases of scurvy had been reported from this region of Pakistan and India, scurvy is highlighted as a recurrent problem in such situations. Outbreaks have occurred in Somalia (1982, 1985), Sudan (1984, 1991), Ethiopia (1989), Nepal (1992) and Kenya (1994) (WHO, 1999b, 2006b).

THE ROLE OF VITAMIN C

Vitamin C has a number of valuable roles in the body, including allowing the maintenance of collagen formation (Jaffe, 1972). The exact role that vitamin C plays in collagen synthesis is still debated (as discussed by Ortner, 2003:383) but defects in synthesis have a number of important consequences including; haemorrhage, delayed wound healing, petechiae, purpura and lack of bone formation in juveniles (Bourne, 1942a; Pimentel, 2003). Petechiae are individual dark spots caused by bleeding into the skin and purpura is the term applied to a rash of such spots. Pain and weakness in the limbs reported by those with scurvy could be related to reduced vitamin C dependent cartine synthesis, which is required to produce energy in muscles. Vitamin C also has a role in a number of other biochemical pathways (Akikusa et al., 2003). It is closely linked to blood formation and the metabolism of iron and folate (Pangan and Robinson, 2001; Cheung et al., 2003) and as a result, individuals with vitamin C deficiency are more likely to develop anaemia. Haemorrhage that occurs as a result of scurvy may also cause sufficient blood loss to contribute to the development of anaemia (Pimentel, 2003). A number of studies have demonstrated that anaemia frequently occurs when scurvy is present (e.g. Levy, 1945; Cutforth, 1958; Clark et al., 1992; Fain, 2005).

Vitamin C has an important role in maintaining adequate immune function and this is discussed in detail by Jacob and Sotoudeh (2002). The mechanisms behind this role are still poorly understood, but vitamin C has two broad functions: assisting in neutralising or destroying pathogens, and producing various protective antioxidants. A number of studies have demonstrated that individuals who are deficient in vitamin C are more prone to infectious conditions and recovery is impaired (e.g. Bartley et al., 1953).

Vitamin C Requirements

The recommended daily allowance (RDA) is widely quoted as being 30 mg/day of vitamin C, although a review by Cheung et al. (2003:248) found that scurvy could be prevented with an intake of between 6.5 and 10 mg/day. However, an individual's requirement will differ for a variety of reasons. For example, pregnancy may increase the required amount of vitamin C to 70 mg/day and lactation may further raise it to 95 mg/day (Pimentel, 2003). The exact amount of vitamin C required to prevent the development of scurvy is variable and depends on factors such as age and the type of physical activities undertaken. It has been suggested that those undertaking physically demanding work will have higher vitamin C requirements due to increased catabolism of tissue proteins and ascorbic acid (Norris, 1983).

Healing of scurvy is relatively rapid once vitamin C is re-introduced into the diet, and clinical improvement can be present in 48 hours (Greenfield, 1990:237). Many of the symptoms of scurvy can be resolved in three to five days and most physical signs within two weeks (Pimentel, 2003). In children, there may be complete obliteration of histological signs of scurvy after three months of receiving vitamin C treatment (Follis et al., 1950).

CONSEQUENCES OF SCURVY

Scurvy has a number of effects but the severity of the symptoms experienced will depend on the age of the affected individual and the length of the deficiency. Some of the first noticeable symptoms are tiredness and lethargy with development of musculoskeletal pain and weakness (Cutforth, 1958; Estes, 1997; Akikusa et al., 2003; Ratanachu-Ek et al., 2003). Such symptoms will be found in both adults and children.

The principal defects in scurvy are haemorrhagic manifestations. These arise because of damage to blood vessels, which have a greatly weakened structure due to impaired collagen formation, leaving them susceptible to damage even following very minor trauma. Sub-periosteal haemorrhages can occur in a variety of locations throughout the skeleton and can result in stripping of the periosteum from bones.

Scurvy also causes depressed osteoblastic activity, resulting in reduced or arrested deposition of osteoid (Ortner and Ericksen, 1997; Ortner et al., 2001; Brickley and Ives, 2006), the framework upon which mineralised bone is formed. As a result, osteopenia is frequently reported in individuals with scurvy (Fain, 2005). The exact expression of such changes will be slightly different in adults and children, as in adults there will only be remodelling, but in children modelling will also be taking place (see Chapter 3 for further information on these processes). Experimental work undertaken in guinea pigs by Bourne reported that there was a delay in the differentiation of cells into osteoblasts in animals with scurvy (Bourne, 1942b). It has since been reported that in children there is an inability of osteoblasts to produce the osteoid seam (Fain, 2005).

Consequences for Adults

Reddening and swelling of the gums, have long been regarded as one of the most characteristic and easily recognisable changes associated with scurvy. As the condition progresses, gums can turn a purple colour with the surface degenerating into a slimy film that may have some ulceration. At this stage the gums also sag away from the teeth (Bartley et al., 1953), which become loose due to degeneration of the connective tissues that hold them in place. The end result of these processes can be ante-mortem tooth loss (AMTL), and the single-rooted teeth are particularly susceptible (Danzeiser Wols and Baker, 2004). However, not all individuals with scurvy will display these changes, and gum changes are unlikely to develop in edentulous individuals (Cutforth, 1958). It has been suggested that gum changes may be more common in individuals who had gingivitis before developing scurvy (e.g. Pangan and Robinson, 2001), but more research is required to establish if there is a clear link between these factors.

Scurvy can result in bleeding into the joints causing haemarthrosis; swelling and pain at the joint. Trauma is one of the most frequent causes of haemarthrosis (Pangan and Robinson, 2001, give a full discussion of the range of causes), but in scurvy bleeding can occur in the absence of trauma. The hips, knees and ankles, have been noted to be especially susceptible in the modern population (Fain, 2005). Blood leaking into surrounding soft tissues can cause pain. Swelling around limb joints may occur, and this is caused by blood leaking into the surrounding soft tissue. In severe cases, these changes may result in limping or an inability/refusal to walk (Ratanachu-Ek et al., 2003). Such changes are also observed in children. Cutforth (1958) reported observing a number of individuals with scurvy who experienced pain, which is described as being 'rheumatic' (1958:454). Such a description could have contributed to some cases of scurvy being overlooked by medical practitioners in the past.

Consequences for Children

Changes of the gums can also occur in children and haemorrhages can be very apparent around newly erupted teeth. Sub-periosteal haemorrhage may be particularly marked in children as the periosteum is less firmly attached than in adults, but in children blood rarely enters the joint cavities (Jaffe, 1972:455). As a result of inflammation associated with bleeding, bone surface porosity may develop in a number of skeletal areas (Ortner and Ericksen, 1997:213; Brickley and Ives, 2006).

In addition to the development of osteopenia of existing tissue owing to defects in osteoid synthesis, in children there will also be a virtual cessation of new bone formation (also found in adults). Weakening of bone structures in juveniles can lead to the development of metaphyseal fractures (Milgram, 1990:844).

SCURVY IN THE MODERN PERSPECTIVE

Cases of scurvy are comparatively rare in the modern population, although the occurrence of sub-clinical and clinical vitamin C deficiency is reportedly increasing (Akikusa et al., 2003; Fain, 2005). Early in their development many conditions can have a sub-clinical stage, when no clinically recognisable signs are present. Although foodstuffs containing vitamin C are now widely available in countries such as the United States, 30% of the adult population have less than the recommended daily intake (RDI) of vitamin C (Akikusa et al., 2003). Many elderly individuals, people with alcohol dependencies, men who live alone and children who adopt very restricted diets have sub-clinical vitamin C deficiency and may even develop scurvy.

Outbreaks of scurvy can become common when socio-agricultural systems have been disrupted due to conflict. There have been a number of reports of scurvy occurring in refugee populations (e.g. Toole, 1992; WHO, 1999b; Stevens, 2001; Prinzo and de Benoist, 2002). Most health workers have little experience of the condition and so there can be a delay in the diagnosis or underestimate of scurvy (e.g. Shorbe, 1953; Akikusa et al., 2003; Ratanachu-Ek et al., 2003). A recent report by Cheung et al. (2003) highlighted the problems faced by refugees living in the mountains of Afghanistan where scurvy has a prevalence of 6.3%. Outbreaks of scurvy arose because of a complex series of emergencies precipitated by recent conflict in the region. Cases of scurvy are highest at the end of the winter in January and February and declined in the spring when the availability of various vegetables, fruits and plants increased. Prolonged 'winter periods' lasting five or six months limit the availability of fresh foods and result in frequent outbreaks of scurvy unless there is supplementation of the diet. In the report by Cheung et al. (2003:251–252) it was stated that *seech* a green plant that grew in the mountainous regions of Afghanistan became available in March, but for the majority of people in this region the last time that fresh foodstuffs had been available was October. Many of those who had suffered scurvy but survived had eaten *seech* when it became available. The individuals found to be most at risk of developing scurvy were women and elderly individuals. Discussion with the population revealed that under these conditions scurvy was present every winter. Most local inhabitants thought that *sialengi*, the local word used for scurvy, was in fact caused by the cold weather and some individuals thought it was contagious (Cheung et al., 2003:253).

A lack of understanding of scurvy and its causes was probably almost universal until very recent times, and it is likely that in past communities scurvy was attributed to a wide range of causes. For example, air quality was often considered a cause of sickness amongst crews (Carpenter, 1986:2–7), and was also blamed in past outbreaks of scurvy amongst land-based peoples (Crellin, 2000). Although treatments for scurvy were worked out independently at a number of times and places as discussed by Maat (2004), it took many centuries before there was sufficient understanding of the condition to enable specific planning to prevent scurvy. There has been a clear gap between knowledge and systematic actions to prevent cases of scurvy developing. In the case of modern refugee communities it could be argued that this situation has continued (see further discussion of health in displaced peoples in Box Feature 9.1).

ANTHROPOLOGICAL PERSPECTIVES

An improved awareness of scurvy in past societies has considerable potential to contribute to a better understanding of a variety of issues in anthropology. In terms of interpreting socio-cultural activities in past societies, important information could be gained from considering a wider range of circumstances in which scurvy occurred. For example, it has been suggested that seasonal-related nutritional stress is very common in developing countries (Ulijaszek and Strickland, 1993). The prevalence of scurvy through time was probably closely related to socio-cultural practice, for example farming methods and ideas relating to food production and storage. In hunter-gatherer groups or band societies storage of food does not occur and food is eaten very soon after it is obtained (Cohen, 1989:17). However, changes in subsistence practice may have brought about an increase in cases of scurvy. The potential for this to be demonstrated during the transition to agriculture is discussed in Box Feature 4.2 and further health issues linked to adaptive and transitional diets are outlined in Box Feature 7.3.

Population movements and the arrival of people in new areas may result in changes to available foods as another likely cause of scurvy in past communities. Without a nutritional knowledge of a balanced range of foods that could be eaten, early settlers may well have consumed a diet that did not include the full range of micronutrients, leaving them open to deficiency diseases. Early colonies established by the French in Canada and the British in America suffered from scurvy, particularly in winter when there was limited availability of fresh foods (Carpenter, 1986:12). Scurvy was further present in several of the new communities moving into North America, such as amongst the early settlers in Newfoundland during the early seventeenth century (Pimentel, 2003). An earlier explorer to the region, Jacques Cartier, used a recipe from native inhabitants of the area to treat scurvy in 1543 (Crellin, 2000). Episodes of this disease occurring amongst individuals moving into areas with which they were unfamiliar appear to have been a frequent problem since modern humans first evolved (e.g. Gandevia and Cobley, 1974; Houston, 1990; Feeney, 1992; Forsius, 1993; Hampl et al., 2001).

Individuals who have their dietary resources constrained and/or who rely on others for provision of sustenance are also likely to be at risk of nutritional diseases. For example, research by both Carpenter (1987) and Hampl et al. (2001) demonstrated that scurvy has frequently occurred in prisoners.

Reference to Probable Scurvy in Early Texts

Some of the earliest written evidence for scurvy comes from Egypt and dates to 1550 BC (Rajakumar, 2005), but various ancient texts have references to diseases that are probably scurvy. Mogle and Zias (1995) report that references to scurvy are present in Babylonian texts, the Old Testament and the Talmud. In each of these sources reference is made to pathological changes in the

Box Feature 4.2. Subsistence Change and the Development of Scurvy: The Origins of Agriculture

The adoption of agriculture is widely accepted as being one of the most important changes in human history (Ulijaszek and Strickland, 1993). This change in subsistence patterns would have resulted in complex social and cultural changes that probably had a significant impact on nutritional stress. Following the work of Cohen and his co-workers (e.g. Cohen and Armelagos, 1984), it has become widely accepted by bioarchaeologists that the transition to agriculture may not necessarily have brought improvements in human health and nutrition. This would have been particularly marked in latitudes where greater seasonality resulted in limited availability of fresh plant foods for part of the year. Although the adoption of agriculture may have resulted in a greater availability of foodstuffs, in many instances the diversity and range of plants grown would have been more limited in some nutritional aspects. Research by Cordain (1999) has suggested that this transition might have resulted in people adopting diets containing a far more restricted range of plants. People could have moved from a diet in which as many as 200–300 species of plant foods to one where in addition to a couple of domesticated cereals, probably only 20–50 plant foods were utilised (Cordain, 1999:23).

A good diet requires more than just an adequate number of calories. Sedentism may have had benefits in terms of greater food production, but it may have had disadvantages in terms of the range of nutrients gathered. As pointed out by Cordain (1999:24) 'all cereal grains have significant nutritional shortcomings'. One important nutrient that cereals lack is vitamin C (see Table 4.1).

The adoption of agriculture or horticulture may have required a significant input of working hours and meant reduced time was available for other activities. Over time some cultural activities such as gathering a wider range of wild resources may have been lost. Fehrsen (1974) has discussed the problems of communicating the concept of nutritional disease in rural areas, and has highlighted the difficulties that exist even amongst modern populations regarding understanding the causes of disease. In past communities that lacked knowledge on micronutrients and the link between diet and disease, it may have been even more difficult for communities, especially ones that had recently undergone changes, to understand the cause of diseases that occurred. These examples illustrate that information provided by nutritional anthropology can provide important tools with which to view patterns of disease in past communities. To date relatively few cases of scurvy have been identified in individuals from these important transition periods (see Table A1). However, recent advances in the osteological identification of scurvy may increase the number of cases identified and help to understand the impact on health during the transition to agriculture.

mouth and the condition is referred to as the stinking disease (Mogle and Zias, 1995). One of the earliest references to what later became known as infantile scurvy was that made by Glisson (1650), but following this there were few references to the condition until the mid-nineteenth century. Infantile scurvy was first specifically identified by Sir Thomas Barlow in the early 1880s, and as a result has been referred to as Barlow's disease (Aspin, 1993). Carpenter (1986) provides a full discussion of the development of knowledge relating to scurvy.

Popular perceptions of scurvy in past communities have tended to focus on events, such as sea voyages as a major causative factor, although we would argue that this focus is misplaced. There are a number of reasons why scurvy in mariners has attracted considerable attention. The severely restricted diet on many longer voyages would probably have produced quite severe manifestations of the condition. The individuals involved would often have been otherwise healthy young males, so an outbreak of an incapacitating disease, which in some cases contributed to their death, would have been remarked upon. During the historic period at least one literate individual who could document the voyage would have been on board many ships. Outbreaks of scurvy would have notably disrupted activities associated with warfare, international relations and trade, including journeys linked to highly desirable and economically valuable spices. The interruption of such activities would have attracted widespread attention.

Longer sea voyages became more frequent from the fifteenth century onwards, and by the late eighteenth/early nineteenth century effective treatments for scurvy, such as citrus fruits, were known and widely used. However, specific strategies on prevention and treatment took longer to become widespread in the merchant navy than in organisations such as the British Royal Navy (Cook, 2005). As a result, scurvy continued to be a problem in the merchant navies until the end of the nineteenth century.

Whilst they occurred less frequently, earlier voyages were also likely to have been affected by scurvy. Evidence documenting the journey between the Indus and the Persian Gulf undertaken by the fleet of Alexander the Great in 329 BC indicates that those on board suffered from scurvy (Tickner and Medvei, 1958). Symptoms reported included leg pains, and problems in the mouth including bad gums and tooth loss. Although reports on scurvy in mariners do contain useful information, it should be remembered that they occurred within a relatively short period in human history and few individuals undertook such activities. An important point made previously by Holck (1984) is that scurvy was probably present as a sub-clinical condition in numerous inhabitants of northern Europe in the past. Many sailors, like the rest of the population, probably had sub-clinical scurvy before they ever got on a ship and this is why once at sea scurvy developed so rapidly and frequently.

Times of conflict and the health of soldiers have also attracted significant interest from those studying scurvy in past populations. The writings of Hippocrates have descriptions of men in the army who suffer from a condition that caused changes of gums and loss of teeth. Tickner and Medvei (1958) report that in 1555 Olaus Magnus (Archbishop of Uppsala) referred to a disease which was probably scurvy affecting soldiers and prisoners across northern Europe. During the Crimean and American Civil Wars, troops were also affected (Pimentel, 2003). The contribution of a poor diet to the development of scurvy in soldiers during the American Civil War has been discussed by a number of authors (e.g. Bollet, 1992; Sledzik and Sandberg, 2002; Danzeiser Wols and Baker, 2004). Although soldiers were supplied with foods such as dried vegetables to prevent scurvy (as scurvy and its causes were recognised), cooking practices used would have destroyed much of the vitamin C (Danzeiser Wols and Baker, 2004:67). It is likely that many of the cases of scurvy that occurred during the American Civil War went undiagnosed (Bollet, 1992) as access to medical care in which a diagnosis could be made was limited, and many individuals were killed in fighting or succumbed to other conditions.

A More Recent View of Scurvy in the Past

Scurvy was likely to have been far more widespread and linked to a much greater range of activities than the recent focus on sea voyages and wars would suggest. Vitamin C deficiency, at various levels, probably occurred relatively frequently and it is likely that many individuals were periodically deficient in vitamin C. However, there can be complications in deriving adequate diagnosis of this condition. An early clinical study by Follis et al. (1950) reported findings of scurvy from autopsies that were undertaken on children who died between 1926 and 1942. Of 69 cases of scurvy identified at autopsy, only six cases had been recognised clinically during life (Follis et al., 1950). In many cases, scurvy was at an early stage and not sufficiently advanced to be identified clinically. There were a number of cases in the children autopsied where the physical signs of scurvy had in fact been thought to relate to rickets by medical staff treating the child (Follis et al., 1950:582). Rickets has long been noted to be frequently present in individuals who have scurvy (Follis et al., 1950) and early medical writings have noted the co-occurrence of the two conditions (e.g. Glisson 1650, cited by Jaffe, 1972; Barlow, 1883). However, the study by Follis et al. (1950) also found that 65.4% of the children with scurvy they examined had an acute illness and most of these were infectious diseases. Milgram (1990:843) further observed that scurvy was often identified during post-mortems of children who had died of infectious diseases.

PALEOPATHOLOGICAL CASES OF SCURVY

Whilst there is quite a lot of written evidence relating to scurvy in the past, direct evidence of the disease process provided by paleopathological investigations contribute significant additional information. For example, it was suggested by Rajakumar (2001) that the lack of reference to infantile scurvy in the literature between 1650 and 1850 probably reflected a decline in the condition during this period. However, the discovery of evidence of infantile scurvy dating to the nineteenth century from the site of St. Martin's, Birmingham, England (Brickley and Ives, 2006) demonstrates that infantile scurvy was present at this time. It is likely that many cases occurring at this time were sub-clinical, and that many of the affected individuals died of acute infectious conditions before obvious symptoms became apparent. Lomax (1986) has also suggested that scurvy may have been present after 1650 but was not being recognised by contemporary individuals.

Research conducted by Ortner and his various co-workers between 1997 and 2001 (Ortner and Ericksen, 1997; Ortner et al., 1999; Ortner et al., 2001) has improved the awareness of the range of manifestations of scurvy in juvenile remains. As a result of these works the number of reported cases of scurvy in paleopathology has increased (see Table A1). Information contained in Table A1 shows that the first reported cases of scurvy were mainly in adults. However, following Ortner's work there was a big shift, and now most cases reported are in juveniles. Paleopathological cases of scurvy reported to date, demonstrate that the condition was evident across a wide geographical area, as well as across a broad temporal range. Research currently being undertaken on the WORD project by the Centre for Human Bioarchaeology, Museum of London has identified four juveniles and one adult with probably scurvy from two Medieval sites, and 41 juveniles from 516 analysed to date from post-Medieval sites in London. There are however, gaps in our current understanding of scurvy in past communities, and it is not yet clear whether this trend reflects a true lack of occurrence of the condition, or a lack of research in some areas.

Whilst, the skeletal manifestations of scurvy in adults are difficult to discern from infectious, traumatic or inflammatory conditions, there is still potential for this disease to be better recognised in future research. One example that had not been fully published at the time of writing came from adults buried near the mining community of Kimberly, South Africa, where there would have been limited access to adequate nutrition and medical treatment (abstract by Van der Merwe and Steyn, 2006). Several of these adults had convincing evidence of scurvy and clear publication of the features observed would be very useful. With better diagnosis in adults it would be possible to establish a more complete picture of the pattern of the condition within past communities.

A number of possible causes are suggested for the development of vitamin C deficiency in past individuals. Various studies have considered environmental stress, for example that undertaken on human bone from Polynesia by Buckley (2000). Specific forms of environmental stress, such as earthquakes, can lead to limited access to fresh foods, as was considered possible for scurvy recorded in human remains from a proto-Byzantine (sixth–seventh centuries AD) site at Eleutherna, Greece (Bourbou, 2003a, 2003b). In countries such as Tonga and Greece fresh fruits and vegetables are generally widely available and scurvy would not normally be a common problem (Buckley, 2000; Bourbou, 2003b). In more temperate regions outbreaks of scurvy may be a seasonal occurrence. Research by Maat (2004) demonstrated that in the past scurvy was endemic amongst people living in the area known as the Low Countries, frequently appearing at the end of the winter and during the early spring (similar to the pattern reported in modern Afghanistan earlier). Following the examination of cases from across sites in North America, it was clear that scurvy was likely to have been a serious problem for many past Native American communities (Ortner et al., 2001). Variations in prevalence rates recorded from different regions were probably due to differences in food resources, cultural practice and other socio-economic factors (Ortner et al., 2001).

DIAGNOSIS OF SCURVY IN ARCHAEOLOGICAL BONE

When considering features that can be used to suggest a possible diagnosis of scurvy, it is important to take the age at death of the individual into account. Therefore information on diagnosis of the condition has been divided between juveniles and adults.

Macroscopic Features of Infantile scurvy

Key macroscopic features used in the diagnosis of scurvy in archaeological bone are listed in Table 4.2. As demonstrated by many of the studies included in Table A1, less severe cranial manifestations including those at the sphenoid, orbits, hard palate and mandibular ramus (Figures 4.3 and 4.4) have recently proved to be very important in suggesting a diagnosis of scurvy (e.g. Ortner et al., 2001; Ortner, 2003:111; Brickley and Ives, 2006; Ortner et al., 2007; Mays, 2008).

Variations in rates of growth between skeletal areas will produce a range of effects in juveniles who acquire the condition at different ages. For example, changes may be particularly marked following periods of rapid bone growth (Rajakumar, 2005). It should be remembered that ossification of new bone in areas of sub-periosteal bleeding would only become apparent when vitamin C is ingested and osteoid formation has begun. As discussed in Chapter 3, it is only at this stage that bone can form (Scott and Helton, 1953; Weinstein et al., 2001). Research by Bourne (1942b) demonstrated that only very small amounts of vitamin C are required to produce a considerable reaction and new bone formation at the periosteum. Complete exclusion of vitamin C from the human diet is quite rare, and it is likely that small amounts would occasionally be ingested. Individuals who show some new bone formation, such as the child from St. Thomas's hospital, London, shown in Figure 4.4, will have obtained some vitamin C following a period of deficiency.

Ratanachu-Ek et al. (2003) report that sub-periosteal new bone formation following bleeding was apparent (using radiological examination) in 33% (7/28) of cases reviewed. In these clinical cases researchers were not able to directly observe the bones, and it is likely that periosteal new bone formation may have been present in an even higher proportion of the cases. However, in a study involving autopsies Follis et al. (1950) noted that haemorrhagic phenomena were present in just two or three of the 69 cases of scurvy studied. This difference could be related to the stage of the disease in each of these two groups. Ratanachu-Ek et al. (2003) were reporting on individuals that were still alive, and following diagnosis may have received some vitamin C. Many of the individuals reported on by Follis et al. (1950) were undiagnosed prior to death and may not have ingested any vitamin C, without which no bone formation would have occurred.

TABLE 4.2 Macroscopic Features of Scurvy in Juveniles

Bones affected	Features	Code	Differential diagnosis	Sources
Cranium	Abnormal porosity of cortex (pores should penetrate the cortex and be <1 mm across) – sphenoid, mandible, maxilla and orbits	D	Anaemia Rickets Trauma Infection	Ortner & Ericksen (1997), Ortner et al. (1999), Ortner (2003)
	New bone formation – orbits and vault	D		Sloan et al. (1999), Brickley & Ives (2006)
Dentition	Loosening of teeth and AMTL	G	Periodontal disease	Wolbach & Howe (1926), Jaffe (1972), Hillson (1996:165–166)
	Hypoplastic defects	G	Conditions causing 'stress'	
Vertebrae	–			
Ribs	Fracture of bone adjacent to costochondral junction	D	Trauma	Ortner (2003:386)
	Enlargement of ribs adjacent to costochondral junction – scorbutic rosary	D	Rickets	
Pelvis	Rare – but sub-periosteal haemorrhages can occur. Porosity and new bone formation.	G	Infection	Ortner (2003:387)
Scapulae	Abnormal porosity of cortex – particularly supraspinous and infraspinous areas	D	Infection Trauma Normal growth	Ortner et al. (2001), Brickley & Ives (2006)
Long bones	New bone formation – particularly common towards ends	D	Infection Trauma Normal growth	Wolbach & Howe (1926), Ratanachu-Ek et al. (2003)

Notes: As discussed in the text new bone formation will only occur when some vitamin C is obtained. The features listed demonstrate the skeletal changes useful for pathological identification. Diagnosis can be determined on the strength of each feature: –, no significant features recorded for this region; S, strongly diagnostic feature; S features are normally required for a diagnosis; D, diagnostic feature, the presence of multiple D features are required for a diagnosis; G, general changes, can occur in many of the metabolic bone diseases, as well as in other conditions; G features can aid a diagnosis together with S or D features but cannot be used alone to suggest a diagnosis. Commonly occurring conditions that display these pathological features are listed to aid differential diagnoses, although this list is not exhaustive.

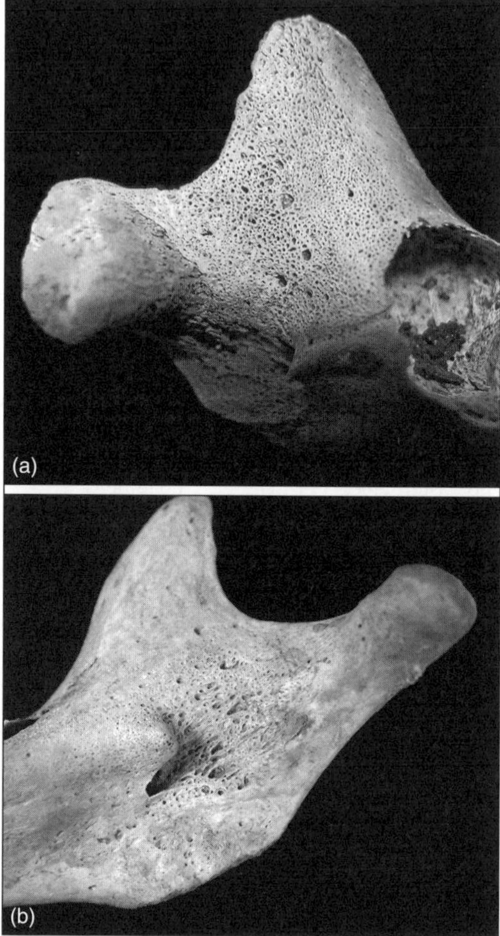

FIGURE 4.3 (a) Marked porosity above the mandibular foramen in a juvenile from London. Photograph by Rachel Ives, courtesy of the Museum of London. (b) Porosity at mandibular foramen in a 1–2 year-old child from an Iron Age (radiocarbon dates c.200 BC–AD 100) site from Kirby, North Yorkshire, England. Photograph by Anwen Caffell for York Osteoarchaeology Ltd., on behalf of MAP Archaeological Consultancy Ltd.

Sub-periosteal haemorrhage has also been reported to occur in the orbits (Sloan et al., 1999) and so new bone formation can also occur on the roof of the orbits (see Figure 4.5). In cases of scurvy increased vascular fragility in this area results in bleeding following very limited tissue movement, but trauma and venous anomalies should also be considered as a possible cause of bone formation in the orbits (Zilva and Still, 1920; Krohel and Wright, 1979; Jacobsen et al., 1988). Figure 4.6 illustrates the formation of bone in the orbit of a juvenile thought to have had scurvy. Such changes could develop in both adults

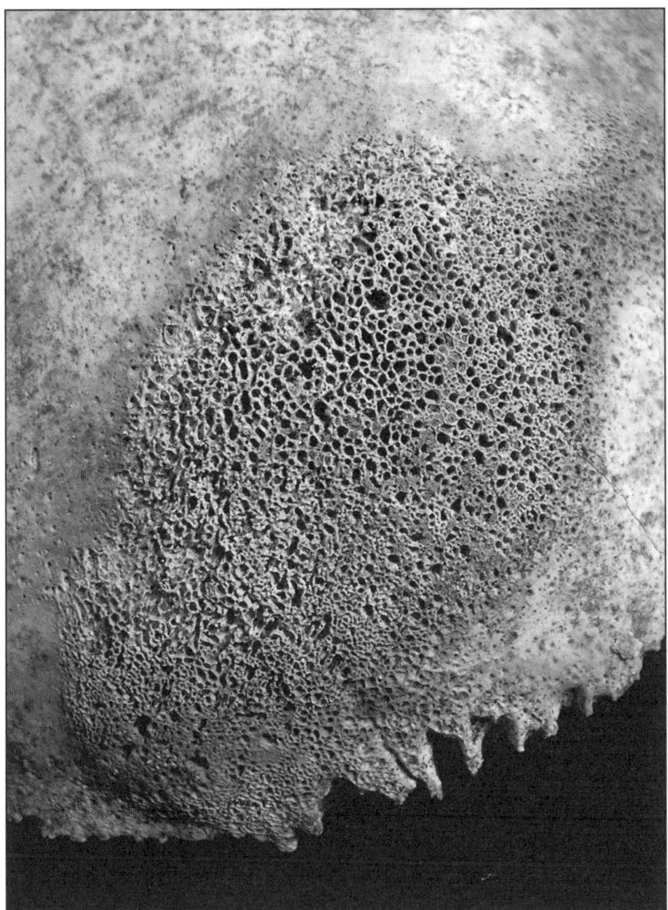

FIGURE 4.4 New bone formation on the skull of a child aged 1–5 years from the post-Medieval site of St. Thomas's hospital, London. These changes are not diagnostic, but the pattern of changes across the skeleton suggested that this individual had scurvy. Photograph by Rachel Ives, courtesy of the Museum of London.

and juveniles. Although these lesions are not commonly reported in modern clinical cases of scurvy they do occasionally occur (e.g. Sloan et al., 1999).

Parrot's swellings, deposits of woven bone on the cranial bosses, have been suggested to be formed by similar mechanisms to those that produce bone formation on the long bones of individuals with scurvy (Ortner and Ericksen, 1997). Care is required however, because as discussed in Chapter 2, many individuals whose remains were kept in pathology museum collections, where such changes have been observed, could have had more than one condition. Barlow's report (1883) on the autopsy of one of a child with scurvy describes evidence of sub-periosteal haemorrhages in the position where Parrot's swellings

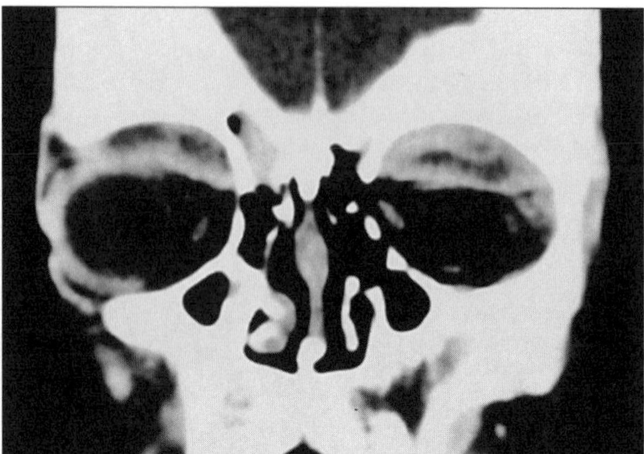

FIGURE 4.5 Coronal CT scan showing bilateral superior sub-periosteal orbital haematomas in a 13-year old girl with scurvy. Photograph published in Sloan et al. (1999).

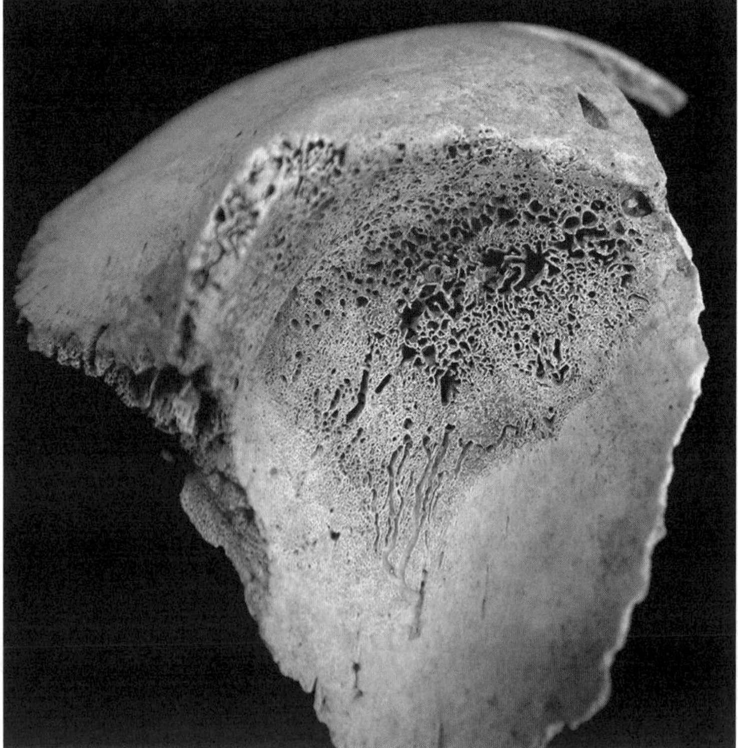

FIGURE 4.6 Extensive new bone formation on the right orbital roof of a 9–12 month old child from an Iron Age (radiocarbon dates c.200 BC–AD 100) site from Kirby, North Yorkshire, England. Photograph by Anwen Caffell for York Osteoarchaeology Ltd., on behalf of MAP Archaeological Consultancy Ltd.

develop, but these changes were also noted in cases of congenital syphilis and rickets (Barlow, 1883:192). Barlow (1883) states that Parrot originally described these skeletal changes in individuals who had congenital syphilis, although they may have had another condition as well. Therefore care is required when linking such changes to scurvy. At present it is probably not safe to regard these features as evidence of scurvy, but continued careful recordings of such features may allow clearer suggestions as to their cause in the future.

Macroscopic Features of Adult Scurvy

In adults, where growth has ceased, the skeletal manifestations of the condition will take a different form and will be very subtle; none of the changes listed in Table 4.3 are pathognomic. During the rapid analysis of human remains excavated from St. Martin's Churchyard, Birmingham, no cases of adult scurvy were identified. However, as discussed in Brickley and Buteux (2006) documentary sources, and pathological changes in juveniles (Brickley and Ives, 2006)

TABLE 4.3 Macroscopic Features of Scurvy in Adults

Bones affected	Features	Code	Differential diagnosis	Sources
Cranium	New bone formation – orbits	D	Trauma	Sloan et al. (1999)
Dentition	AMTL single-rooted teeth particularly susceptible Inflammation in alveolar bone	G G	Periodontal disease	Wolbach & Howe (1926), Hillson (1996), Pangan & Robinson (2001), Ortner (2003:387)
Vertebrae	Osteopenia and possible bi-concave compression (Figure 4.10)	G	Osteoporosis Trauma	Joffe (1961)
Ribs	Transverse fractures at osteocartilaginous junction of ribs	D	Trauma	Ortner (2003:387)
Pelvis	–			
Long bones	New bone formation – particularly common towards ends	D	Trauma Infection	Walbach & Howe (1926), Ortner (2003), Fain (2005)

Notes: As discussed in the text new bone formation will only occur when some vitamin C is obtained. –, no significant features recorded for this region; AMTL, *ante-mortem tooth loss. See* Table 4.2 *for definition of codes used in diagnosis.*

indicated that scurvy was present in Birmingham at this time. It is likely that any bone changes that were present were too subtle to be identified during the rapid recording undertaken to produce the site report.

The lack of clear bone changes in cases of adult scurvy is demonstrated by the detailed study undertaken by Maat (1982) on the Dutch whalers buried at Spitsbergen. Preservation of the human remains from Spitsbergen was remarkable with traces of blood and soft tissues being present. Features such as subperiosteal haematomas were recorded in many of the men. However, most of these cases were identified through the presence of remnants of bloodstains, and only one individual actually had any new bone formation. This person must have ingested some vitamin C and this had enabled bone to form at the locations of haematomas (Maat, 1982). In some individuals there were infractions of the cortex, but it is possible that these were made more noticeable due to freezing of blood within these lesions (Maat, 1982). Infractions of cortical bone have not been noted in clinical reports of scurvy. There was evidence for in vivo tooth loss in a small number of individuals excavated from Spitsbergen. Retraction of the alveolar bone was recorded in quite a few of the individuals, but as pointed out by Maat (1982), this could have been caused by periodontal disease.

The most important thing to emerge from this study is that the number of bone changes linked to possible cases of scurvy was very small. Many adults who have scurvy will not have bone changes that enable clear diagnosis of the condition. Blood clots, in which bone formation had taken place were one of the features used by Stirland to suggest possible cases of scurvy in an adult from the ship the *Mary Rose* (Stirland, 2000).

Radiological Features of Infantile Scurvy

A range of the radiological changes that can appear in children with scurvy are illustrated in Figures 4.7 and 4.8, and the diagnostic features are given in Table 4.4. Features such as a metaphyseal 'white line', which have been referred to as 'white lines of Frankel' can appear due to the preservation of areas of provisional calcification at the ends of the metaphyses (Tamura et al., 2000). In some cases a fine white line of more mineralised bone may be apparent outlining the epiphyses and this has been termed a 'Wimberger ring' (Tamura et al., 2000), or a 'pencilled effect' (Grewar, 1965). As no new matrix is formed, calcification is intensified in a limited area. Adjacent to these white lines are lucent areas referred to as the 'scurvy line' (Grewar, 1965) or the 'Trümmerfeld zone'. Radiolucent lines may be present in juveniles although once individuals receive adequate amounts of vitamin C these can be resolved in two to six weeks (Ratanachu-Ek et al., 2003). Because of their relatively rapid resolution, such features will be of limited use in cases of healed scurvy. At the edge of the white lines at the lateral edge of the metaphyses, spurs of more mineralised bone can be observed, known as 'Pelkan spurs' that result from healing of micro-fractures (Tamura et al., 2000). The term 'corner sign' or 'corner sign of

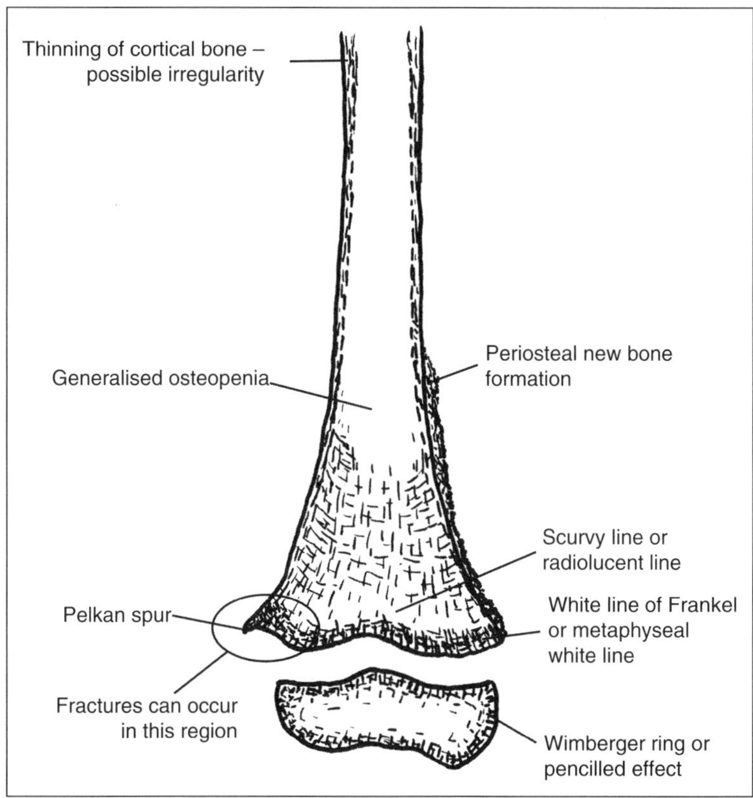

FIGURE 4.7 Radiological changes that can be observed in infantile scurvy.

Park' has also been applied to fractures of trabecular elements below the epiphysis. Fractures occur due to minor trauma as pre-existing bone becomes brittle.

It is noted by Follis et al. (1950:569) that skeletal lesions must be 'marked' before they can definitely be identified using X-ray, and Joffe (1961) reported that in around half of those with severe cases of scurvy no radiological features were apparent (see also Akikusa et al., 2003). A number of early signs of scurvy that could be identified radiologically were reported by Park et al. (1935), readers are referred to his published radiographs. Not all of the features described here and shown in Figures 4.7 and 4.8 are present in all cases of scurvy (Tamura et al., 2000), and no one feature alone is pathognomic for scurvy (Grewar, 1965).

Radiological Features of Adult Scurvy

As illustrated by the features listed in Table 4.5, many of the radiological features of scurvy in adults are non-specific. Generalised osteopenia will extend

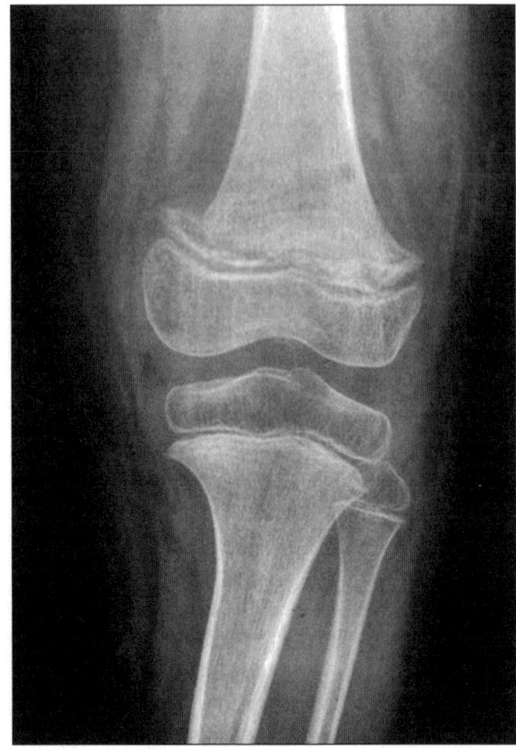

FIGURE 4.8 Antero-posterior radiograph of the left knee showing generalised osteopenia, metaphyseal lucent bands and irregularity of the medial cortex of the distal femoral metaphysis due to fractures on the left side. Picture courtesy of Michael Weinstein and the American Academy of Pediatrics, Pediatrics (2000), 108:E55.

throughout the skeleton, not just to the bones listed in Table 4.5. An example of to the uneven bone loss around the knee is illustrated in Figure 4.9.

Loss of bone in the vertebrae may also contribute to osteoporotic fractures in these bones (see Figure 4.10) (Joffe, 1961). However, as discussed by Fain (2005) the reported cases of this type of fracture occurred in elderly female patients and so age-related bone loss (as discussed in Chapter 6) was almost certainly also involved. Elderly individuals frequently have multiple deficiency diseases, and the possibility that this is occurring in the elderly has been largely overlooked (Fain, 2005). There is also the potential for the deliberate mismanagement of foods as a manifestation of elder abuse and this is considered in Box Feature 5.2.

To date none of the published cases of scurvy in archaeological human bone have used radiological features of the condition in suggesting a diagnosis. This is probably because none of the radiological features of the condition are pathognomic and many are transitory and not always present even in clinically diagnosed cases of scurvy. However, radiological analysis is worth conducting

TABLE 4.4 Radiological Features of Juvenile Scurvy

Bones affected	Features	Code	Differential diagnosis	Sources
Cranium	–			
Dentition	–			
Vertebrae	Osteopenia (across the skeleton this has been referred to as 'ground-glass' osteoporosis)	G	'Ground-glass' appearance also described in fluorosis	McCann (1962)
Ribs	Osteopenia	G		
Pelvis	Osteopenia	G		
Long bones	Osteopenia – of cortex and trabeculae	G		McCann (1962), Grewar (1965), Chatproedprai & Wananukul (2001)
	Irregularity and thinning of the cortex	D	Healing rickets	
	White line of Frankel or dense metaphyseal line	D		
	Radiolucent or 'scurvy' line – band of diminished density beneath band calcified cartilage	D	Trauma	
	Wimberger ring	D		
	Corner fractures or sign – defect in spongiosa and cortex below provisional zone of calcification (see Figures 4.7 and 4.8 for illustration of these features)	D		

Note: – denotes to no significant features recorded for this region. See Table 4.2 for definition of codes used in diagnosis.

as it may provide additional observations that could help suggest a diagnosis particularly in juveniles.

Histological Features of Infantile Scurvy

Key features linked to infantile scurvy that can be identified histologically are listed in Table 4.6. Investigations by Follis et al. (1950) suggest that

TABLE 4.5 Radiological Features of Adult Scurvy

Bones affected	Features	Code	Differential diagnosis	Sources
Cranium	–			
Dentition	–			
Vertebrae	Generalised osteopenia (may contribute to compression fractures, see Figure 4.10)	G	Osteoporosis See Chapter 7	Joffe (1961), Greenfield (1990), Tamura et al. (2000), Fain (2005)
Ribs	Generalised osteopenia	G	See Chapter 7	Greenfield (1990), Tamura et al. (2000)
Pelvis	Generalised osteopenia	G	See Chapter 7	Greenfield (1990), Tamura et al. (2000)
Long bones	Generalised osteopenia	G	See Chapter 7	Greenfield (1990), Joffe (1961)
	Uneven areas of osteopenia around knee or ankle joints Figure 4.9	D		

Notes: – denotes to no significant features recorded for this region. See Table 4.2 for definition of codes used in diagnosis.

scurvy is characterised by fractures within the trabecular bone at the epiphyseal plate in individuals in whom endochondral bone formation should be taking place. This leads to an area with limited trabecular development, as illustrated in sections of a scorbutic and normal guinea pig rib shown in Figure 4.11. These investigators observed fragments of mineralised trabecular bone lying at various angles in individuals who had scurvy, or who had recently started to recover (but importantly not all individuals with scurvy showed such changes). Fractures caused by scurvy may be present for a couple of months after recovery, but, this type of fracture is not pathognomic (Follis et al., 1950). Histological changes can be used to help suggest a diagnosis in archaeological bone, but other features (macroscopic and radiological) would need to be present for a firm diagnosis to be considered.

An important point noted by Follis et al. (1950) was the difficulty of identifying features linked to scurvy using histology if severe rickets was present (see Figure 4.12). Schultz (2001) also states that some aspects of the histological appearance of scurvy and anaemia are very similar, although the cases discussed

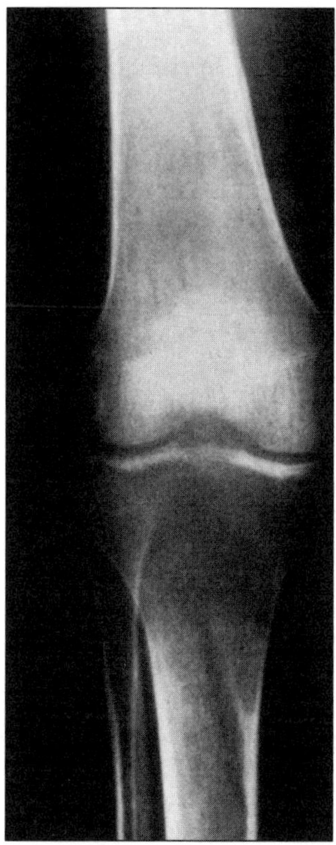

FIGURE 4.9 An anteroposterior radiograph of the right knee of an adult male with scurvy. Para-articular osteopenia can be observed around the knee joint. A smooth narrow layer of periosteal reaction was observed on the medial border of the femur and this extended along the length of the shaft. This individual did receive some vitamin C prior to the radiograph. Joffe N. 1961. Some radiological aspects of scurvy in the adult. British Journal of Radiology 34:429–437.

in this paper are all archaeological so an exact diagnosis is not known. However, Schultz does provide information in differentiating between changes caused by scurvy and anaemia (Schultz, 2001:134). There are difficulties with the histological analyses of bone sections especially where diseases co-exist, but microscopic analysis is worth considering where invasive investigative techniques are possible, as it may provide additional information that can assist with a diagnosis. As shown by Brickley and Ives (2006) microscopic analysis of bone surfaces, which is relatively easy to undertake, is particularly useful for characterising the extent and type of porosity abnormal surface present in early stages of the condition.

Histological Features of Adult Scurvy

As can be seen in Table 4.7, there are no clear histological features that can be identified in adult individuals with scurvy. Changes such as the widespread

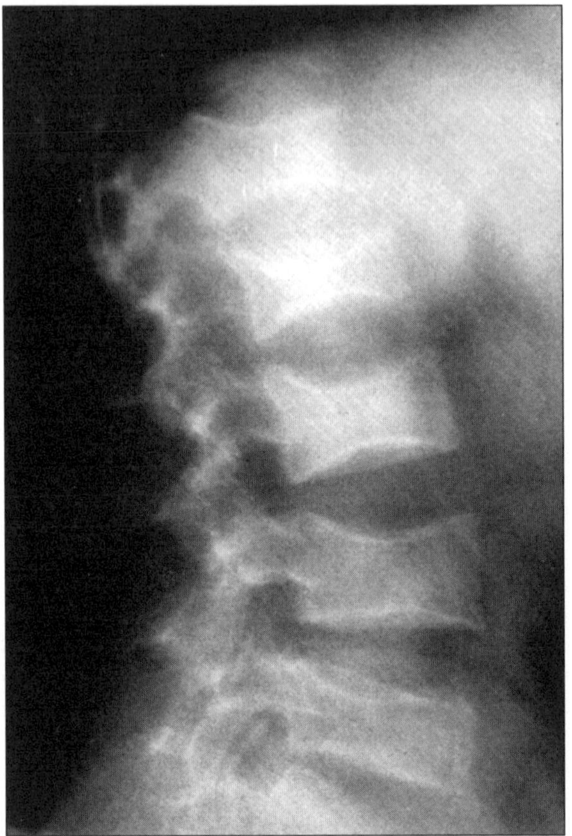

FIGURE 4.10 Lateral radiograph of the lumbar spine of an adult male with scurvy, showing generalised osteopenia and bi-concave compression of the vertebrae. Joffe N. 1961. Some radiological aspects of scurvy in the adult. British Journal of Radiology 34:429–437.

TABLE 4.6 Histological Features of Juvenile Scurvy

Tissue type	Features	Code	Differential diagnosis	Sources
Cortical bone	Metaphyseal fractures	D	Trauma	Milgram (1990)
Trabecular bone	Fractures in trabecular bone elements	D	Trauma	Bourne (1942b), Follis et al. (1950), Milgram (1990)
	Broken or irregular ossified cartilage columns	D	Leukaemia	

Note: See Table 4.2 for definition of codes used in diagnosis.

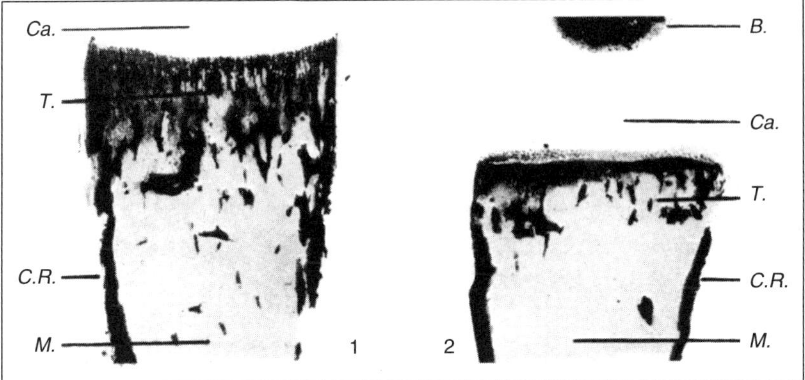

FIGURE 4.11 1. Un-decalcified section of costochondral junction of rib of guinea pig receiving ample diet of greenstuff. Numerous calcified trabeculae may be seen extending from the junction. 2. Un-decalcified section of costochondral junction of rib of guinea pig on a scorbutic diet for two weeks. Number of trabeculae greatly reduced (Ca., cartilage; T., trabeculae; C.R., cortex of rib; M., marrow; B., deposit of bone salt in cartilage). Courtesy of Geoffrey H. Bourne. 1943. Some experiments on the possible relationship between vitamin C and calcification. Journal of Physiology, Wiley-Blackwell Publishing Ltd.

osteopenia will be observable, but as discussed in Chapter 7, the range of conditions and circumstances that can lead to the development of secondary osteopenia are extensive.

DIFFERENTIAL DIAGNOSIS

Increased knowledge about scurvy, and especially its appearance in juvenile bone, has prompted many investigators to consider scurvy when considering a differential diagnosis for pathological changes in human skeletal remains (e.g. Fairgrave and Molto, 2000; Lambert, 2002; Lewis, 2002; Danzeiser Wols and Baker, 2004; Blom et al., 2005).

The importance of checking for multiple deficiency diseases is discussed by Ortner et al. (2001:349) and here one of the individuals analysed from Native American skeletal collections is also suggested to have anaemia due to the skeletal changes present. The likely co-existence of a number of pathological conditions was also reported in the human bone from Tonga (Buckley, 2000), where there was evidence for anaemia and infectious diseases (possibly due to conditions such as weanling diarrhoea and yaws) (Buckley, 2000). The presence of acute infectious diseases may have resulted in malabsorption of vitamin C (Buckley, 2000:499).

Although periosteal new bone formation frequently occurs with scurvy, a wide range of other pathological and traumatic conditions can cause such new bone formation (see Table 4.4) and all of these should be considered before making a diagnosis.

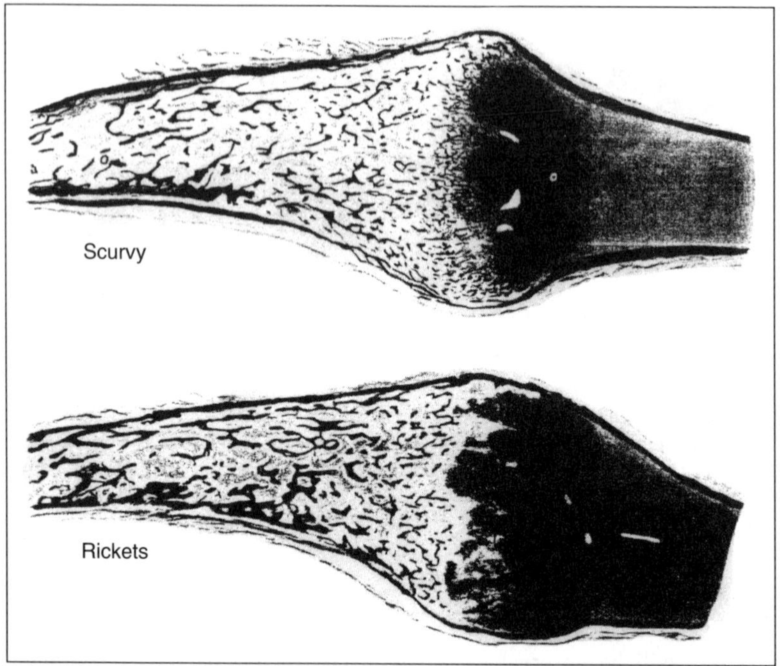

FIGURE 4.12 Camera lucida drawing of transverse sections of a typical scorbutic and rachitic rib. The upper borders correspond to the outer surfaces of the costochondral junctions. In rickets the sharp angle of junction between cartilage and shaft is not apt to be present because of the deep rachitic intermediate zone interposed between the proliferative cartilage and the shaft. Reproduced with permission from the BMJ Publishing Group ©. Archives of Disease in Childhood, Park et al. 1935. 10:265–294, Figure 37.

TABLE 4.7 Histological Features of Adult Scurvy

Tissue type	Features	Code	Differential diagnosis	Sources
Cortical bone	Generalised osteopenia	G	See Chapter 7	Bourne (1942a)
Trabecular bone	Generalised osteopenia	G	See Chapter 7	Bourne (1942a)

Note: See Table 4.2 for definition of codes used in diagnosis.

The diagnosis of scurvy can be complicated by the frequent co-occurrence of other deficiency diseases, such as vitamin D deficiency (Pimentel, 2003). The presence of metaphyseal cupping of long bones, a feature frequently associated with rickets (discussed in Chapter 5), can in rare cases also occur as a result of infantile scurvy (Hallel et al., 1980). As stated by Ortner and

Box Feature 4.3. Scurvy in Non-Human Primates: A Result of Human Actions

All primates lack the ability to synthesise vitamin C (Akikusa et al., 2003), but it is unlikely that many non-human primates would experience scurvy in their natural habitat with unrestricted access to their normal foods. Spontaneous scurvy will be rare in non-human primates and the condition will almost certainly always relate to inappropriate feeding by humans, or dietary restrictions that occur as a result of human actions. A reported case of scurvy occurred in young rhesus macaques held at a research centre (Morgan and Eisele, 1992). Scurvy developed as a result of feeding the macaques a commercially available diet, which turned out to be deficient in vitamin C. In this case the problem was spotted and rectified, and all but one of the animals regained full health. An older case of scurvy is reported from mummified baboons from ancient Egypt, which would have been kept in captivity (Nerlich et al., 1993). These authors found that pathological investigations contributed significantly to understanding life and living conditions of baboons at this time.

It is unclear if the research suggested by Stone (1965) regarding further testing of all non-human primates to see which if any animals have the gene that is responsible for synthesis of ascorbic acid has ever been undertaken. An answer to this question would provide important information that would contribute to a clearer understanding of human evolution and the relationship of humans to other primates.

Ericksen (1997) care should be taken that skeletal changes caused by scurvy are not confused with those such as infections and anaemia. It has been suggested that in children, the bone pathology associated with scurvy could be confused with the types of changes that occur during child abuse (Rajakumar, 2005), but it is possible that neglect of a child could result in them developing scurvy due to a nutritionally inadequate diet. These issues are discussed in Box Feature 5.2.

CONCLUSIONS

The recognition that vitamin C deficiency does, and certainly in the past, did exist quite widely means that anthropologists will be able to use information on this condition in a wide range of current debates. With the publication of the many cases of scurvy that have been recognised from a range of archaeological contexts over the next couple of years, bioarchaeologists in particular are well placed to contribute to a diverse range of fundamental debates in anthropology, including wider recognition of the health impacts of displaced peoples as well as culturally mediated practices within infant feeding and weaning. Information on scurvy in both past and present societies is also likely to be an important source of information in future studies of disease interactions on human health and the life course.

APPENDIX: SUMMARY OF PUBLISHED ARCHAEOLOGICAL EVIDENCE FOR VITAMIN C DEFICIENCY

TABLE A1 Reported Published Paleopathological Cases of Scurvy

Location	Date	Total burials	Age/Sex, bones affected	Methods	Attributed cause	Sources
Germany	3500–2900 BC (EN)	NS	3 J. Cranial	Mac. Rad. Mic.	NS	Carli-Thiele (1995, 1996)
Greece	3500–2000 BC (N)	NS	J? Cranial	Mac.	Link to adoption of agriculture	Papathanasiou (2005)
Austria	2200–1800 BC (N)	5	1 J. NS	Mac. BiT.	NS	Schmidt-Schultz & Schultz (2004)
Austria	2200–1800 BC (EBA)	110	7 J. NS	Mic.	Seasonal occurrence with adoption of cereal grains	Schultz (2001)
Jordan	3300–3200 BCE (EBA)	153	J. (2 clear, 2?). Cranial, Long	Mac.	Episodes of malnutrition in the community	Ortner et al. (2007)
Slovakia	2200–1800 BC (EBA)	89	15 J. NS	Mic.	Seasonal occurrence with adoption of cereal grains	Schultz (2001)
England	2200–1970 BC (BA)	14	1 J. Cranial, Long	Mac.	Suggested lack of breastfeeding	Mays (2008)
Turkey	2200–1800 BC (EBA)	123	17 J. NS	Mic.	Seasonal occurrence with adoption of cereal grains	Schultz (2001)
Israel	ca. 2200 BC (MiBA)	1	1 J. Cranial Mandible	Mac. Rad.	NS	Mogle & Zias (1995)

England	650 BC–AD 100 (IA-ER)	NS	1 J. NS	Mac. Rad.	NS	Roberts (1987)
England	AD 43–410 (R)	52	2 J. Cranial	Mac.	Seasonal availability of foods	Melikian & Waldron (2003)
Peru	200 BC–AD 1530	NS	J. (10%). Cranial	Mac.	NS	Ortner et al. (1999). Additional data Ortner (2003:390)
England	AD 410–1050 (A–S)	NS	A. Cranial and Long	Mac.	NS	Wells (1964)
Greece	AD 500–600 (PB)	NS	2 J. Cranial Mandible Long	Mac.	Natural event e.g., earthquake restricting fresh food supplies	Bourbou (2003a, 2003b)
Portugal	AD 300–500 (R)	8	1 J. Cranial Scapula	Mac.	NS	Ferreira (2002)
England	AD 750–1550 (M)	52	2 J. Cranial	Mac.	Seasonal availability of foods	Melikian & Waldron (2003)
Tonga	AD 800–1700 (PE)	NS	J. and A. Cranial Long	Mac. Rad.	Possible crop destruction following natural disaster. Co-occurring with a number of other diseases.	Buckley (2000)
Guatemala	800 BC–AD 300	15?	J. and A. Cranial Long	Mac.	Seasonal availability of fruit, food storage and preparation practices	Saul (1972)
Denmark	AD 750–1550 (M)	800	28 J. and A. Cranial and Long	Mac.	NS	Møller-Christensen (1958)

Continued

TABLE A1 (*Continued*)

Location	Date	Total burials	Age/Sex, bones affected	Methods	Attributed cause	Sources
North America	AD 1300–1850	NS	J. (2–38% by site). Cranial Long Scapula	Mac. SSEM.	Variation in food resources and cultural practices	Ortner et al. (2001). Additional data Ortner (2003:390)
England	AD 1485–1603 (PM)	NS	A. Cranial and Long	Mac.	NS	Stirland (2000).
Germany	500–1400 AD	159	57 J. NS	Mic.	Seasonal occurrence	Schultz (2001)
The Netherlands	AD 1600–1700 (PM)	50	39 A and J. Cranial Long	Mac. Mic.	Endemic during winter periods – seasonal availability of foods	Maat (1982), Maat & Uytterschaut (1984), Maat (2004).
Alaska	AD 1800–1900	NS	1 J. Cranial	Mac.	Change in diet post-contact	Ortner (1984)
England	AD 1800 (PM)	164	6 J. Cranial Mandible Scapula	Mac. SSEM.	Socio-economic status likely restricting food available, and potato crop destruction	Brickley & Ives (2006)
Peru	Not specified	19	3 J. Cranial	Mac.	NS	Melikian & Waldron (2003)

Notes: *Dates given are only approximate to provide a time line. Time periods given by original authors are indicated in brackets*: E, *Early*; Mi, *Mid.*; N, *Neolithic*; BA, *Bronze Age*; IA, *Iron Age*; R, *Roman*; A-S, *Anglo Saxon*; M, *Medieval*; PM, *Post Medieval*; PE, *pre-European*; PB, *proto-Byzantine*; J, *juveniles*; A, *adults*; NS, *not specified*; Long, *Long bone*; Mac., *macroscopic examination*; Mic., *histological examination*; SSEM, *surface* SEM; Rad., *radiological examination*; BiT, *biochemical test*.

Chapter 5

Vitamin D Deficiency

Vitamin D is a pro-hormone that is vital for skeletal health and which plays a significant role in many bodily functions including immune reaction, mineral metabolism, cell growth and maintaining cardiovascular health (Holick, 2005:2746S, see also Holick and Adams, 1998; Hochberg, 2003; Pettifor, 2003). Clinical research into the actions and effects of vitamin D and interruptions in its synthesis and utilisation is developing rapidly and the next few years are likely to yield significant advances in our knowledge of this element in health. Similarly, investigations into the role of vitamin D in health in past populations are developing, and will enable a greater insight and wider interpretations of the range of factors affecting life in the past.

Vitamin D synthesis is dependent on the exposure of skin to ultraviolet rays in sunshine or dietary intake of vitamin D from eggs, fortified milk, liver and oily fish such as salmon, tuna and mackerel (Francis and Selby, 1997; Holick, 2006). At the skeletal level, vitamin D is essential for proper mineralisation of osteoid formed during bone growth and remodelling (Pitt, 1988; see Chapter 3 for these processes), and plays a significant role in calcium homoeostasis. The evolutionary development of an inter-relationship between vitamin D and calcium has recently been considered by Holick (2003).

THE SKELETAL REQUIREMENT OF VITAMIN D

Figure 5.1 illustrates how vitamin D is required during the mineralisation of newly formed bone osteoid. Vitamin D is particularly important in aiding intestinal mineral absorption and regulating renal re-absorption and excretion to ensure that sufficient calcium and phosphorous exist in the blood serum to enable bone mineralisation (Pitt, 1988; Heaney, 1997a; Holick and Adams, 1998).

The impact of reduced amounts of vitamin D on calcium metabolism is demonstrated in Figure 5.2. The principal feature is an increase in bone resorption to release calcium back into the blood serum. If severe, a vitamin D deficiency can lead to skeletal osteopenia (Chapuy and Meunier, 1997; Heaney, 1997a; Bilezikian and Silverberg, 2001; Holick, 2002a). During a prolonged deficiency of vitamin D, the subsequent shortages of serum calcium and phosphorous will prevent or delay the mineralisation of newly formed organic bone matrix leading

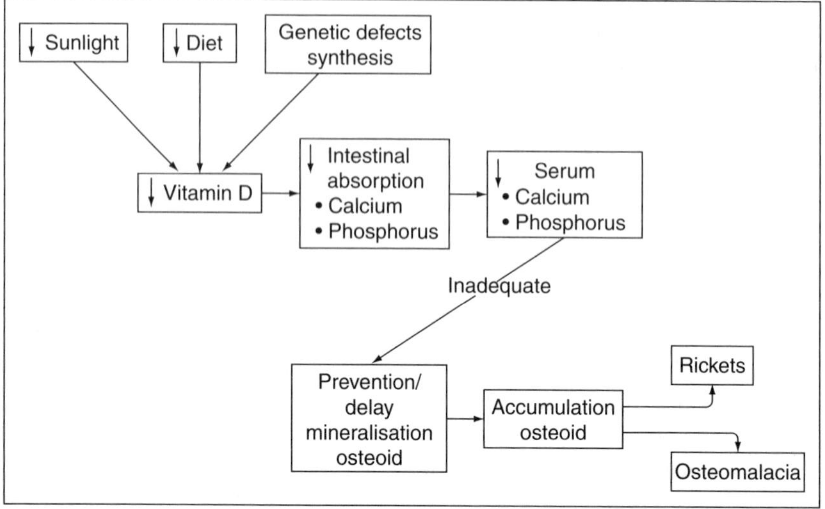

FIGURE 5.1 Flow diagram illustrating the effects of insufficient attainment of vitamin D.

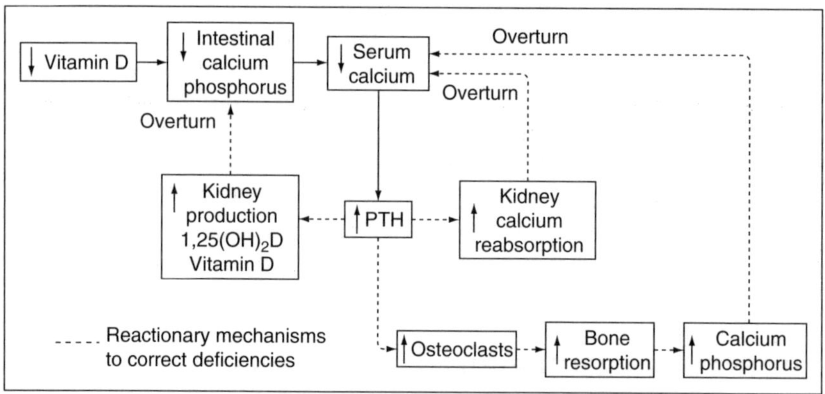

FIGURE 5.2 Physiological effects and reactions to vitamin D insufficiency.

to the accumulation of unmineralised patches of osteoid throughout the skeleton (St-Arnaud and Glorieux, 1997:295; Holick, 2002a:733; Holick, 2003). These predispose to sites of weakness and stress in the skeleton (Chalmers, 1970 in Parfitt, 1986a:268; Mankin, 1974:118–119). Skeletal deformities arising from poorly mineralised bone comprise rickets in children and osteomalacia in adults (discussed further below).

The specific actions of vitamin D on bone cells are to date unclear. Osteoblasts contain a vitamin D receptor, and respond to stimulatory signals in differentiation and osteoid synthesis and mineralisation (St-Arnaud and Glorieux, 1997:297). Vitamin D appears to have an effect on increasing production of factors such as alkaline phosphatase, osteocalcin and osteopontin, which are necessary for osteoid mineralisation (Favus, 1999:617). Under a severe vitamin D deficiency, osteoblast formation and activity are impaired further preventing osteoid mineralisation (Parfitt, 1997).

TERMINOLOGY

Vitamin D is a pro-hormone rather than a traditional vitamin, as it requires synthesis or transformation to become actively utilised in the body (Mankin, 1974; Holick, 2003). Different terminology is used to refer to the status of vitamin D at each stage of formation. As these terms are used interchangeably throughout literature concerning vitamin D, a summary is presented in Table 5.1.

CAUSES OF VITAMIN D DEFICIENCY

Vitamin D deficiency is primarily caused by a prolonged lack of exposure to sunlight and/or dietary deficiency of foodstuffs containing vitamin D (see Holick, 2003; Pettifor, 2003). There are a number of rare hereditary and other conditions that can result in severe skeletal manifestations of vitamin D deficiency and these changes are discussed later in this chapter. It is likely that high levels of vitamin D deficiency in past populations will be attributable to deficiencies of sunlight and dietary practices (see Mays et al., 2006a). As such, the presence of this condition in past populations, whether manifest in adults or juveniles, can provide an important indicator of socio-economic, cultural, environmental and nutritional conditions in the past (see Ortner and Mays, 1998; Mays et al., 2006a; Brickley et al., 2007).

Sunlight

The method of vitamin D synthesis from ultraviolet radiation is detailed in Figure 5.3. It is the final form of vitamin D which is metabolically active and is required for cellular, muscle and skeletal functions. Exposure of the skin to sunlight is therefore extremely important in the synthesis of vitamin D (Hess, 1930:107–117; Mankin, 1974; Holick, 2003).

Holick recently estimated that between 5 and 15 minutes exposure to sunlight per day of adult arms and legs or hands, face and arms between the spring, summer and autumn should generate a minimum required amount of 1000 International Unit (IU) of cholecalciferol (2005:2746S). Fully clothed infants but without a hat, need sunlight exposure for 2 hours a week to maintain normal levels of calcifediol (25(OH)D) (Pettifor and Daniels, 1997:665). Sunlight exposure varies between the geographic latitude of countries and also fluctuates with seasonal alterations in sunlight (see Table 5.2). These factors can significantly alter the amount of vitamin D synthesis.

TABLE 5.1 Summary of Terminology Relevant to Different Aspects of Vitamin D Metabolism

Terminology	Additional terms	Descriptions
7-Dehydrocholesterol	N/A	Exists in the skin and is required to turn into pre-vitamin D_3
Vitamin D_3	Cholecalciferol (also referred to as calciferol)	Formed from irradiation of pre-vitamin D_3 in the skin
Vitamin D_2	Ergocalciferol (also referred to as calciferol)	Produced by irradiation of food sources
25-Hydroxyvitamin D 25(OH)D	Calcifediol	Produced via hydroxylation of vitamin D_3 in the liver
1,25-Dihydroxyvitamin D $1,25(OH)_2D$	Calcitriol	The most biologically active form of vitamin D produced via a hydroxylation of 25(OH)D in the kidney
Vitamin D insufficiency	N/A	Decrease of vitamin D in the serum and in metabolism May initiate secondary hyperparathyroidism If worsened will result in deficiency
Vitamin D deficiency	Rickets – children* Osteomalacia – adults*	A level of vitamin D at which pathological changes will occur Changes in intestinal and renal (kidney) handling of calcium and phosphorous If prolonged, pathological skeletal changes will occur

Source: Reid (2001:554)

Discussed further in this chapter.

Table 5.2 clearly demonstrates that there is substantial variation in yearly sunshine hours throughout various countries based on data published by Hess (1930:43). The growing accumulation of air pollution and continued concentration of urban architecture (such as taller buildings) are likely to have further impacts on sunlight availability in various locations in more recent times.

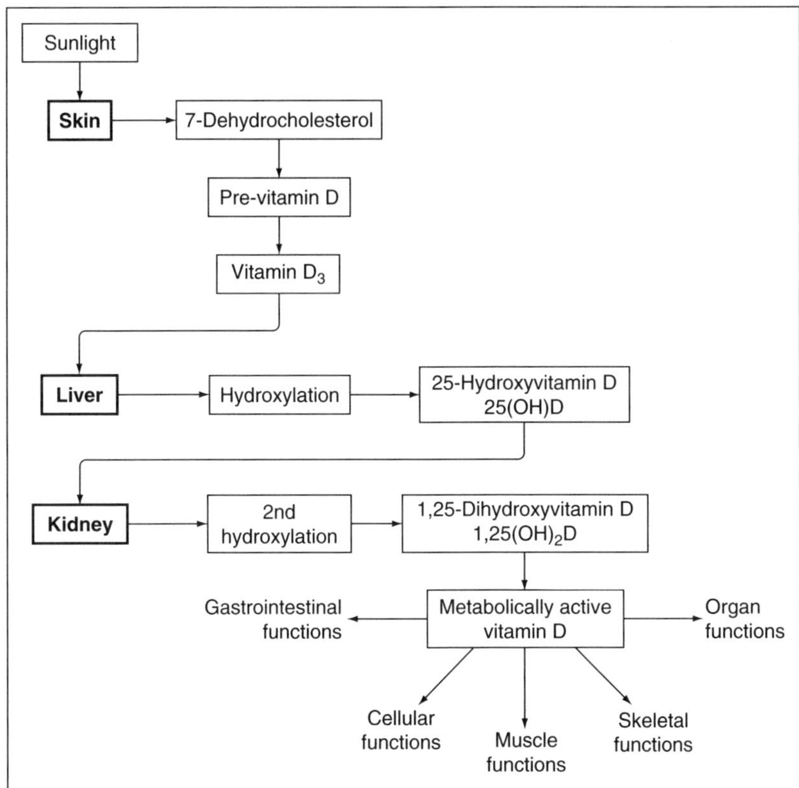

FIGURE 5.3 Attainment and synthesis of vitamin D from sunlight.

The lower levels of sunlight in areas of the United States and Northern Europe are attributable both to greater latitude as well as to the effect of seasonal climatic variation. Changes in the solar zenith angle alters the amount of ultraviolet B radiation that can penetrate through the ozone layer and reach the earth (Holick and Adams, 1998:128; Holick, 2003). Individuals living cities in the northeastern parts of the United States (such as Boston) produce no vitamin D during winter months even when sunlight exposure is prolonged (Holick and Adams, 1998). Whilst age-related changes can increase the susceptibility to vitamin D deficiency in the elderly (see later description), various studies have demonstrated that younger adults are also at risk. For example, a study of young adults living in Finland identified 86% of women and 56% of men aged 31–43 years as vitamin D insufficient at the end of the winter (Lamberg-Allardt et al., 2001:2070; see also Tangpricha et al., 2002). Young adults living in Ireland, England, Italy and France are similarly affected during the winter (see Chapuy and Meunier, 1997:684).

Seasonal vitamin D insufficiency may not result in skeletal changes of rickets or osteomalacia, but may result in bone loss resulting from underlying

TABLE 5.2 Yearly Average Number of Hours of Sunlight in Various Cities (Arranged by Greatest Number of Hours)

City (country)	Yearly average sunlight hours	City (country)	Yearly average sunlight hours
Phoenix (US)	3752	Tsing-tao (China)	2202
Cairo (Egypt)	3238	Tokyo (Japan)	2171
Kingston (Jamaica)	3169	Sofia (Bulgaria)	2145
Port-au-Prince (Haiti)	3056	Sydney (Australia)	2125
San Diego (US)	3049	Portland (US)	2095
Tampa (US)	2948	Toronto (Canada)	2048
Denver (US)	2946	Seattle (US)	2022
Madrid (Spain)	2909	Budapest (Hungary)	1963
San Francisco (US)	2878	Oslo (Norway)	1949
Honolulu (Hawaii)	2840	Vienna (Austria)	1852
San Juan (Porto Rico)	2720	Cracow (Poland)	1733
Athens (Greece)	2655	Zurich (Switzerland)	1693
Chicago (US)	2632	Berlin (Germany)	1672
Washington (US)	2598	Paris (France)	1663
New York (US)	2557	Brussels (Belgium)	1570
New Orleans (US)	2519	Utrecht (Holland)	1469
Buenos Aires (Argentina)	2396	Leningrad (Russia)	1427
Rome (Italy)	2362	Stockholm (Sweden)	1418
Palermo (Sicily)	2261	London (England)	1227
Port of Spain (Trinidad)	2245	Glasgow (Scotland)	1086
Bucharest (Romania)	2238		

Source: Hess (1930:43).

secondary hyperparathyroidism caused by the secondary effects of lowered vitamin D on calcium metabolism (Chapuy and Meunier, 1997). Where deficiencies become more marked and prolonged, without means of additional intake of vitamin D from the diet or artificial supplementation, pathological skeletal changes can develop. The effect of seasonality on sunlight exposure is significant in the pathogenesis of vitamin D deficiency. Behaviour related to seasonal climate (e.g. warm layers, and hats and gloves) can limit skin surfaces available for pre-vitamin D synthesis.

Cultural Practices and Sunlight Exposure

Customs exist in many population groups, which dictate the adoption of specific clothing practices. Cultural practices may be specific to certain occasions or circumstances and there may be a different practice adopted between public and private living (Groen et al., 1964; Groen et al., 1965). However, individuals who follow such cultural practices may receive extremely limited sunlight exposure minimising vitamin D synthesis, which can be exacerbated without extensive consumption of dietary sources of vitamin D (see Berlyne et al., 1973; Güllü et al., 1998). The health consequences of some of these practices are becoming more widely recognised with the onset of detailed investigation in affected regions, and may vary depending throughout the life course. Elderly females, or those experiencing pregnancy and lactation may be affected more than other demographic groups (see later description).

Many modern developed countries strongly advocate the use of sunscreens, and skincare products routinely incorporate protective sunscreen lotions to prevent harmful over-exposure to sunlight and consequent risk of skin cancer. By preventing the absorption of ultraviolet light by the skin, these factors also limit the means of dermal production of vitamin D (Holick, 2003). These developments may have a significant impact in the risk of heightening the occurrence of low levels of vitamin D in many demographic groups within affected countries.

Skin Pigmentation and Genetic Adaptations

Skin pigmentation is caused by melanin, a granular substance that can absorb and filter components of ultraviolet light in the skin. Melanin is responsible for skin and hair colour (Jablonski and Chaplin, 2000; Jobling et al., 2004:407–409). Variation in skin colour derives from differences in the number, size and distribution of the vesicles which contain melanin in the skin (melanosomes). Dark skin contains a number of large, dark melanosomes, whereas lighter skin contains smaller and less dense melanosomes (see Jobling et al., 2004:408). The chemical structure of melanin may also vary between individuals of different skin colour, but such differences are incompletely understood (see Jobling et al., 2004:408).

There are differences in vitamin D concentrations between individuals of different ancestry, which act to vary the risk of developing rickets and osteomalacia (Mitra and Bell, 1997). The interaction between the degree of skin pigmentation and risk of vitamin D deficiency is complex, but is suggestive of greater risk with darker skin. Research has indicated that black population groups living away from the equator, tend to experience lowered levels of 25(OH)D (Holick, 2003). It is thought that increased melanin in the darker skin pigmentation reduces the dermal production of vitamin D_3, which is combined with reduced sunlight exposure evident in countries with greater latitude. In consequence, Holick (2005:2742S) has argued that 'African-Americans who are heavily pigmented require at least five to ten times longer [sunlight] exposure than whites to produce adequate cholecalciferol in their skin' (see also Clemens et al., 1982; Nesby-O'Dell et al., 2002).

Low levels of vitamin D in black individuals tend to be correlated with increased concentrations of parathyroid hormone, in an attempt to increase vitamin D synthesis and improve calcium metabolism and availability. Genetic differences, for example, in enzymes required for vitamin D hydroxylation, may also exist but require further study (Mitra and Bell, 1997). Much of what is currently known regarding vitamin D metabolism derives from research into population groups in North America and Europe. Future research across a wider geographical range will provide a better understanding into population differences of vitamin D amounts and utility, together with the implications for health.

Population groups with increased skin pigmentation who immigrate into more northerly latitudes have been recently documented as experiencing vitamin D insufficiency and deficiency (e.g. Meyer et al., 2004; De Torrenté de la Jara et al., 2006; Roy et al., 2007). It is likely that decreased sunlight exposure due to seasonality and latitude are significant causative factors, together with increased skin pigmentation compared to European white groups. Co-occurring dietary factors may play an important role in increasing the risk of vitamin D deficiency in Asian groups within northern latitudes (see later description).

Food Sources

Vitamin D can be obtained from various food sources, although in the developed world these largely consist of foods that have been artificially fortified to boost vitamin D intake (Mankin, 1974; Holick and Adams, 1998:13; Bilezikian and Silverberg, 2001:72). Figure 5.4 demonstrates the process of vitamin D synthesis from food sources. Natural foods containing vitamin D are primarily eggs and oily fish (Holick, 2003). Table 5.3 summarises the vitamin D content of various foods. Individuals who lack sunlight exposure are dependent on large doses of vitamin D in the diet or on modern artificially created supplements (see the requirements of submariners discussed by Preece et al., 1975 cited by Boyle, 1991:59).

In 1997 the Institute of Medicine recommended that the minimum intake of vitamin D should be 400IU in middle aged adults and over 600IU in older adults (70+ years) (Holick, 2002a:4). However, Holick has argued that these levels are too low and that in the absence of sunlight exposure, dietary intake

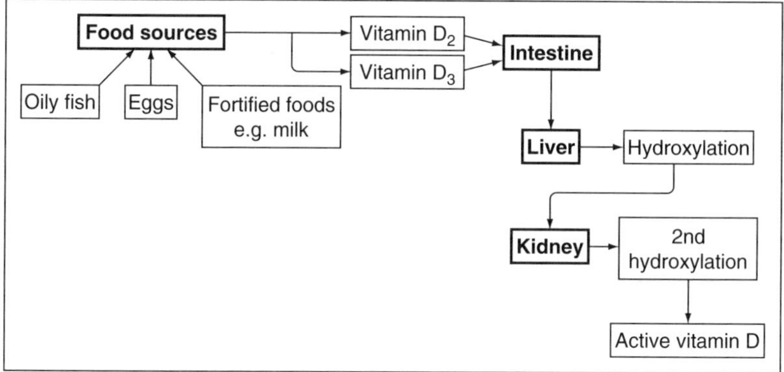

FIGURE 5.4 Attainment and synthesis of vitamin D from food sources.

TABLE 5.3 Vitamin D Content of Various Modern Foodstuffs

Food source	Vitamin D content	Per portion size	Food source	Vitamin D content	Per portion size
Milk*	100IU	8 oz	Liver, beef cooked	30IU	3.5 oz
Orange juice*	100IU	8 oz	Tuna canned	236IU	3.5 oz
Yoghurts*	100IU	8 oz	Mackerel canned	~250IU	3.5 oz
Butter*	56IU	3.5 oz	Sardines canned	~300IU	3.5 oz
Margarine*	429IU	3.5 oz	Salmon canned	~300–600IU	3.5 oz
Cheese*	100IU	3 oz	Salmon fresh	~400–500IU	3.5 oz
Egg yolk (fresh or lightly cooked, not dried)*	~20IU	Yolk	Eel cooked	200IU	3.5 oz
Shiitake mushroom (fresh)	100IU	3.5 oz	Oysters cooked	272IU	3 oz
Shiitake mushroom (sun-dried)	1600IU	3.5 oz	Cod-liver oil	400IU 1360IU	5 ml 15 ml

Sources: Hess (1930:76) and Holick (2006:2065). The recommended daily amount of vitamin D is 1000IU (see text).
**Contain artificially fortified forms of vitamin D that are unavailable to populations in many developing countries and in the past.*

of vitamin D should be increased to 1000IU per day (2002a:4). It is recommended that infants receive 400IU per day and females lactating should receive up to 2000IU per day to maintain infant levels of calcifediol (25(OH)D) and also benefit maternal calcium metabolism (see Pettifor and Daniels, 1997:665, 670).

Modern estimates suggest that between three and four portions of oily fish, such as salmon are needed per week to maintain the minimum adequate vitamin D status (Holick, 2003, 2004). Whilst milk can be a valuable source of vitamin D and calcium in modern western societies, the vitamin D component is only apparent following artificial irradiation (Bilezikian and Silverberg, 2001:72).

Dietary intake of vitamin D is not as efficient as dermal synthesis following sunlight exposure. For example, Holick (2005:2743S) states that an adult white individual exposed to sufficient sunlight to cause a slight pinkness of the skin during one tanning, will produce a comparable amount of vitamin D to an individual ingesting approximately 10,000–20,000IU of vitamin D from food intake.

Of considerable importance in modern and across past populations, is consideration of the potential for various foods to affect the utility of vitamin D and its interaction with calcium metabolism. For example, high levels of phytate from cereal grains can bind to calcium preventing its absorption and use in the body. The resulting calcium imbalance interferes with the normal vitamin D synthesis (see Dunnigan and Henderson, 1997; Pettifor and Daniels, 1997:674). Low-calcium diets increase the concentrations of active vitamin D ($1,25(OH)_2D$), which in turn heightens the rate of clearance of calcifediol (25(OH)D). This reduces the supply of vitamin D available for hydroxylation in the kidney, requiring greater vitamin D_3 production (Pettifor and Daniels, 1997:673; Holick, 2003). High-fibre diets and intestinal malabsorption also reduce the half-life of 25(OH)D.

Vegetarian diets which include sources of calcium (e.g. milk, cheese) but which do not contain oily fish, meat or eggs may pose a risk to health if sunlight exposure is minimal. The low protein levels of such a diet may contribute to a lack of vitamin D through reducing oxidation required in the transformation of pre-vitamin D to active vitamin D (Dunnigan and Henderson, 1997). Low dietary protein is a factor implicated with the onset of osteopenia (see Chapter 7), and this risk will be exacerbated if there is insufficient vitamin D to efficiently regulate calcium absorption and utility.

Pregnancy and Lactation

Foetal development requires substantial calcium accumulation through placental transfer during pregnancy (Prentice, 2003:249; Chapter 6). Adequate vitamin D is essential to increase the intestinal absorption of calcium and for provision of binding proteins (e.g. calbindin$_{9k}$-D) to enable calcium utility (Care, 1997;

Prentice, 2003:250). Additional synthesis of active vitamin D can occur in the placenta and oestrogen, prolactin and growth hormone can enhance the renal production of vitamin D during pregnancy (Holick and Adams, 1998:137). Pregnancy, and lactation in particular, have been associated with an increase in bone remodelling, with temporary losses of bone tissue occurring where extra serum calcium is required (see Prentice, 2003:252; Chapter 6). Prolonged deficiencies of vitamin D during this period will not only impact on calcium availability and absorption from the diet, but may exacerbate bone loss through secondary hyperparathyroidism. This will occur together with a failure to mineralise bone that may be formed during the increased remodelling. Recent clinical evidence has demonstrated the presence of vitamin D insufficiency in reproductive aged females (Holick, 2002b; Nesby-O'Dell et al., 2002).

Severe vitamin D deficiency occurring in childhood and adulthood can affect the structure and morphology of the pelvis, particularly in the narrowing of the pelvic outlet (see Figure 5.5). These changes have severe implications for obstructing normal delivery following pregnancy. Hess (1930:227) has discussed the frequency with which caesarean section operations have been

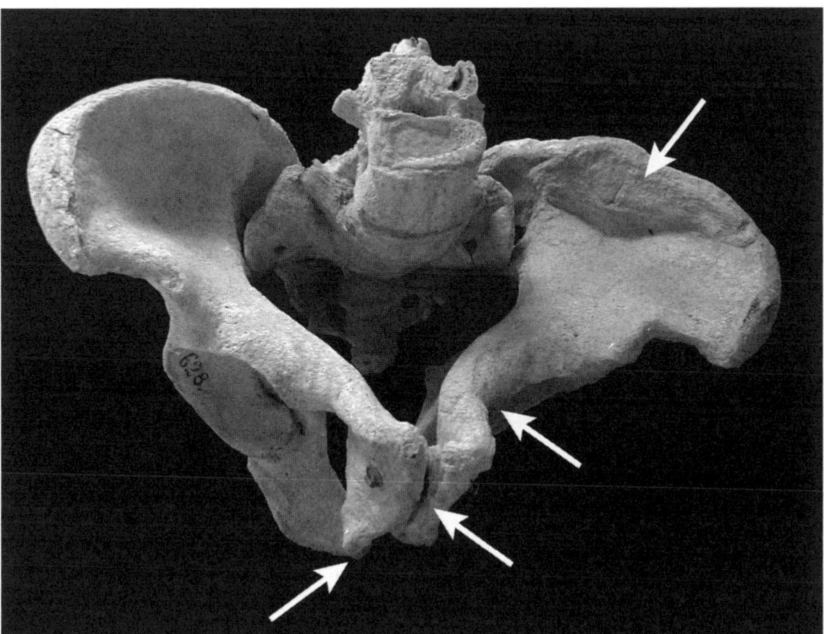

FIGURE 5.5 Severe effects of vitamin D deficiency osteomalacia on an adult pelvis. Note the buckling of inferior pubic rami, displacement of pubic symphyses, fracture of superior pubic rami with protrusion into pelvic inlet, buckling and collapse of ilium and angulation with anterior protrusion of the sacrum. The range of these defects in cases of osteomalacia is likely to have had implications for childbirth. Photograph by authors, courtesy of the Federal Museum for Pathological Anatomy, Vienna, Austria.

required in cases of vitamin D deficiency (see also Mankin, 1974; Herm et al., 2005; and Ortner and Mays, 1998; Ortner, 2003; Brickley et al., 2005 for illustrations of skeletal changes). It is likely that the undertaking of this type of operation even in the relatively recent past was a significant concern for maternal health.

Cultural clothing practices which limit sunlight exposure can particularly exacerbate the risk of vitamin D deficiency during pregnancy and lactation. Increases in bone remodelling combined with insufficient dietary intake of calcium and limited sources of vitamin D can result in severe consequences for the skeleton (see examples in Serenius et al., 1984; Henry and Bowler, 2003:329; Herm et al., 2005). Further differences in socio-economic status, food availability (such as cost of milk products), housing conditions (degree of outside space in rural and urban settlements) and air pollution, are factors contributing to vitamin D deficiency particularly during pregnancy (e.g. Sachan et al., 2005).

Maternal stores of calcifediol (25(OH)D) can be passed to the foetus via the placenta and to the infant through breast milk, but these stores are rapidly utilised within three to four weeks, necessitating additional sources of vitamin D from a young age (Hess, 1930:90–91; Pettifor and Daniels, 1997; Holick, 2003). These sources are severely limited by maternal vitamin D deficiency, and like other metabolic conditions (e.g. scurvy in Chapter 4), can result in infantile expression of the maternal pathology (Care 1997; Mitra and Bell, 1997; Pettifor and Daniels, 1997).

Unfortified cow's milk contains little vitamin D and a relatively equal ratio of calcium to phosphorous. Human milk contains more calcium and vitamin D, with the latter ranging between 20 and 60IU per litre (Pettifor and Daniels, 1997:664). Pettifor (2003:545) has calculated that if an infant consumed 600–700 ml of breast milk a day, this would confer a vitamin D amount of less than 40IU, which is considered insufficient to protect against vitamin D deficiency. Maternal supplementation with 1000–2000IU of vitamin D per day can increase the amount of vitamin D within breast milk. Alternatively, increased infant sunlight exposure is needed to provide adequate vitamin D status.

Increased Age

Clinical evidence indicates that vitamin D concentrations in older individuals are less than those found in younger adults with the same diet and sunlight exposure (Bilezikian and Silverberg, 2001; Mosekilde, 2005). Dermal levels of 7-dehydrocholesterol decline with age (Bilezikian and Silverberg, 2001) and both skin thickness and capacity to produce pre-vitamin D_3 are also limited (Chapuy and Meunier, 1997:681; Halloran and Portale, 1997:542; Holick and Adams, 1998). There is currently no evidence to indicate age-related variation in the capacity of a healthy liver to produce calcifediol (25(OH)D). However, kidney function decreases with age, potentially reducing active vitamin D ($1,25(OH)_2D$) synthesis (Halloran and Portale, 1997; Reid, 2001). Decreased kidney response to parathyroid hormone may also contribute to a reduction of vitamin D (Bilezikian and Silverberg, 2001). It is not yet clear whether there are age-related changes in the intestinal absorption of dietary vitamin D (Chapuy and Meunier, 1997:681; Holick and Adams, 1998:131; Bilezikian and Silverberg, 2001:72). Further research is needed to extrapolate whether there are different effects of aging on vitamin D synthesis between the sexes, as well as between groups of different ancestry (Halloran and Portale, 1997).

Age-Related Osteoporosis

Vitamin D insufficiency has been increasingly noted in elderly individuals with osteoporosis and post-menopausal females (Chapuy and Meunier, 1997; Holick and Adams, 1998; Lips, 2001). Oestrogen deficiency following the menopause can exacerbate secondary hyperparathyroidism contributing to low levels of vitamin D (Holick and Adams, 1998:137). These changes can increase the rate of bone loss, which together with the accumulation of poorly mineralised bone will prevent efficient load bearing and contribute to the risk of fracture. Individuals with age-related osteoporosis have demonstrated markedly low concentrations of vitamin D (Bilezikian and Silverberg, 2001), likely exacerbated where dietary calcium intake is also low (Malabanan and Holick, 2003). A correlation between vitamin D deficiency and osteoporosis-related fracture of the femoral neck has been suggested in several early clinical studies (e.g. Aaron et al., 1974; Parfitt, 1986a; Chapter 6). Several more recent studies have demonstrated that a reduction in the number of osteoporosis-related fractures with supplementation of vitamin D does occur (Chapuy and Meunier, 1997; Feskanich et al., 2003), but such results are not consistent across studies and additional lack of calcium is likely to prevent supplementations of vitamin D from being effective (see recent reviews in Eastell and Riggs, 1997; Favus, 1999; Lips, 2001; Reid, 2001; Malabanan and Holick, 2003). Furthermore, it remains difficult to extrapolate the effects of increased age acting on vitamin D concentrations and synthesis from effects specifically attributable to age-related osteoporosis (Bilezikian and Silverberg, 2001; Reid, 2001).

Additional Causes of Vitamin D Deficiency with Effects on Mineral Metabolism

Lack of sunshine exposure remains a significant cause of vitamin D deficiency. However, as mentioned briefly above, there are a large number of additional conditions that can result in the skeletal manifestations of rickets and osteomalacia, principally via defects in the synthesis of vitamin D, alterations in mineral metabolism or defects within remodelling such as hypophosphatasia (see Chapter 9; Arnstein et al., 1967; Parfitt, 1998; Berry et al., 2002). These conditions are relatively rare today but are gaining increasing clinical interest due to attempts to identify many of the genetic changes responsible (Holick, 2003; Holick, 2006). Table 5.4 highlights some of the principal conditions likely to result in rickets and osteomalacia.

TABLE 5.4 Summary of the Various Forms of Inherited and Acquired Rickets and Osteomalacia

Condition	Features
25-OHase-deficiency rickets	Defect in hydroxylation of vitamin D in the liver Inability to form 25(OH)D Limits the amount of vitamin D passed to kidney for final hydroxylation Rare condition as various enzymes can undertake equivalent of liver hydroxylation of pre-vitamin D
Hereditary, vitamin-D-dependent rickets Type I	Pseudo-vitamin D deficiency rickets Defect in conversion of 25(OH)D to active vitamin D in the kidneys Rare hereditary disorder Condition is often manifest before 2 years
Hereditary, vitamin-D-dependent rickets Type II	Vitamin D resistant rickets Mutations exist in vitamin D receptor gene Tends to have early onset in children (under 2 years) but can occur in some adults
Hereditary, vitamin-D-dependent rickets Type III	Abnormal expression of hormone response element binding protein Binds to vitamin D responsive elements preventing linkage with vitamin D active receptors Rickets or osteomalacia occurs despite normal intake of vitamin D Completely resistant to the action of $1{,}25(OH)_2D$ Severe skeletal deformities
Fibroblast growth factor 23 (FGF23)	Problems in phosphorous homeostasis Includes regulating factors (phosphatonins), such as fibroblast growth factor 23 (FGF23)
Hypophosphatemia	Decreased renal re-absorption of phosphorous Reduction in intestinal absorption of calcium and phosphorous Levels of $1{,}25(OH)_2D$ decrease Deficiencies of vitamin D, calcium and phosphorous may occur

Continued

TABLE 5.4 *(Continued)*

Condition	Features
	May include failure of or excessive production of FGF23 Rickets and osteomalacia can occur If present, normal calcium levels will prevent secondary hyperparathyroidsism
Autosomal dominant hypophosphatemic rickets	Mutation of the FGF23 gene Increased levels of growth factor production Increased phosphorous excretion Decreased phosphorous absorbed from the diet
X-linked hypophosphatemic rickets	Unclear cause Hypophosphatemia and decreased intestinal absorption of calcium and phosphorous Mutation may affect the phosphate regulating endopeptidase homolog X-linked gene May increase expression of FGF23 affecting phosphorous metabolism
Tumour-induced osteomalacia	Small benign tumours can secrete FGF23 Imbalance in the amount of phosphorous absorbed and excreted
Renal tubular disorders	Includes Fanconi syndrome Rickets and osteomalacia secondary to inherited and acquired defects in renal function
Hypophosphatasia	Mutations in the gene for non-specific alkaline phosphatase Results in rare heritable osteomalacia and rickets (see Chapter 9) Incidence of one per 100,000 live births in severe forms, with 300 modern cases (see Whyte, 1999:337) Infantile cases can occur aged 6 months May include craniosynostosis and ante-mortem tooth loss Condition can be fatal in about 50% of the cases
Fibrogenesis imperfecta ossium	A rare bone disorder associated with abnormal bone mineralisation and defective collagen fibres, resulting in features of osteomalacia

Sources: *Glorieux* (1999), *Liberman and Marx* (1999), *Whyte* (1999), *Berry et al.* (2002), *Reginato and Coquia* (2003:1074) *and Holick* (2006).

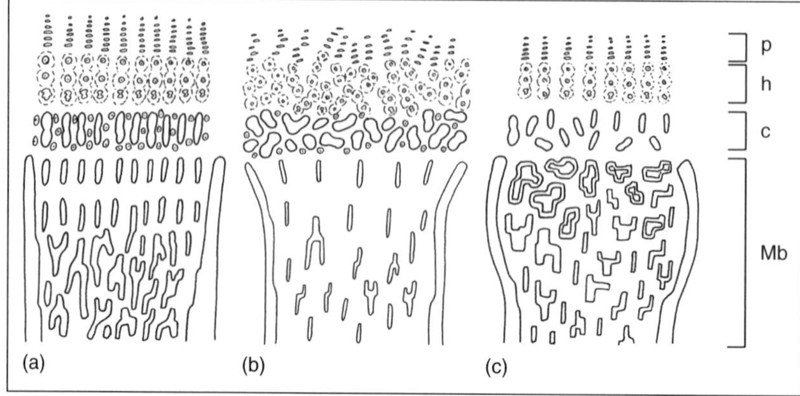

FIGURE 5.6 Schematic illustrations of normal and pathological changes in endochondral development in rickets. p, zone of proliferating cartilage cells; h, hypertrophic zone of cartilage cell maturation and degeneration; c, zone of cartilage mineralisation and osteoblast synthesis of osteoid on cartilage template; Mb, zone of mineralised bone incorporating the growth plate, cortical bone in the metaphyses and internal trabeculae. (a) Shows normal arrangement, (b) shows early changes of rickets with cortical flaring, osteopenia of the trabeculae and disorganised arrangement of forming cartilage and (c) shows recovery from the vitamin D deficiency, with disordered ossification of cartilage formed during the deficiency at the growth plate and metaphyses. The newly developing cartilage is regaining normal columnar structure.

RICKETS

Vitamin D deficiency in growing children prevents calcium from being deposited in the developing cartilage as well as in newly formed bone osteoid, impeding bone mineralisation (Mankin, 1974; St-Arnaud and Glorieux, 1997:295; Chapter 3). As such, deformation particularly occurs in the cartilaginous growth plate between the mineralised metaphysis and the epiphysis, with significant effects on the forming bone structure (Pettifor, 2003).

Lack of bone mineralisation is prominent at sites of endochondral growth (e.g. costocartilage rib junctions and long bones). In these regions, lack of mineralisation promotes continual cartilage development by removing the mechanism needed to cause cartilage cell death and allow penetration of the cartilage template by osteoblasts. The normal, tightly ordered columnar arrangement of cartilage cells fails, and as such disordered cartilage is formed. This poor cartilage arrangement is exacerbated by mechanical forces acting on affected bones (e.g. crawling, walking), which further spreads the cartilage into a wider, splayed, horizontal form, than the vertical orientation which occurs in normal growth. Whilst these changes derive from alterations in the cartilage growth plate, when bone mineralisation occurs following attainment of small amounts of vitamin D or with complete recovery, endochondral bone can only be formed on the pre-existing cartilage template. Therefore, deformities in the metaphyses will become manifest in ossified bone before remodelling attempts to rectify the changes. These pathological changes are illustrated in Figures 5.6

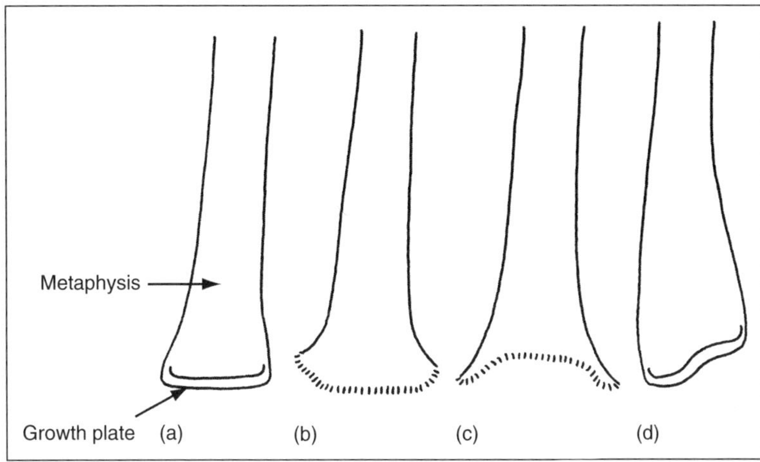

FIGURE 5.7 Schematic overview of pathological changes at the growth plate and metaphyses in rickets. (a) Normal outline of the metaphyses and growth plate in a juvenile radius, (b) increased porosity of the growth plate in active rickets, (c) increased porosity and cupping of the metaphysis in rickets, similar to changes in scurvy and (d) grossly disordered swelling of the metaphysis and angulation of the growth plate in severe deficiency. Severe cupping and swelling changes are likely to continue to be evident in healed cases. Porosity will be rectified with recovery and normal mineralisation of osteoid (no porosity of the growth plate).

and 5.7 and are discussed further in older literature by Park (1923) and Hess (1930) and more recently by Ortner and Mays (1998) and Mays et al. (2006a).

Defects in endochondral mineralisation are specific to growing juveniles and are called rickets. Juveniles also experience a failure to mineralise bone that has been laid down on previously formed bone surfaces and this defect is referred to as osteomalacia (Mankin, 1974; Pettifor, 2003). Therefore, juveniles with vitamin D deficiency experience both rickets and osteomalacia. Adults affected by vitamin D deficiency will only demonstrate skeletal changes relating to incomplete mineralisation of bone on pre-existing surfaces (osteomalacia) with growth at the epiphyseal plates having finished, preventing the manifestation of endochondral defects (see Pettifor, 2003:541; Brickley et al., 2007). In keeping with terminology adopted previously in paleopathology, we will refer to rickets as vitamin D deficiency in childhood, to residual rickets as evidence of the healed childhood condition detectable in adult skeletal remains, and to osteomalacia as vitamin D deficiency which occurred during adulthood. The macroscopic, radiological and histological features of each condition are detailed at the end of this chapter.

CONSEQUENCES OF RICKETS

Rickets has a peak incidence in modern cases between 3 and 18 months, particularly affecting children before they are able to freely move around in the sunshine and when dietary sources of vitamin D are inadequate (Pettifor, 2003:544).

The age of onset may vary considerably across other contexts, depending on maternal health and socio-cultural practices concerning the diet and mobility of juveniles within a population group.

The physical effects of rickets can include a temporary cessation of breathing (apnoea), involuntary muscle contractions and bodily contortions (convulsions) or spasms particularly affecting the face, hands and feet (tetany) (see Hess, 1930; Pettifor, 2003:548). Increased muscle weakness is a significant concern in rickets and adult onset osteomalacia, and can be significant enough to prevent movement (Hess, 1930:243; Parfitt, 1998; Pettifor, 2003:550).

Bone deformation in rickets depends on the age at which the disease occurs and behaviour relevant to the age of occurrence. In severe cases, softening and weakening of the bone caused by the accumulation of unmineralised osteoid and poorly formed bone, causes bending deformities in long bones in response to weight-bearing, including in the forearm during crawling and legs during walking. Young infants that are tightly swaddled may exhibit bowing, angulation deformities and even metaphyseal fractures in long bones and ribs, independent of mechanical force from weight-bearing (see macroscopic features in later description). Apart from causing marked changes in physical appearance (see later in this chapter), these changes may prevent efficient mobility, may cause eventual increased stress on joints and may complicate events such as pregnancy (see above).

Historical Recognition of Rickets

Accurate descriptions of the pathological changes in the condition were reported on by Whistler in 1645 and Glisson in 1650, both noting the occurrence in children aged six months to two and a half years. Trousseau, writing during the 1800s, related rickets to nutritional disturbances and lack of sunlight. The use of cod-liver oil for the treatment of rickets existed since the early 1700s but became widespread after 1920, together with sunlight exposure for treatment. Developments in the artificial irradiation of foodstuffs were originally conducted by McCollum, Park and Mellanby. The history of the recognition and treatment of rickets has been previously discussed by Hess (1930), Aufderheide and Rodríguez-Martín (1998:309), Mays (2003), Pettifor (2003), O'Riordan (2006) and Lewis (2007:119–126).

RICKETS IN THE MODERN PERSPECTIVE

Surveys of infant health within tropical counties have demonstrated that cases of rickets are geographically widespread, occurring throughout the Eastern Mediterranean, Palestine, Egypt, Syria and Lebanon, Pakistan, India, China, Africa including Nigeria, the Ivory Coast and Sierra Leone, as well as Brazil in South America (Jelliffe, 1955; see also Hess, 1930; Pettifor, 2003:545; Pettifor, 2004; Holick, 2006; Thacher et al., 2006).

Various factors may result in the onset of infantile rickets throughout these countries, most notably maternal malnutrition causing vitamin D deficiency during pregnancy and lactation, as well as an inadequacy of vitamin D in infant foods used in partial weaning and complete weaning (Jelliffe, 1955:94). Diets low in calcium and often high in cereals containing phytates can exacerbate vitamin D deficiency (Jelliffe, 1955:95; see also Pettifor and Daniels, 1997:665; Pettifor, 2004; Chapter 7 and above). Jelliffe (1955:29–30, 34) particularly observed subsistence practices countries such as Egypt, Lebanon, Syria and Morocco, of excluding foodstuffs which contain vitamin D (e.g. eggs and fish) from young infant weaning diets, likely contributing to vitamin D deficiency where sunlight exposure was also minimal. Poor nutrition and the onset of many illnesses can lead to diarrhoeal diseases in young infants which will also prevent the dietary intake and absorption of calcium and vitamin D, exacerbating ill-health.

Cultural practices can be particularly important in governing sunlight exposure for very young children. For example, rickets occurred in the Dode !Kung, observed between 1967 and 1968, despite adequate nutrition but caused by excessive protection of infants from sunlight exposure. Spontaneous resolution of the condition occurred when sunlight exposure became a normal part of infant social maturation (Cohen, 1989:82). Young infants in Ibadan, western Nigeria, who were consistently swaddled and carried on their mother's backs (see Figure 5.8) thus receiving limited sunlight exposure experienced infantile rickets (Jelliffe, 1955:93). The rickets often resolved itself in older children, who could spend more time in the sunlight independent of their mother's movements (see also discussion in Pfeiffer and Crowder, 2004:24). Dense forest canopy provides marked shade from the sun and was likely to be a contributing factor to rickets cases along the Ivory Coast observed by Jelliffe (1955:93), as well as in the Congo (Barnes, 2005:327) (see also discussion of light density in tropical rainforests across Mesoamerica, Africa and Southeast Asia by Keesing and Strathern, 1998:90).

Overcrowding of living space, particularly by increasingly tall buildings creating dark alleyways and courtyards and minimising sunlight availability in outside spaces, are factors frequently associated with the living conditions and the onset of the Industrial Revolution during the post-Medieval period in England (see later description, and Mays et al., 2006a). However, the creation of walled cities across many urban centres from a range of present and past contexts are likely to have created similar limitations on building spaces and dark spaces (see e.g. Jelliffe, 1955:94; and Wells' discussion of Lahore, India, 1975:752). Jelliffe (1955:92) also identified urban children as more affected by rickets more than those in rural villages in for example, Syria and Lebanon, with living conditions in urban locales and differences in child weaning foods thought responsible (Jelliffe, 1955:30). Jelliffe (1955:30) noted a socio-economic divide in some area affected by rickets, with poorer individuals in towns in Syria and Lebanon unable to afford milk, with potential complications

FIGURE 5.8 Cultural practices of consistently swaddling and carrying infants on the back of the mother was observed to be a contributory factor to infantile rickets in Ibadan, western Nigeria by Jelliffe (1955). Figure courtesy of the World Health Organization, and taken from Jelliffe DB. 1955. Infant feeding in the subtropics and tropics. Geneva: World Health Organization, Plate 2.

for calcium balance and vitamin D metabolism. Seasonal changes in climate and sunlight exposure may further contribute to rickets occurrence in countries such as Israel (see Costeff and Breslaw, 1962).

ANTHROPOLOGICAL PERSPECTIVES: RICKETS

Investigations into the effects of urban living in the past are frequently related to contexts of the Industrial Revolution. However, historical evidence suggestive of rickets occurring in urban life was documented in Rome, as discussed by Soranus of Ephesus, dating to the second century AD (see Steinbock, 1993:978), as well as from China dating to the seventh and eighth centuries AD (Lee, 1940 discussed in Steinbock, 1993:978).

Various effects of urban living, including factors such as overcrowding and variable living conditions, sanitation, food availability and quality and employment prospects may affect health, and increase the risk of rickets. Heightened understanding of the interactions between people, environment and health is developing (see Howe, 1997; Gage, 2005). Analysis of rickets from urban contexts in the past has enabled better understanding of its range of manifestations, socio-economic prevalence and effects on juvenile growth (see Lewis, 2002; Mays, 2003; Mays et al., 2006a; Pinhasi et al., 2006) as well as seasonal episodes of disease frequency (see Mays et al., 2006a and in press).

Air pollution from coal smog was likely to have been a particularly important factor in contributing to rickets and poor health from the later Medieval period in England (Brimblecombe, 1987; O'Riordan, 2006). Pollution worsened in urban centres during the Industrial Revolution, with increased factory smog, leading to the frequent co-occurrence between many infectious and respiratory diseases (Hardy, 1993). The identification of co-existing conditions, such as between rickets and evidence of pulmonary infection could be reflective of these living and working conditions. Rickets can be an important secondary factor underlying other illnesses, particularly where individuals are too sickly to venture outdoors, as has been attributed to a number cases of affected young infants excavated from the Medieval rural village of Wharram Percy, England (Ortner and Mays, 1998).

There are a wide range of contexts in which vitamin D deficiency can be demonstrated and might be further investigated in past communities. For example, investigation of infant nutrition and health is increasing in bioarchaeology (see Lewis, 2007; Chapter 4), and rickets in infant remains can highlight potential deficiencies in quality of breast milk or the transition to milk replacement foods. It is likely that research will need to incorporate evidence from historical, archaeological and potentially enthnohistorical sources to make accurate interpretations regarding this transition and its effects in past populations (e.g. Fitzgerald et al., 2006; see also Box Features 4.1 and 7.3). The potential for multiple conditions to co-exist and indicate poor nutritional quality and health in infancy is important in the accurate recognition of rickets (Box Feature 5.1).

The most frequently identified traits of vitamin D deficiency in the physical remains of past populations are bending deformities that occurred during childhood. Bone deformation can be severe in healed cases despite subsequent bone remodelling attempts. The social implications of disease are inter-related with concepts of disability and associated factors including provision of aid or food (Box Feature 7.1; Roberts, 2000). Integration of archaeological evidence with historical documentation may be able to better determine the nature of such effects on life in the past, both in terms of social attitudes, manifestations of care and residual effects (see discussion in Sledzik and Sandberg, 2002:190–191). Circumstances under which care may be deliberately withheld can result in various pathologies including rickets (see Box Feature 5.2; Blondiaux et al., 2002). It is of further importance to question whether those affected by physical deformities, such as bowed limbs in rickets, manifest different treatment within native funerary cultures (e.g. discussion in Formicola and Buzhilova, 2004).

Box Feature 5.1. Beyond Fighting: The Physiological Impact of Warfare

Warfare has a significant impact on individuals affected by conflict but who may be removed from direct physical aggression. Conflict can result in the destruction of crops and food stores, may involve the stealing or killing of livestock and can result in damage to farming equipment, as was recently reviewed by Bizzari (2004) (see also Johansson and Owsley, 2002:540). War in Rwanda in 1995 displaced three out of every four farmers resulting in harvest volumes being subsequently reduced (Bizzari, 2004). The consequences of modern warfare may render land unusable for cultivation, for example, containing landmines (Bizzari, 2004). Pathological conditions are likely to occur from food shortages, including many of the metabolic bone diseases, in similar pathways as those affecting peoples displaced from natural disasters (Moss et al., 2006; see Chapter 4 and Box Feature 9.1).

Clinical evaluation of large numbers of Jewish individuals deported into Warsaw during World War II identified physical manifestations of disease together with behavioural and psychological health problems (Braude-Heller et al., 1979). Markedly reduced levels of physical activity and muscle strength occurred in some children causing dependency of daily care. Some female adolescents displayed malnutrition-induced amenorrhoea (cessation of menstruation) (Braude-Heller et al., 1979:48, 55) which, together with inactivity, may have exacerbated the risk of osteopenia (see Chapters 6 and 7). Overcrowded conditions in which Jewish deportees were housed invoked the spread of tuberculosis, particularly prevalent in children (Braude-Heller et al., 1979:55). Vitamin D deficiency was evident in infants and children under two years of age, despite attempts to make preventative treatment widely available (Braude-Heller et al., 1979:51). Poor nutrition, and illness-preventing mobility and ability to move outside in the sunshine are likely responsible factors, in a similar manner as postulated for very young children affected by rickets from the Medieval site of Wharram Percy, England (Ortner and Mays, 1998; see also Jelliffe, 1955:93). Whilst the study of warfare in the past is developing (e.g. Frayer and Martin, 1997; Sledzik and Sandberg, 2002; Mitchell, 2004), increased consideration of how such events may impact on health is required to interpret the full range of effects experienced by past and present communities.

PALEOPATHOLOGICAL CASES OF RICKETS

Table A1 summarises the paleopathological evidence for rickets recorded in juvenile remains. Whilst there are a number of cases linked to increasing urbanisation and industrialisation from the post-Medieval period, the occurrence of this condition in earlier periods may be linked to living conditions, as well as dietary or cultural practices. Recording of human burials in the Centre for Human Bioarchaeology, Museum of London, has so far confirmed 1/18 juvenile cases in Roman burials with 2/324 Medieval and 28/206 post-Medieval cases.

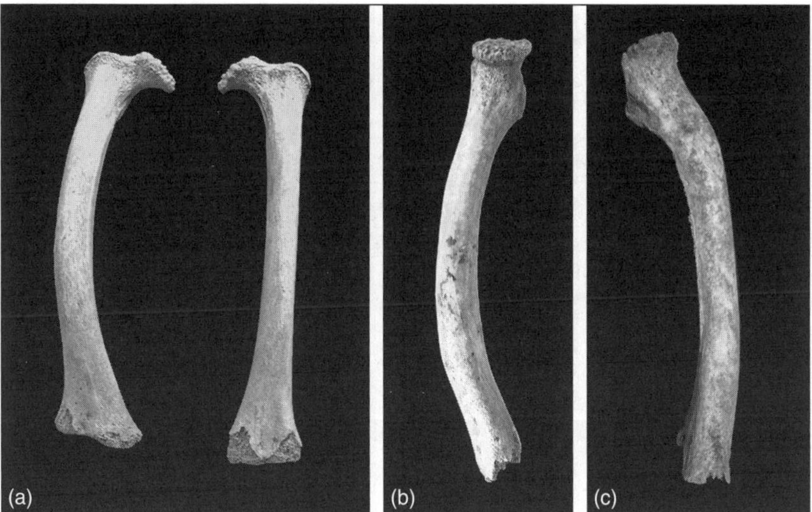

FIGURE 5.9 Bending deformities in juvenile femora resulting from vitamin D deficiency rickets. Note unilateral difference in severity of deformity in (a) and different apex of anterior bend at distal shaft in (b) and at proximal shaft in (c). There is also distal metaphyseal flaring and *coxa vara* of the femoral neck in (a). Femora: (a) (b) from Medieval East Smithfield Black Death cemetery, London. Juvenile aged approximately 5 years (MIN11415). Photograph by Rachel Ives, courtesy of the Museum of London, (c) from post-Medieval St. Martin's Churchyard, Birmingham. Photograph by authors, child aged approximately 2 years (SMB100).

DIAGNOSIS OF RICKETS IN ARCHAEOLOGICAL BONE

Macroscopic Features of Rickets

Long bone bending deformities and metaphyseal swelling are characteristic and well-recognised skeletal changes of rickets, although there are a number of conditions where one or the other feature can occur (see differential diagnosis in later description) (see Figures 5.9, 5.10 and 5.11). There can be considerable variation in the manifestations of bending between affected limbs and between individuals as demonstrated in tibiae shown in Figure 5.10, with post-Medieval cases from St. Marylebone, Westminster, London (Miles et al., in preparation) and St. Martin's Churchyard (Mays et al., 2006a; Figure 5.12).

Early stages of the condition may not include bending. Instead the initial changes at the metaphyses will include fraying and flaring of the growth plate margins and metaphyseal junction as the cartilage structure starts to lose the organised vertical arrangement (see Figures 5.6, 5.7, 5.10(d) and 5.12). These changes may progress to marked swellings in a severe deficiency. Active rickets can be recognised by increased porosity of bone surfaces, particularly the cranium (see Figure 5.13) and the growth plates, where unmineralised osteoid was present during life. Detailed discussion of these features and methods for

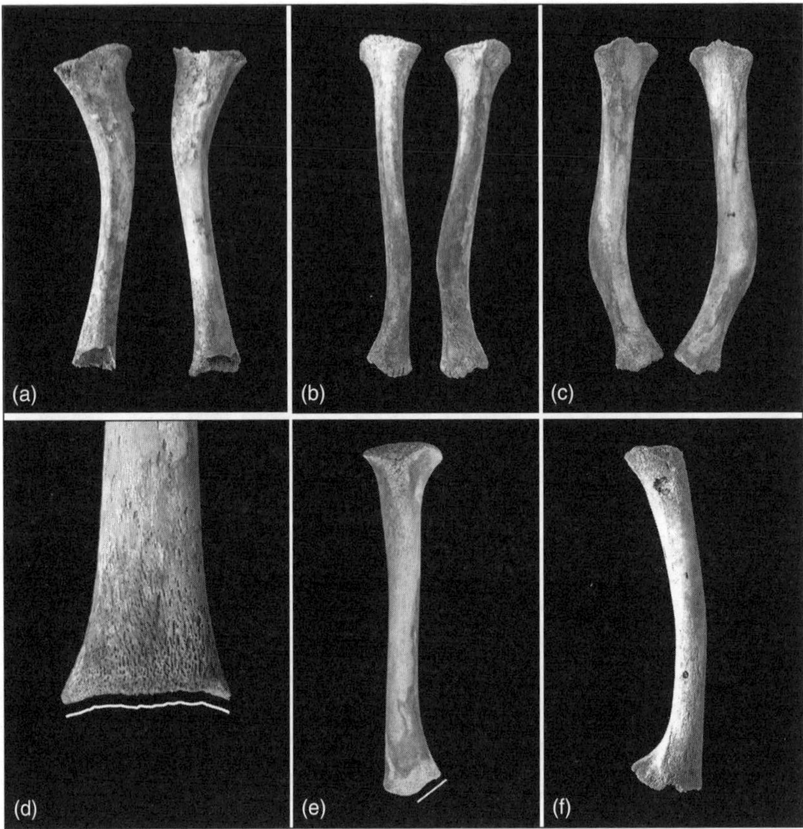

FIGURE 5.10 (a–c) Range of bending deformities in juvenile tibiae due to vitamin D deficiency rickets, (d) fraying and slight cupping of the distal tibial growth plate with outline emphasized, (e) medial tilting of the distal tibial growth plate with outline emphasized and (f) posterior swelling and flaring of the distal tibial growth plate. Figures (a–c) and (f) post-Medieval examples from St. Marylebone, Westminster, London (see Miles et al., in preparation). Photographs by Rachel Ives, courtesy of the Museum of London Archaeology Service. Figures (d) and (e) are post-Medieval examples from St. Martin's Churchyard, Birmingham. Child aged approximately 2 years (SMB100).

osteological recording are provided by Ortner and Mays (1998) and Mays et al. (2006a) (see Figure 5.14; Table 5.5). Where a child has overcome the vitamin D deficiency, but skeletal defects are still clearly visible, this is considered healed rickets with evidence of porosity limited to the concave surface of severe bending deformities due to compensatory mechanical remodelling.

Two forms of rickets have been noted by previous researchers report edly based on the nutritional health of the affected juvenile (see Silverman, 1985; Stuart-Macadam, 1989; Ortner, 2003). The atrophic form is linked

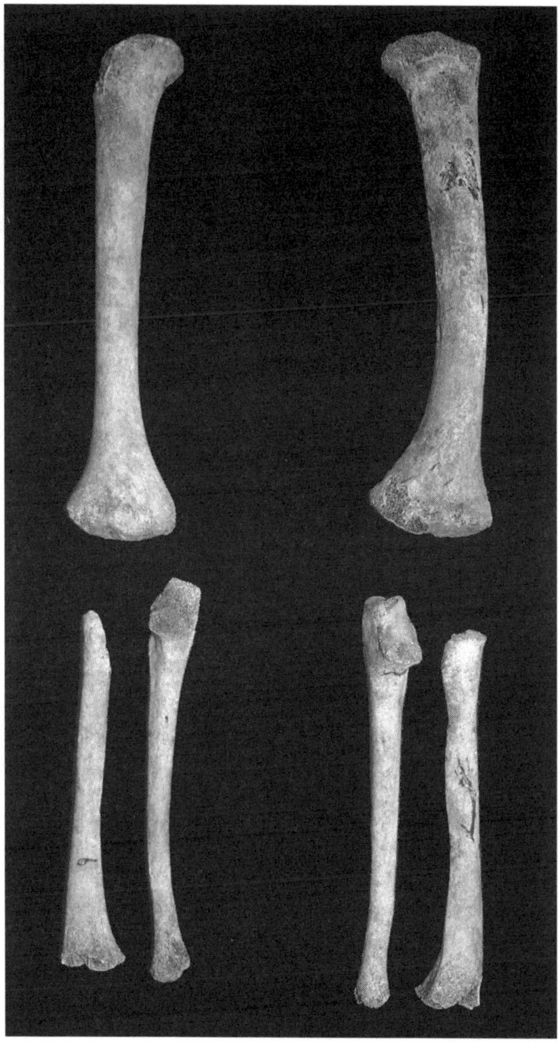

FIGURE 5.11 Arms from a juvenile affected by severe, but healed vitamin D deficiency rickets. All arms bones are thickened, with curvature and flaring of distal left humerus, and curving and flaring of the distal ulnae and radii. Post-Medieval archaeological example from St. Martin's Churchyard, Birmingham. Child aged approximately 2 years (SMB100).

to malnourished children and the hypertrophic form to well-nourished children. There is however, significant overlap between skeletal features seemingly attributed to each form (e.g. discrepancies between presence of cupping deformities and osteopenia present in both types in Silverman, 1985:668, 669). More importantly, the stage of the disease will determine the presence pathological manifestations (e.g. active or healed) together with the age of the child.

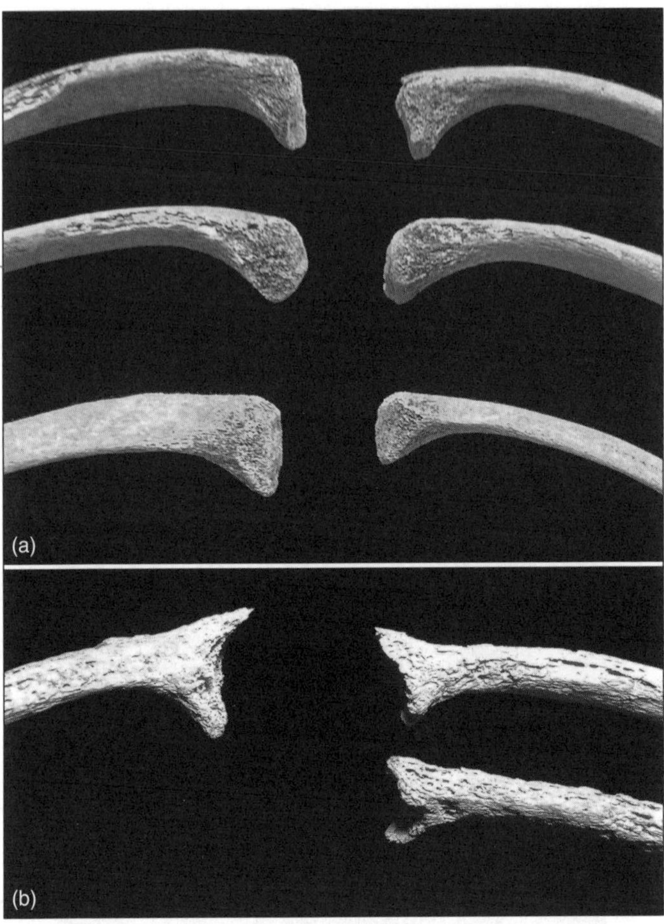

FIGURE 5.12 Pathological changes in the costocartilage junctions of juvenile ribs in vitamin D deficiency rickets. (a) Shows rounded swellings compared to sharp flaring in (b). Archaeological examples: (a) post-Medieval example from Cross Bones burial ground, London, juvenile aged 7–11 months (REW 109A), (b) from post-Medieval St. Benet Sherehog, London, juvenile aged 1–5 years (ONE616). Photographs by Rachel Ives, courtesy of the Museum of London.

Recent researchers have noted difficulties in utilising these divisions (see Littleton, 1998), and Mays et al. (2006a) present further discussion of these issues.

Radiological Features of Rickets

A detailed discussion is presented in Table 5.6 on radiological features of vitamin D deficiency rickets.

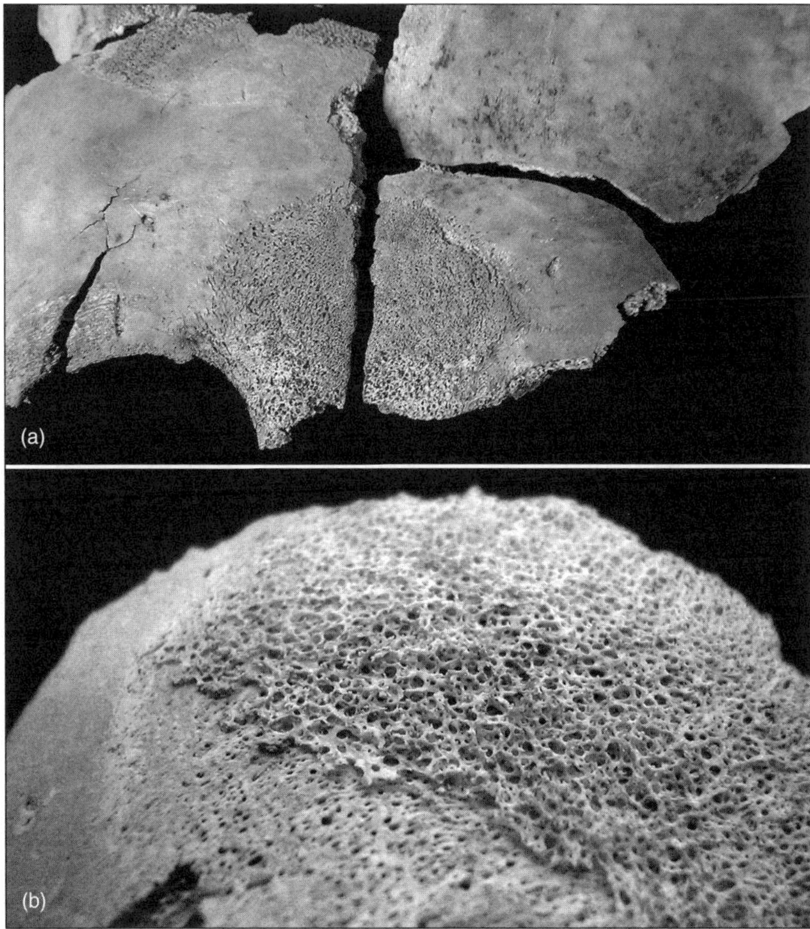

FIGURE 5.13 Plaques of porous, spiculated bone in juvenile cases of vitamin D deficiency rickets. (a) Porous bone affecting the glabella from post-Medieval St. Benet Sherehog, London. Photograph by Rachel Ives, courtesy of the Museum of London, juvenile aged 1–5 years (ONE616). (b) Juvenile parietal bone from Medieval Wharram Percy of an infant aged 6–8 months (NA236). Photograph by Rachel Ives, courtesy of Simon Mays, English Heritage.

Histological Features of Rickets

Table 5.7 shows the histological features of vitamin D deficiency rickets. Figure 5.15 presents a BSE-SEM image of a section of juvenile bone with pathological changes of vitamin D deficiency rickets.

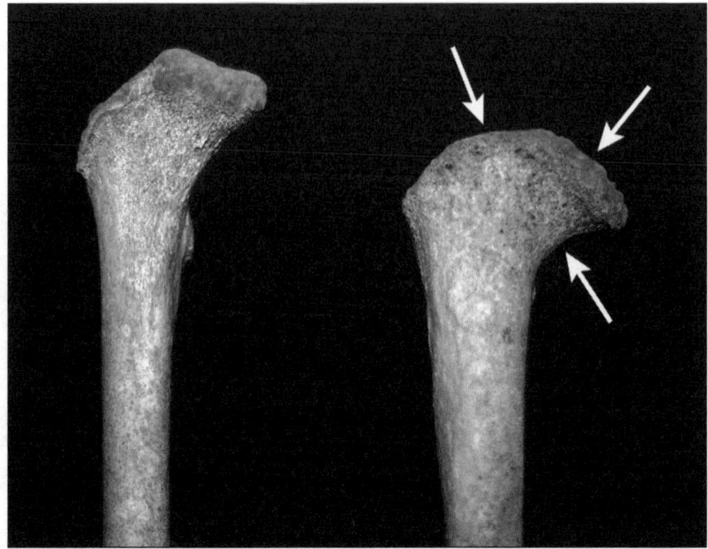

FIGURE 5.14 Pathological changes at the juvenile proximal femur (arrowed) due to vitamin D deficiency rickets compared to normal. Increasingly acute angulation of the neck (*coxa vara*), depression and flattening of the head and proximal epiphyses. Post-Medieval examples from St. Martin's Churchyard, Birmingham, juvenile aged 2–3 years (SMB108).

RESIDUAL RICKETS IN THE ANTHROPOLOGICAL PERSPECTIVE: ADULT EVIDENCE OF CHILDHOOD VITAMIN D DEFICIENCY

It is a significant challenge to the reconstruction of many aspects of health in the past that illness affecting juveniles cannot be construed according to a sex basis when based on the examination of the juvenile skeleton (Lewis, 2007). Residual rickets deformities in adults may provide a broad insight into any potential sex-related differences that could indicate variation in health in juveniles. Research has suggested that males may be more commonly affected than females, although the mechanisms for this are unclear (Hess, 1930:91; Pettifor and Daniels, 1997:664). Patterns of disease onset are also likely to vary depending on cultural practices throughout the world. Inferences regarding sex-related disease trends from past populations require caution given the composition of the archaeological record (see discussion in Chapter 2) and the potential for bone remodelling to obliterate pathological changes over time. However, a greater consideration of these factors could offer additional insights into reconstructions of health in past populations.

Investigations into the health of various slave populations in the past have demonstrated a range of skeletal health insults. Slaves working on plantations in the Caribbean have shown evidence of chronic and likely seasonal vitamin and mineral deficiencies (Corruccini et al., 1987:183; Hutchinson, 1987:235–237; Lambert, 2006). Slaves living within the rural South of North

TABLE 5.5 Macroscopic Features of Rickets

Bones affected	Features	Code	Differential diagnosis	Sources
Cranium	Delayed closure of fontanelles	G	Hydrocephalus, Congenital or developmental conditions	Hess (1930), Pettifor & Daniels (1997), Pettifor (2003)
	Cranial bone thinned	G	Osteopenia	
	Frontal and parietal bossing	G	Scurvy, Anaemia	Ortner & Mays (1998), Mays et al. (2006a)
	Craniotabes (softening of bone behind ears over occipital region and adjacent to lambdoid suture)	G	Normal variant, Prematurity, Congenital syphilis, Osteogenesis imperfecta (Chapter 9)	
	Formation of large, square-shaped head	G	Normal variant, or swaddling/binding	
	Layers of spiculated, irregular, porous bone formation can occur during healing when osteoid is mineralized	D	Normal growth (endocranial), Scurvy, Anaemia	
Dentition and mandible	Delayed eruption of deciduous and permanent dentition	G	Various illness and childhood stress, Generalized nutritional defects, Scurvy	Hess (1930), Hillson (1996), Berry et al. (2002)
	Enamel hypoplastic defects	G	Influence of sex	
	Dental caries	G		
	Mandibular ramus may show medial angulation	D		
Vertebrae	Kyphosis or scoliosis (generally T9-L3)	D	Congenital or developmental conditions	Hess (1930), Pettifor & Daniels (1997), Pettifor (2003)

Continued

TABLE 5.5 (*Continued*)

Bones affected	Features	Code	Differential diagnosis	Sources
Ribs and sternum	Alteration in rib neck angle	D	Age*	Pettifor & Daniels (1997),
	Lateral straightening of shaft (thorax narrowing)	D	Age*	Scheuer & Black (2000a)
	Enlargement of costochondral rib junctions; flaring to swelling and beading ('rachitic rosary')	D	Scurvy	Pettifor (2003)
	Protrusion of sternum together with rib angulation ('pigeon-chested')	D		
Pelvis and sacrum	Exaggerated medio-lateral curvature of ilium	D		Hess (1930), Ortner & Mays (1998)
	Acetabulae pushed dorsally into pelvic cavity and angled anteriorly (*protrusio acetabulae*)	D	Developmental conditions	
	Retarded disproportionate growth of pelvis	G		
Long bones	Flaring and swelling of distal metaphyses	D	Scurvy	Hess (1930), Pettifor (2003), Pettifor & Daniels (1997),
	Fraying bone margins growth plate	D	Post-mortem damage	Ortner & Mays (1998), Mays et al. (2006a)
	Porosity growth plate	D		
	Cupping deformities of growth plate and metaphyses	D	Post-mortem damage. Trauma if severe, likely unilateral, scurvy.	Various. See Table 5.10. Prolonged habitual posture (e.g. sitting, lying). TPBD.

Bending (*genu varum*): forearms crawling, legs with walking	D	
Bending may be variable between limbs (e.g. *valgus* and *varus* changes)	D	
Angulation of femoral neck due to weight-bearing on weakened bone (*coxa vara*)	D	Osteopenia, Developmental (e.g. dislocation)
Angulation of knees ('knock-knees')	G	Blount's disease
Shortening with gross deformity and growth stunting	G	Various childhood illnesses and Trauma
Fractures (e.g. tight swaddling – ribs, distal radius, ulna, fibula or weight-bearing)	G	Osteogenesis imperfecta (Chapter 9), Variant forms of rickets, Congenital syphilis, Scurvy
Thickening – healed rickets or variant form (especially with bending)	D	Scurvy, Congenital syphilis, Non-specific infection, Trauma
Periosteal porosity concave of bending compensatory remodeling	G	

Notes: *The features listed demonstrate the skeletal changes useful for pathological identification. Diagnosis can be determined on the strength of each feature:* S *strongly diagnostic feature,* S *features are normally required for a diagnosis.* D *diagnostic feature, the presence of multiple* D *features are required for a diagnosis.* G *general changes can occur in many of the metabolic bone diseases, as well as in other conditions.* G *features can aid a diagnosis together with* S *or* D *features but cannot be used alone to suggest a diagnosis. Commonly occurring conditions that display these pathological features are listed to aid differential diagnosis, although this list is not exhaustive.* TPBD *denotes traumatic plastic bending deformity.*

**Changes may appear in a healthy infant as a manifestation of normal development at young age* (<3 *months*). *For differential conditions to bending deformities see also* Table 5.10.

TABLE 5.6 Radiological Features of Vitamin D Deficiency Rickets

Bones affected	Features	Code	Differential diagnosis	Sources
Cranium	Generalised osteopenia.	G		Hess (1930)
	Healing: increased periosteal apposition on ectocranium, irregular trabecular spurs	D	Scurvy	
Dentition	–			
Vertebrae	Possible osteopenia	G		
Ribs and sternum	Loss of integrity of the growth plate at costochondral junction	D	Scurvy	Hess (1930), Silverman (1985)
Pelvis and sacrum	Possible osteopenia	G		
Long bones	*Active changes*:			
	Porosity of bone along growth margin	D	Scurvy	Hess (1930), Silverman (1985), Reynolds & Karo (1972), Pettifor & Daniels (1997), Berry et al. (2002)
	Fraying of growth plate; brush-like, fringed bone projections	D		
	Osteopenia of trabeculae along growth plate and in metaphyses	D		
	Coarsening remaining trabeculae	D		
	Osteopenia of cortex, loss of clear distinction in margins	D		
	The epiphyses may be ill-defined and osteopenic	D		

	Healing changes:			
	New trabecular structure regained along metaphysis	D	Scurvy	
	Some continuation of disorganised trabecular orientation with mineralisation of previously formed cartilage defects	D		
	Vertical trabeculae may be coarsened with mineralisation, in distal metaphyses (e.g. femur)	D		
	Margin growth plate regains definition (solid white line), lack fraying or porosity	D		
	Cortical outline sharpened	D		
	Cortex may be thickening with mineralisation of periosteal apposition	D		
	Morphological changes (bowing, flaring of metaphysis) will remain if severe	D		
Other	Fractures and Looser's zones (pseudofractures) may be evident in severe cases	D		Hess (1930)

Notes: *– denotes no significant features at this location.* See Table 5.5 *for definition of codes used in diagn osis.*

TABLE 5.7 Histological Features of Vitamin D Deficiency Rickets

Tissue type	Features	Code	Differential diagnosis	Sources
Cortical bone	If active, loss of new bone formation	G	Scurvy, Osteopenia, Genetic forms of rickets Defects as in osteomalacia	Mankin (1974), Pitt (1988)
	May be increased number of sites of resorption	G		
	Increased cortical porosity, large Haversian channels Irregular Haversian systems	G		
	Bone formed will be poorly mineralised	D		
	Mineralisation defects may be adjacent to cement lines	G		
	Osteocyte lacunae may be enlarged	G		
Trabecular bone	Resorption may be increased. If severe may show perforation of trabeculae	G	Scurvy, Osteopenia Defects as in osteomalacia	Mankin (1974), Pitt (1988), Mays et al. (2007)
	Separation of recently formed bone from more mature sections	D		
	With secondary hyperparathyroidism 'bite-like' defects and tunnelling resorption	D		
	If mineralisation continues, new bone will be poorly formed	D		
	Defects may be localised adjacent to cement lines	D		
	Remodelling in healed cases may show increased resorption at metaphyses to remove cupping deformities	G		

Note: See Table 5.5 for definition of codes used in diagnosis.

America, including Maryland, Virginia and the Carolinas, endured deficiencies of calcium and vitamin C in the diet. Residual rickets were identified in 50% of females and 75% of males examined by Kelley and Angel (1987:206). A multi-factorial cause including skin pigmentation and decreased vitamin D synthesis, poor nutritional quality, and high rates of parasitic infestations may have adversely affected intestinal calcium metabolism and contributed to the onset of vitamin D deficiency (Kelley and Angel, 1987).

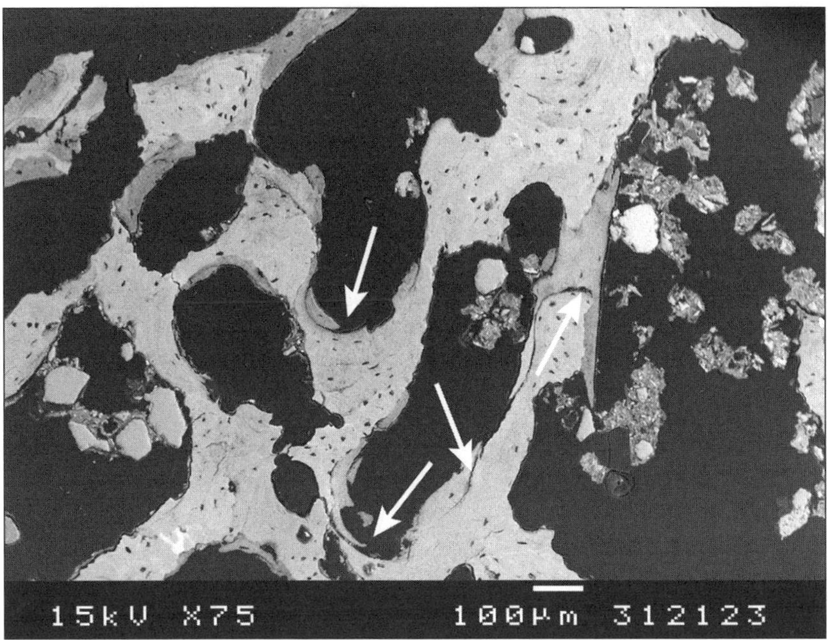

FIGURE 5.15 BSE-SEM image of a section of juvenile bone with pathological changes of vitamin D deficiency rickets. Incomplete layers of bone formation and defective mineralisation of bone adjacent to cement lines are arrowed. Post-Medieval sample of juvenile bone from St. Martin's Churchyard, Birmingham.

In a contrasting pattern to that above, urban slaves in New Orleans displayed no evidence of dietary insufficiency (Owsley et al., 1987), and disease presence may vary depending on the environmental as well as socio-economic status of the relevant individuals examined (Kelley and Angel, 1987; Rathbun, 1987). Martin et al. (1987) successfully demonstrated that poor health and diet had significant impacts on bone quality in adults and juveniles in the Cedar Grove, Arkansas African-American sample, which dated 1890–1927, by combining macroscopic and histological analyses. Table A2 details the paleopathological evidence for residual rickets in bioarchaeological analyses. On-going recording of a large skeletal collection from London has confirmed 5/1393 Medieval cases of residual rickets and 13/406 post-Medieval cases (Centre for Human Bioarchaeology, Museum of London).

DIAGNOSIS OF RESIDUAL RICKETS IN ARCHAEOLOGICAL BONE

Macroscopic Features of Residual Rickets

Table 5.8 outlines the macroscopic changes in adult skeleton indicative of residual rickets in childhood.

TABLE 5.8 Macroscopic Changes in Adult Skeletons Indicative of Residual Rickets in Childhood

Bones affected	Features	Code	Differential diagnosis	Sources
Cranium	Residual frontal or parietal bossing	G	Congenital/developmental, Normal variant, Prematurity	Hess (1930), Ortner & Mays (1998), Mays et al. (2006a)
	Formation of large, square-shaped head	G	Normal variant, Cultural (e.g. swaddling/binding) Age-related changes in cranial shape	
	Mandibular ramus may show medial angulation	G	Influence of sex	
Dentition	Enamel hypoplastic defects	G	Various childhood illness, Generalised nutritional quality	Hillson (1996)
	Dental caries	G		
Vertebrae	Kyphosis or scoliosis (generally T9-L3)	D	Congenital/developmental Osteomalacia	Ortner (2003)
	Vertebral body collapse	G	Osteoporosis, Buckling in osteomalacia	
Ribs and sternum	Alteration in rib neck angle	D	Osteomalacia	Pettifor (2003), Pettifor & Daniels (1997), Brickley et al. (2005)
	Lateral straightening of shaft (thorax narrowing)	D		
	Protrusion of sternum together with rib angulation ('pigeon-chested')	D		

Pelvis and sacrum	Lateral narrowing of pelvis	D	Osteomalacia	Hess (1930), Ortner & Mays (1998), Ortner (2003), Brickley et al. (2005)
	Bulging at pubic symphysis			
	Ventral projection of sacrum	D	Osteomalacia	
	Narrowed pelvic inlet	D	Osteomalacia	
	Acetabulae pushed dorsally into pelvic cavity anteriorly (*protrusio acetabulae*)	D		
	Curvature/abnormal shape of ilia (see Ortner and Mays, 1998 for expression in childhood)	D	Osteomalacia	
	Anterior angulation/bending of sacrum	D	Osteomalacia	
Long bones	Residual bending (*genu varum*), typically of legs rather than arms	D	See Table 5.10. Prolonged habitual posture, TPBD, Mechanical adaptation, Osteomalacia, Paget's disease	Hess (1930), Pettifor & Daniels (1997), Ortner (2003), Brickley et al. (2005)
	Heightened angulation of femoral neck (*coxa vara*)	D		
	Angulation of knees ('knock-knees')	G	Blount's disease Childhood stress, Trauma Infection, Paget's disease	
	Shortening	G		
	Thickening, especially filling in concavities of bends	D		
	Medio-lateral widening of proximal femora (sub-trochanteric)	D	Osteomalacia	

Notes: For differential conditions to bending deformities, see also Table 5.10. TPBD *denotes traumatic plastic bending deformity. See* Table 5.5 *for definition of codes used in diagnosis.*

TABLE 5.9 Radiological Changes in Adult Skeletons Indicative of Rickets in Childhood

Bones affected	Features	Code	Differential diagnosis	Sources
Cranium	–			
Dentition	–			
Vertebrae	–			
Ribs and sternum	–			
Pelvis and sacrum	–			
Long bones	Prominent thickening of cortical bone on concave of bending deformities, as mechanical remodelling (see Figure 5.16) Long-standing recovery from active childhood condition is likely to have restored cortical and trabecular structure to near normal	D	Paget's disease, Healed osteomalacia, Prolonged habitual posture, TPBD, Mechanical adaptation	Hess (1930), Mankin (1974)

Notes: For differential conditions to bending deformities, see also Table 5.10. TPBD, traumatic plastic bending deformity; –, no significant features at this location. See Table 5.5 for definition of codes used in diagnosis.

Radiological Features of Residual Rickets

Radiological features of residual rickets are presented in Table 5.9. Figure 5.16 shows a medial radiograph of an adult femur with residual bending deformities due to childhood vitamin D deficiency rickets.

Histological Features of Rickets

Adequate recovery of vitamin D deficiency suffered in childhood, will result in the attainment of normal bone tissue structure at the histological level (see Chapter 3 for features). It is extremely unlikely that histological evaluation will be able to detect evidence of healed childhood rickets.

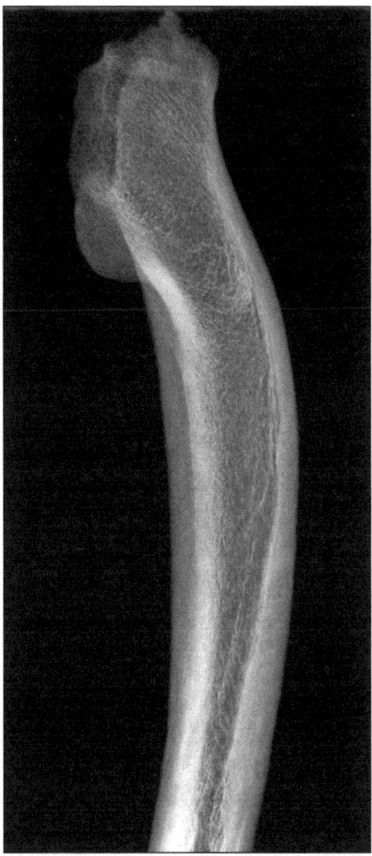

FIGURE 5.16 Medial radiograph of an adult femur with residual bending deformities due to childhood vitamin D deficiency rickets. Marked anterior bending of the proximal shaft, with anterior and posterior cortical thickening into the midshaft. Post-Medieval archaeological example from St. Martin's Churchyard, Birmingham.

CO-MORBIDITIES

Rickets can be associated with an increased risk of both respiratory and gastrointestinal infections, which can increase the risk of mortality (Hess, 1930:235–236). Rickets has been linked to the onset of anaemia and malnutrition and risk of dying (Pettifor, 2003:550). Iron deficiency in particular can impair intestinal absorption of minerals, potentially limiting vitamin D and calcium availability. Malnutrition may result in a number of co-occurring conditions, including anaemia, rickets and scurvy (see Stuart-Macadam, 1989).

As discussed in Chapter 4, there are close links between rickets and the co-occurrence of vitamin C deficiency. A deficiency of vitamin C will prevent osteoid from being secreted and may prevent the typical skeletal manifestations of rickets from developing. Several of the skeletal changes in each condition may appear similar, including swelling at the rib costochondral junctions and metaphyseal cupping (see Table 5.5; Chapter 4 and Figure 4.12). The range of skeletal changes apparent in each condition will need to be considered to enable differentiation of macroscopic lesions.

Vitamin D has a complex role in aiding immune function (Holick, 2003). Continued research is likely to clarify the extent to which vitamin D may be interlinked with the prevention of many illnesses, including cancers and tumour conditions (Pettifor and Daniels, 1997:667; Mensforth, 2002 and pers.com; Holick, 2002b, 2004). Vitamin D may enable improved immune response to tuberculosis (TB) infection (see McMurray et al., 1990; Strachan et al., 1995; Ustianowski et al., 2005). Research has questioned whether a genetic defect in the vitamin D receptor can be linked with cases of TB (Lewis et al., 2005), although extrapolating the independent effects of lifestyle or diet on vitamin D status may be difficult. Increases in serum vitamin D have been observed in hypercalcaemia in some cases of TB and synthesis of this hormone may also be undertaken by granlomas, potentially complicating changes in vitamin D status (Cadranel et al., 1994; Bell, 1998).

DIFFERENTIAL DIAGNOSIS

Bending deformities may often be the most obvious evidence of vitamin D deficiency, particularly in skeletal remains, and can occur in rickets, as evidence of residual rickets in adults and also in osteomalacia. However, there are a number of other conditions that can result in long bone bending deformities, and these conditions are summarised in Table 5.10.

Determining the presence or absence of bending deformities tends to be subjective. Quantitative analyses of variation in bending morphology have been attempted (e.g. Ivanhoe, 1994; Shackelford and Trinkaus, 2002). However, such indices can be affected by differences in habitual patterns of mobility and weight-bearing, the physical environment, morphological differences in robusticity between the sexes, across age and between different population groups (Ruff et al., 2006; Wescott, 2006). These factors necessitate modifications within quantitative analyses (e.g. Ivanhoe, 1994) and may further complicate what is deemed 'normal' and what constitutes a 'pathological' change especially across different population groups.

VITAMIN D DEFICIENCY OSTEOMALACIA

Vitamin D deficiency osteomalacia occurs where insufficient vitamin D prevents mineralisation of osteoid. The range of macroscopic deformities observable in the skeleton are listed in Table 5.11. In the adult the effects of vitamin D deficiency can be quite non-specific. Generalised or local muscle pains and weakness may exist, and in prolonged cases muscle and bone deformity in

TABLE 5.10 Summary of the Most Frequently Occurring Differential Conditions with Long Bone Bending Deformities

Condition	Features	Sources
Normal long bone bowing; bow-leg and knock knee	Normal exaggeration of bone shape	Silverman (1985:515, 813, 814)
Pre-natal bowing of long bones	Faulty fetal positioning, congenital malformations, congenital dislocations Tibia or femur anterior bowing Rarely affects arms; humerus may show bending and thickening Unilateral or bilateral Typically corrects by 2 years. If severe can remain visible in adults	Silverman (1985:515–518)
Birth defects/trauma	Varus deformity long bones, e.g. tibia, femur, humerus May be unilateral or bilateral	Silverman (1985:515–518), Molto (2000)
Trauma: Incomplete acute plastic bowing fractures, greenstick fractures. Entrapment neuropathy, muscle trauma, joint instability	Varus deformities various upper and lower limbs. Usually unilateral. May be associated with fracture in paired bones, e.g. forearm or lower leg. Typically apparent in children as result of falls. Can remain evident in adults	Borden (1975), Crowe & Swischuck (1977), Cail et al. (1978), Komara et al. (1986), Martin & Roddervold (1979), Cook & Bjelland (1979), Orenstein et al. (1985), Churchill & Formicola (1997), Stuart-Macadam et al. (1998), Glencross & Stuart-Macadam (2000)
Osteogenesis imperfecta	Severe bending of long bones Increased brittleness of bones Multiple fractures Joint laxity may increase bowing/angulation deformities	Silverman (1985), Milgram (1990), Whyte (1999), Lewis (2007); Chapter 9

Continued

TABLE 5.10 (*Continued*)

Condition	Features	Sources
Changes in mobility patterns	Marked antero-lateral bending distal femora with increased mobility and temporal changes in long bone robusticity	Shackelford & Trinkaus (2002), Holt (2003), Ruff et al. (2006), Wescott (2006)
Blount's disease	Localised growth disturbance medial proximal tibial epiphysis	Silverman (1985:515–518), Apley & Solomon (1994:199)
	Depressed angle to medial joint surface in adults	
	Unilateral or bilateral	
	Bone spurs medial aspect epiphysis and metaphysic	
	Tibia vara; medial bending of proximal tibia, or swelling of medial cortex at proximal shaft	
	Oblique meeting of femoral condyles with tibia joint surface	
Metaphyseal chondrodysplasia	Congenital, inherited disorder	Silverman (1985:573–580)
	Delayed mineralisation	
	Cupping of metaphyses, metaphyseal widening, long bones, ribs	
	Long bone bowing	
	Proximal femur particularly affected	
	Coxa vara of femoral necks	
	Cortical bone erosions, sub-periosteal bone formation	
	Irregularly calcified bone filling metaphyses	

Infantile cortical hyperostosis (Caffey's disease)	Cause unknown Average age of onset nine months Swelling of soft tissues Diffuse inflammatory reaction within periosteum, periosteal new bone formation Cortical thickening of long bones, typically clavicles, ulnae and affects mandibular ramus Fusion of adjacent bones, e.g. ribs May present expansion of medullary cavity and some endosteal resorption may occur in remodelling phase Complications include bowing deformities of lower limbs	Silverman (1985:841)
Treponemal diseases	Bowing deformities may be associated with bending, particularly of tibia Deformity can be true bend or pseudo-bend created by irregular bone formation (e.g. saber shin)	Webb (1995)

the pelvis and hip may lead to difficulties in walking and a 'waddling gait' (Mankin, 1974; Chapuy and Meunier, 1997; Güllü et al., 1998; Lips, 2001; Reginato and Coquia, 2003).

Recent investigations aimed at improving the identification of this disease in bioarcheaology have been undertaken utilising historical pathological collections (Mensforth, 2002; Schamall et al., 2003b; Brickley et al., 2005) and studies of archaeological remains (Schamall et al, 2003a; Brickley et al., 2007). Further investigation of the range of skeletal manifestations of this disease is continuing (Mensforth, 2002, pers. com.). Recent research by Brickley et al. (2007) determined that severe manifestations of osteomalacia frequently identified in pathology museum collections, were rarely observed in the archaeological cases. Whilst this may have occurred for a number of reasons inherent to the nature of each museum collection (as discussed in Chapter 2), the defective mineralisation that occurs in osteomalacia may have a significant role in increasing the fragility and post-mortem damage of skeletal remains, and hindering recognition of the disease from archaeological contexts (Brickley et al., 2007).

Consistent manifestations of osteomalacia have been identified in recent investigations of skeletal remains from archaeological contexts and correlated with clinical observations, both macroscopically and histologically using scanning electron microscopy (see Schamall et al., 2003a; Brickley et al., 2007).

Pseudofractures

The macroscopic manifestations of vitamin D deficiency osteomalacia are summarised in Table 5.11, and have been discussed by Schamall et al. (2003a), Ortner, (2003), Brickley et al. (2005, 2007) and Ives (2005). Skeletal features of osteomalacia include pseudofractures ('Looser's zones'), accumulations of unmineralised osteoid or poorly mineralised bone, which occur at specific locations throughout the skeleton (Steinbach et al., 1954; Mankin, 1974; Brickley et al., 2005, 2007). Stresses exerted on such defectively mineralised bone result in small, linear fractures (Chalmers, 1970 in Parfitt, 1986a:268; Mankin, 1974).

Clinical literature suggests two causative mechanisms for these lesions; one concerning the pressure or pulsations of arteries on weakened bone tissue (Le May and Blunt, 1949; Steinbach et al., 1954; Mankin, 1974), and the second invoking a fracture response in weakened bone to muscle and soft tissue forces during normal movement (Mankin, 1974; Adams, 1997:625; Francis and Selby, 1997; Berry et al., 2002). It will remain difficult to be certain of the origin of these lesions. Whilst the latter cause may seem more plausible, the pattern and location of some of the lesions, for example affecting the scapulae and ilia bodies do appear to suggest a response to arterial pressure (see Le May and Blunt, 1949; Steinbach et al., 1954; Ives, 2005).

The linear cortical bone features are infractions and so a repair mechanism is initiated. However, the fracture callus formed is disorganised during an active deficiency owing to impairment of the mineralisation process by the lack of vitamin D (Looser, 1920 discussed by Steinbach et al., 1954:388; Le May and Blunt, 1949; Sevitt, 1981:61). With recovery and vitamin D attainment, callus ossification will progress but may remain irregular, especially if excess osteoid has accumulated at the fracture site resulting in irregular and spiculated bone formation around the fractures margins (see Brickley et al., 2007).

Recent investigations have demonstrated the consistent occurrence of pseudofractures, particularly at the base of the scapula spinous process (see Figure 5.17), throughout the ribs (Figure 5.18), and also affecting the forearms, pelvis (see Figure 5.5) and femora (Mensforth, 2002; Schamall et al., 2003a; Ives, 2005; Brickley et al., 2007). The consistent location of these features is likely to play an important role in increasing the identification of future cases of osteomalacia. Prolonged osteomalacia may result in extremely severe skeletal deformities, particularly of the spine and pelvis (see Ortner, 2003; Brickley et al., 2005) and where there is no evidence of recent or healing bone changes bone deformities can be difficult to extrapolate from those of severe residual rickets (Table 5.8).

ADULT VITAMIN D DEFICIENCY IN THE MODERN PERSPECTIVE

Recent clinical investigations have highlighted the diverse range of conditions that will cause osteomalacia, including calcium–phosphorus imbalance, intestinal malabsorption states, renal disorders and failure, hereditary defects in aspects of vitamin D synthesis, as well as various heavy metals (including fluoride, see Chapter 9) and modern treatments such as bisphosphonate (see Holick and Adams, 1998; Reginato and Coquia, 2003). It is clear that careful distinctions need to be made in comparisons of the onset and manifestations of some of these conditions to osteomalacia that occurred in past populations from different causes (see also Mays et al., 2006a).

There is currently increasing awareness of vitamin D insufficiency in many areas of the world. There is particular concern for many elderly individuals in the developed countries, who may develop vitamin D deficiency and if prolonged skeletal evidence of osteomalacia from imbalances in renal function and synthesis of vitamin D with age, together with probable lowered dietary intakes and sunlight exposure (Halloran and Portale, 1997; Bilezikian and Silverberg, 2001; Lips, 2001). Elderly individuals who are institutionalised through illness or provision of care may be unable to go outside, increasing the reliance on dietary sources of vitamin D (Chapuy and Meunier, 1997:681; Lips, 2001; Plehwe, 2003; Mosekilde, 2005).

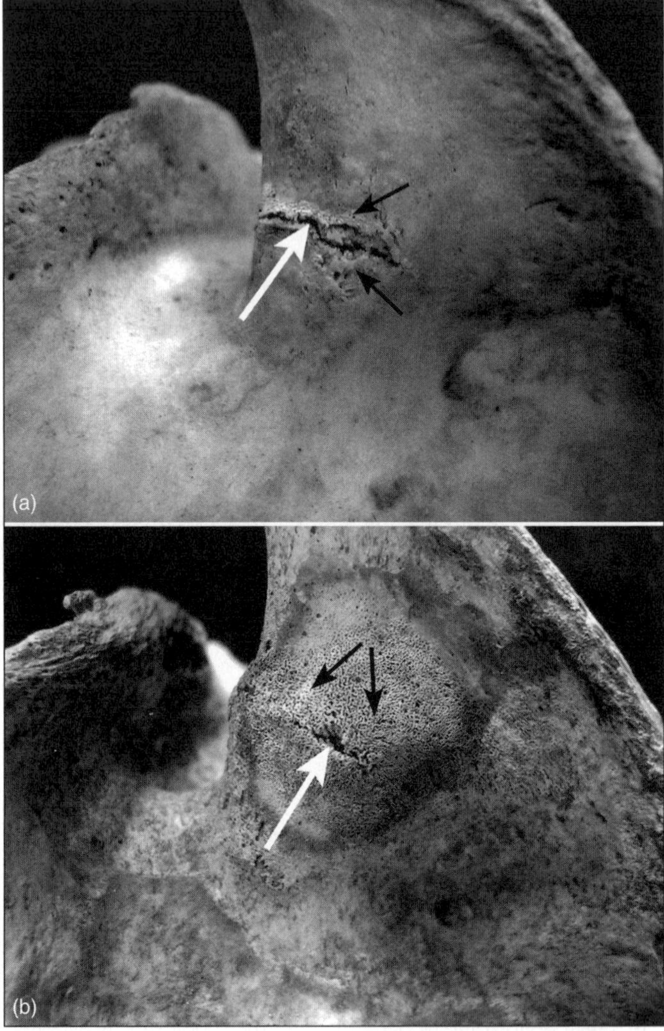

FIGURE 5.17 Linear fractures (pseudofractures) affecting cortical bone surface at the base of the scapula spinous process in adult vitamin D deficiency osteomalacia. (a) Bone callus (black arrows) surrounding the fracture margins (white arrow), (b) large build-up of irregular bone reaction (black arrows) almost covering the fracture (white arrow). Post-Medieval examples, (a) St. Martin's Churchyard, Birmingham, reproduced with permission of Wiley from Brickley et al. (2007), Figure 2g. (b) Chelsea Old Church (OCU615). Photograph by Rachel Ives, courtesy of the Museum of London.

In many modern contexts, elderly individuals tend to increasingly cover the skin when outdoors, limiting dermal conversion of vitamin D_3 from sunlight. In some instances, sunlight exposure may be deliberately denied under regimes of cruelty or neglect with an effect on health (see Box Feature 5.2).

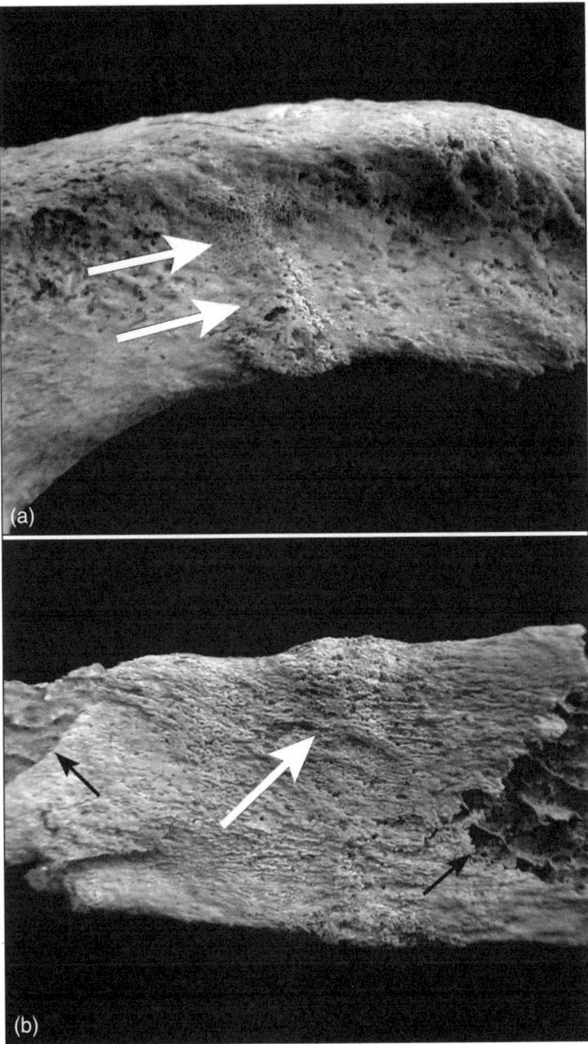

FIGURE 5.18 Rib fragments showing linear ridges of bone spicules indicating underlying pseudofracture in adult vitamin D deficiency osteomalacia. Black arrows demonstrate areas of post-mortem damage in contrast to the pathological changes. Post-Medieval examples from St. Martin's Churchyard, Birmingham.

Research has further considered the extent to which environmental pollutants and toxic metals have detrimental effects on health. The exposure of cadmium to water sources by zinc mining in the Jinzu River basin in Japan, as well as in Norway, Belgium and China, together with pollutant discharges from nickel–cadmium battery production in Southeast Sweden, have been linked with the skeletal effects of osteomalacia (see Låg, 1987; Mensforth,

Box Feature 5.2. Physical and Non-Violent Manifestations of Abuse

The physical manifestations and consequences of child abuse are becoming increasingly evident in modern research (Walker et al., 1997). Investigations of abuse in the past are also increasing with consideration of the physical evidence (Blondiaux et al., 2002; Lewis, 2007:175–183), as well as into theoretical perceptions of cruelty and conceptual changes throughout time (e.g. Baraz, 2003). Accuracy in determination is vital in modern contexts and is important in identifying and determining the social context of abuse in the past. Recent examples demonstrate the similarity in bone expression between child abuse (multiple fractures and new bone formation from haemorrhaging) and a range of pathological conditions, including vitamin K deficiency following birth (Brousseau et al., 2005), spontaneous fractures in disuse osteopenia, vitamin D deficiency, osteogenesis imperfecta, as well as scurvy and congenital syphilis (see Kerley, 1978; Kaplan, 1986; Walker et al., 1997; Torwalt et al., 2002; Rosen et al., 2003).

However, manifestations of abuse may not only take a violent or physical form. Neglect of illnesses or the withholding of food resources can be detrimental to health (see Jelliffe, 1955:93) and where deliberate can be considered as abusive (see Palmer and Weston, 1976; Meade and Brissie, 1985; Kellogg and Lukefahr, 2005; Cundiff and Harris, 2006; Piercecchi-Marti et al., 2006). Dietary deprivation is a significant concern in cases of child abuse, whether occurring as an independent action (neglect) or as an underlying factor secondary to violence (e.g. Anon. 1890:460–461, 1906:1088). Forcible imprisonment with lack of sunlight, together with dietary deprivation may result in vitamin D deficiency.

Physical and non-violent abuse is not restricted towards children. There is growing evidence of neglect and maltreatment towards the elderly in a variety of contexts particularly in the developed world (see review Shields et al., 2004). Failure in adequate provision of care can be as simple as not enabling elderly individuals to go outside, especially where physical deformities may limit individual mobility and may increase the dependence on others. Deliberate dietary deprivation is currently a significant factor in elder abuse (Shields et al., 2004), particularly as elderly individuals require increased dietary intakes of various nutrients (calcium, vitamin D, vitamin C, dietary fibre and magnesium) to maintain skeletal health (Berner et al., 2002; Volkert et al., 2004). Whilst likely complex to extrapolate, multidisciplinary anthropological research into the treatment and health of children and the elderly, could provide valuable social and cultural insights into the various contexts in which abuse may have occurred in the past.

2002; Järup and Alfvén, 2004; Jin et al., 2004; Vahter et al., 2004). Cadmium toxicity is associated with decreased concentrations of vitamin D, potentially via associated kidney damage or via a direct effect on osteoclast differentiation and action (Davies, 1994; Järup, 2003; Vahter et al., 2004:62; Järup and Alfvén, 2004). The skeletal consequences include secondary hyperparathyroidism and manifestations of osteomalacia and osteopenia.

As with many of the metabolic bone diseases, the spectrum of bone defects in rickets and osteomalacia can have equivalent expression in zoological species (see Zongping, 2005). Historical records attest to osteomalacic bone fractures occurring in cattle from Southern Norway in the seventeenth century, at the time linked to consumption of the plant *Narthecium ossifragum*. Subsequent analysis demonstrated that deficiencies of phosphorus in bedrock, soil and plants were the underlying cause of the condition (Låg, 1987:63) although the northerly latitude may have been an additional causative factor.

Both rickets and osteomalacia attributable to phosphorus deficiency in grazing land exacerbated by calcium deficiency currently exist in Bactrian camels in the Badanjiling and Tengeli deserts in China (Zongping, 2005). Skeletal deformities and subsequent lameness have severe impacts on the feeding ability of these animals, which typically need to graze large areas to achieve a general nutritional adequacy. Prolonged conditions resulted in death of animals following exhaustion and starvation (Zongping, 2005). The human implication of these pathological changes is loss of herd members, particularly important if affecting reproductive females, as well as loss of supplies of meat and an important economic and transport provision (see Zongping, 2005). Animals may suffer from vitamin D deficiency if a mother's milk is withdrawn and unsupplemented cow's milk is substituted. For example, a marked case of rickets with bowed leg bones has been documented in a six-month old elephant calf reared on cow's milk (Hochberg, 2003:3).

ANTHROPOLOGICAL PERSPECTIVES: OSTEOMALACIA

Paleopathological Cases of Osteomalacia

World-wide archaeological evidence for vitamin D deficiency osteomalacia in adults is infrequent although few studies to date represent systematic large-scale population-based investigations of this condition (Schamall et al., 2003a; Brickley et al., 2007). A summary of the currently known cases is shown in Table A3.

The various studies that have investigated the presence of adult vitamin D deficiency in many communities have provided valuable insights into the circumstances of in which this disease occurs. For example, vitamin D deficiency can be a notable concern when individuals are institutionalised for a range of underlying causes, unless dietary sources make up for the deficit of sunlight exposure. It is unclear at present to what extent sunlight exposure was limited for individuals who were affected by a range of mobility problems (see Chapter 7). If unable to walk and when lacking additional support, it is likely that individuals may not have been able to venture outside. This situation is often noted in the elderly in many developed countries (e.g. Plehwe, 2003), but it would be useful to explore the additional contexts under which such circumstances may have arisen, particularly in the past. Schamall et al. (2003a) have discussed a case of osteomalacia in a middle adult female from a Roman military hospital in Austria. It is plausible that injury may have resulted in immobilisation and subsequent osteomalacia, although the assessment of dietary adequacy also needs consideration (e.g. Sledzik and Sandberg, 2002). Additional causes in such circumstances could include the age and sex of the individual affected (hormonal status and age-related renal function decline).

Evidence of vitamin D deficiency in bioarchaeological samples may be complex to interpret, owing to the possible range of causative factors that can result in skeletal changes. However, continued investigations will provide the potential to examine the relationship between clothing customs or cultural dietary practices on health in the past. Clothing fashions can be inextricably linked with the individual's position in society and can also be largely interpreted as reflecting a cultural impact on lifestyle and stage in the life course. Informative analyses will need to provoke inter-disciplinary research in drawing on a wide spectrum of anthropological and historical evidence to identify specific social conventions relevant to each period. For example, in Victorian England, dresses that permitted glimpses of a lady's ankle were associated with prostitutes (Picard, 2005:172). It was held as popular fashionable practice for females to remain pale-skinned, limiting sunlight exposure and aided by the use of arsenic in soap products to whiten the skin (see Freeman, 1989; Howe, 1997), although these values principally reflect those held by the middle and upper classes. It would also be of value to consider if clothing customs altered with age and to what extent this varied between social classes and what effects these may have had on health.

Food availability in many present and past populations is no guarantee of food quality, and practices of food adulteration are documented throughout periods in history (see a perspective on this issue across British history in Roberts and Cox, 2003:311–312, 360). During the post-Medieval period, adulteration practices affected foodstuffs containing both calcium (milk and cheeses) as well as vitamin D (fish), particularly in enabling the sale of stale products (Freeman, 1989). Further variation in dietary consumption may exist between the classes or sexes, or across different stages of childhood, and may result in health effects that could demonstrate such cultural patterning. Investigations could be combined with paleodietary analyses for a more integrated understanding of past diets and health impacts.

Dietary practices may have been limited according to social or religious customs both in the present (e.g. Jelliffe, 1955, above) as well as in the past, and normal dietary customs may have altered in times of ill-health. For example, some religious houses and monasteries in late Medieval England, particularly Cistercians, were not allowed to eat fish except when ill, whereas Benedictine monks were allowed to eat fish but not meat (see Roberts and Cox, 2003:244). Dietary restrictions would emphasise the requirements on sunlight exposure to obtain sufficient vitamin D status. Dietary reconstruction through isotope analysis can aid the reconstruction of consumption of marine protein compared to terrestrial protein, which could be correlated with investigations of health (e.g. Mays, 1997). Geographical location and transportation means will affect the availability of food resources and could potentially be correlated with health effects (see Mays, 1997). Evidence of vitamin D deficiency in Britain as recently surveyed by Roberts and Cox (2003) highlights increased occurrence in Roman Britain, with decreases through the Medieval period. Increased prevalences are evident in the post-Medieval period, attributable to urban living and industrial pollution and poor diet and research is continuing to document the occurrence of this condition across the social classes (Mays et al., 2006a; Brickley et al., 2007; Ives, in preparation).

Comparisons of geographical and regional differences in the prevalence of this condition could enable better insight into paleoclimatic interpretations and would facilitate comparisons with known geographic prevalence of this condition in the modern world. Cases of adult osteomalacia from the post-Medieval period in England have been linked with increasing air pollution deriving from coal smoke and factory smog, which together with climatic variability served to filter out available sunlight (see Brickley et al., 2007; Brimblecombe, 1987:109; Luckin, 2003). Continued investigation of skeletal samples during periods of industrialisation and increasing urbanisation may be able to shed light on how widespread the effects of such atmospheric pollution may have been during this period.

DIAGNOSIS OF OSTEOMALACIA IN ARCHAEOLOGICAL BONE

Macroscopic Features of Osteomalacia

Recently occurring osteomalacia will be manifest through pseudofractures, small, linear fractures, within the external bone cortex. These features are frequently bilateral and tend to show irregular fracture callus surrounding the fissure. In severe cases pseudofractures can develop into complete fractures. Healed evidence of osteomalacia may comprise severe deformities in the axial skeleton caused by bone softening (see Schamall et al., 2003b; Brickley et al., 2005) (see Figure 5.19).

Pseudofractures can occur in a number of conditions, including osteogenesis imperfecta, Paget's disease, congenital syphilis and osteopetrosis (Mankin, 1974; Brickley et al., 2007). However, the pattern of skeletal occurrence and trend towards bilateral locations affected, particularly in the scapulae can be correlated with defective mineralisation indicative of vitamin D deficiency (Steinbach et al., 1954; Steinbach and Noetzli, 1964; Boyde et al., 1986; Brickley et al., 2007), and Mankin (1974:119) has further argued that those pseudofractures occurring in other conditions are more likely to progress to complete fractures to a greater extent than those in osteomalacia (Table 5.11).

Radiological Features of Osteomalacia

Radiologically, pseudofractures may appear as small, linear bands of decreased density occurring perpendicular to bone margin. Pseudofractures may demonstrate a sclerotic margin if prolonged and an irregular fracture callus can be formed (Table 5.12).

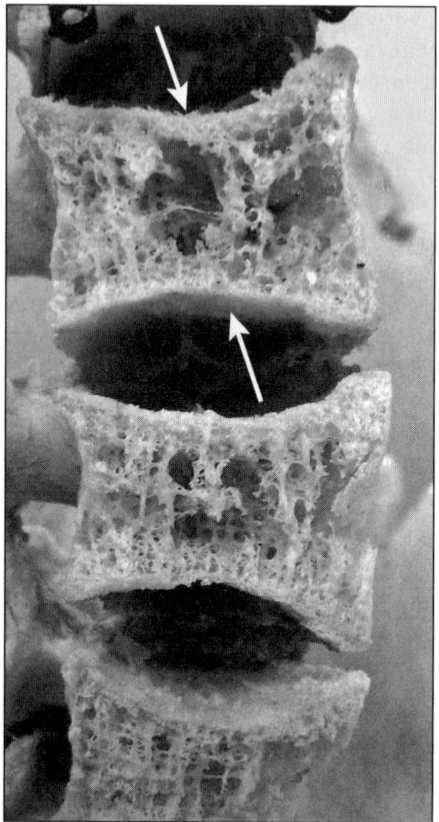

FIGURE 5.19 Section through vertebrae of an adult individual with osteomalacia curated in the Galler Collection, Basel. Note biconcave compression of superior and inferior surfaces in all three vertebrae (e.g. arrowed), unlike osteoporosis which tends to manifest in a single compressed vertebra. Also note the coarsening of vertical trabeculae and loss of horizontal trabeculae. Photograph by the authors courtesy of the Natural History Museum in Basel, Switzerland.

Histological Features of Osteomalacia

Pathological defects in bone mineralisation can be identified histologically using BSE-SEM. In particular, defects in bone mineralisation adjacent to the cement line, together with poor quality of newly formed bone and irregularly sized osteocyte lacunae are bone features indicative of osteomalacia (see further, Boyde et al., 1986; Schamall et al., 2003a; Brickley et al., 2007) and Table 5.13 and Figure 5.20.

Histological changes of osteomalacia can be identified in trabecular and cortical bone tissue. Pathological changes can be quite distinctive but should be demonstrated as consistently occurring throughout a sample for definite diagnosis. It is necessary to consider the potential for various diagenetic and taphonomic alterations to have affected bone quality and structure as distinct from pathological alterations.

TABLE 5.11 Macroscopic Features of Vitamin D Deficiency Osteomalacia

Bones affected	Features	Code	Differential diagnosis	Sources
Cranium	Fine pitting/diffuse porosity of cortical surface, cardboard-like consistency of bones, low weight	G	Paget's disease, Rickets	Mankin (1974), Ortner (2003:398, 400)
	Basilar invagination if severe			
Dentition	AMTL	G	Dental disease, Age, Metabolic, Infectious	Hess (1930)
Vertebrae	Buckling, folding of body	D	Osteoporosis (usually compression not buckling)	Jaffe (1972), Mankin (1974), Pitt (1988), Barnes (1994), Francis & Selby (1997), Ortner (2003), Schamall et al. (2003b), Brickley et al. (2005), Brickley et al. (2007)
	Loss of body height and irregularity of endplates	D		
	Multiple sites of biconcave compression superior and inferior surfaces of vertebral bodies	D	Osteoporosis, usually single site	
	Kyphosis and scoliosis; if lumbar affected may protrude anteriorly to obstruct pelvic inlet	G	Congenital and developmental or Trauma, Metabolic, Infection	
Ribs and sternum	Multiple pseudofractures as linear ridges of irregular, spiculated bone formation	S	Infection, but usually not linear, Trauma	Albright et al. (1946), Mankin (1974), Pitt (1988), Adams (1997), Berry et al. (2002), Ortner (2003), Brickley et al. (2005), Brickley et al. (2007)
	Multiple complete fractures, numerous fractures may appear adjacent	D		
	Irregularly formed trabecular bone spicules occur at fracture margins	D		
	Lateral straightening of rib shafts where extreme may show medial indentations	D	Residual rickets	
	Acute rib neck angle	D	Residual rickets	
	Bending sternum, severe cases	D		

Continued

TABLE 5.11 (*Continued*)

Bones affected	Features	Code	Differential diagnosis	Sources
Pelvis and sacrum	Pseudofractures superior/inferior pubic rami. True fractures or buckling deformities	S	Trauma	Mankin (1974), Adams (1997), Ortner (2003), Brickley et al. (2005), Ives (2005), Brickley et al. (2007)
	Pubic rami lie adjacent rather than opposing, dislocation of the pubic symphysis. Obstruction of pelvic inlet	S		
	Pseudofractures in medial aspect ilium adjacent to greater sciatic notch	S		
	Protrusion of ilia into pelvic inlet	D		
	Narrowing of pelvis and curvature of ilia	D	Residual rickets	
	Anterior facing of acetabulae and protrusion into pelvic inlet	D	Residual rickets	
	Thinning of the ilia, collapse and buckling/folding (anterior) and fracture of the iliac crest	D	Osteopenia, Paget's disease	
	Sacrum extreme ventral angulation, typically at S3, potentially with fracture. Contributes to obstruction of pelvic inlet	G	Residual rickets, Trauma, Extremes of sexual dimorphism	
Long bones	Pseudofractures affecting medial femoral neck or medial shaft adjacent to lesser trochanter	D	Osteoporosis, Paget's disease	Hess (1930), Dent & Hodson (1954), Mankin (1974), Schamall et al. (2003a), Brickley et al. (2005), Ives (2005), Brickley et al. (2007)
	Antero-lateral bending of femoral shafts	D	Residual rickets, CDF	

	Coxa vara of femoral necks	D	Residual rickets, CDF	
	Genu valgum ('knock-knees') angulation	D	Blount's disease	
	Pseudofractures may affect proximal ulna shaft beneath proximal joint surface as well as distal third of shaft	D	Trauma, Activity related development	
	Irregularity of poorly mineralised bone accumulation at the site of muscle attachments, e.g. proximal ulna	D		
	Pseudofractures may affect shaft of radius. Colles' fractures indicative of osteoporosis can co-occur in osteomalacia	G	Osteopenia	
	Pseudofractures clavicles, humeri, tibiae and fibulae, less frequently than other bones	D	Stress fractures	
Other: Scapulae	Pseudofractures (Looser's zones) affecting lateral border and inferior lateral margin of spinous process	S		Dent & Hodson (1954), Steinbach et al. (1954), Mankin (1974), Schamall et al. (2003a), Brickley et al. (2005), Ives (2005), Brickley et al. (2007)
	Increased posterior curvature of scapula blade	D	Residual rickets	
	Buckling/collapse of superior border	D		
Other: Hands and feet	Pseudofractures can affect metacarpals and metatarsals	D	Stress fractures, Trauma	Dent & Hodson (1954), Mankin (1974), Schamall et al. (2003a)

Notes: See Table 5.5 for definition of codes used in diagnosis. CDF, *congenital dislocation of the femur* (Mitchell and Redfern, 2008); AMTL, *ante-mortem tooth loss.*

TABLE 5.12 Radiological Features of Vitamin D Deficiency Osteomalacia

Bones affected	Description of changes	Code	Differential diagnosis	Sources
Cranial	Osteopenia, although can be inconsistent	G	Various metabolic, Inflammatory, Haematogenous, Malignant conditions	Swann (1954), Dent & Hodson (1954), Mankin (1974)
Dentition	AMTL	G	Various	
Vertebrae	Osteopenia of cortical margin and trabeculae	D	Osteopenia, Paget's disease, Malignancy and Infection	Dent & Hodson (1954), Reynolds & Karo (1972), Mankin (1974), Berry et al. (2002), Reginato & Coquia (2003), Schamall et al. (2003b)
Ribs and sternum	Multiple pseudofractures	D	Trauma	Dent & Hodson (1954), Reynolds & Karo (1972), Mankin (1974), Parfitt (1986a), Berry et al. (2002), Reginato & Coquia (2003)
Pelvis and sacrum	Pseudofractures at pubic rami Osteopenia at iliac crest and ilia	D G	Trauma Various conditions (Chapter 7)	Dent & Hodson (1954), Reynolds & Karo (1972), Mankin (1974), Parfitt (1986a), Berry et al. (2002)
Long bones	Pseudofractures medial femoral neck and medial sub-trochanteric level	S	Paget's disease	Dent and Hodson (1954), Reynolds & Karo (1972), Mankin (1974), Parfitt (1986a), Berry et al. (2002)

	Pseudofractures radius, ulna and clavicle	D		
	Long bone curvature. May result in differential thinning and thickening of affected cortices	D	Residual rickets, Paget's disease, CDF	
Other: Scapulae	Pseudofractures in lateral border and inferior margin spinous process	S		Dent & Hodson (1954), Steinbach et al. (1954), Reynolds & Karo (1972), Mankin (1974), Parfitt (1986a), Adams (1997), Berry et al. (2002), Schamall et al. (2003a)
Other: Hands and feet	Periarticular osteopenia and sub-chondral erosions, cortical bone striations or intra-cortical porosity (fine detail required)	G	Secondary hyperparathyroidism, Osteoporosis, Rheumatoid arthritis	Dent & Hodson (1954), Reynolds & Karo (1972), Mankin (1974)
	Pseudofractures in metacarpals and metatarsals			
Other: General	Generalised osteopenia	G	Various metabolic. Osteopenia; outlines tend to remain fairly distinct	Hess (1930), Milkman (1930, 1934), Albright et al. (1946), Dent & Hodson (1954), Steinbach et al. (1954), Steinbach et al. (1959), Steinbach & Noetzli (1964), Mankin (1974), Francis & Selby (1997)
	Loss of definition of cortical margins			
	Thinning trabeculae and coarsening and 'fuzzy' outline to remainder			
	There may be no detectable radiological changes in early stages			

Notes: See Table 5.5 for definition of codes used in diagnosis. AMTL *denotes ante-mortem tooth loss.* CDF, *Congenital dislocation of the femur*

TABLE 5.13 Histological Manifestations of Vitamin D Deficiency Osteomalacia in Skeletal Remains

Tissue type	Features	Code	Differential diagnosis	Sources
Cortical bone	Increased resorption bays in early stages of conditions	G	Secondary hyperparathyroidism, Osteopenia, Fibrous dysplasia	Garner & Ball (1966), Sato & Byers (1981), Boyde et al. (1986), Mankin (1974), Francis & Selby (1997), Parfitt (1998), Corsi et al. (2003), Schamall et al. (2003a), Ives (2005), Brickley et al. (2007)
	Poor mineralised bone quality	S	Fibrous dysplasia	
	Defectively mineralised bone adjacent to osteon cement lines	S		
	Incomplete mineralisation of layers of bone	S		
	Separation of mature bone from newly formed bone by post-mortem disappearance of osteoid	S		
	Enlarged osteocyte lacunae	D	Different to abnormal bone mineralisation in fibrous dysplasia	
	Mineralisation proceeds normally with obtainment of vitamin D, may be excess bone if osteoid accumulation is rapid	D		
Trabecular bone	As above. Changes may be more widespread throughout trabecular structure	S		Garner & Ball (1966), Sato & Byers (1981), Mankin (1974), Boyde et al. (1986), Parfitt (1997:652), Francis & Selby (1997), Schamall et al. (2003a), Brickley et al. (2007)
	May be increased trabecular thinning and perforation with and increased resorption	G	Secondary hyperparathyroidism, Osteopenia	

Notes: *See Table 5.5 for definition of codes used in diagnosis.*

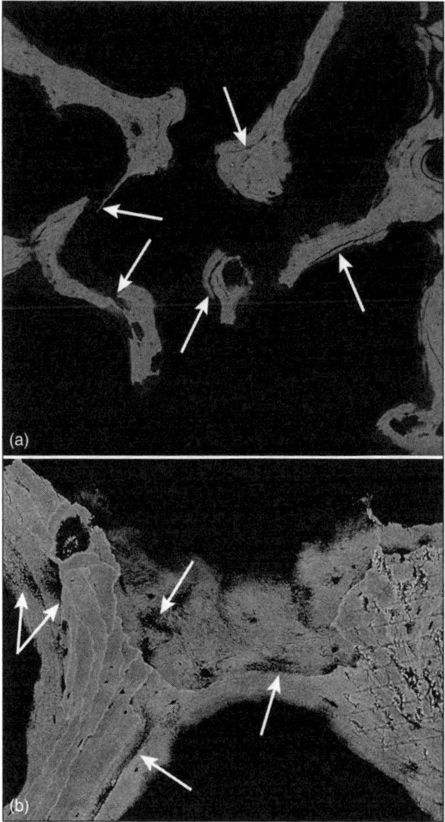

FIGURE 5.20 BSE-SEM images of pathological mineralisation defects in adult vitamin D deficiency osteomalacia. (a) Incomplete layers of new bone formation separated from existing bone by defective mineralisation adjacent to cement lines. (b) Poorly mineralised new bone surrounded by defects adjacent to the cement lines in a clinical sample. Copyright Alan Boyde QMUL.

CONCLUSIONS

The manifestations of vitamin D deficiency have recently received increasing attention and as a result, this condition is becoming more widely recognised from a range of contexts in both present and past communities. The causative factors of vitamin D deficiency can be widespread and recognition of these conditions may facilitate greater interpretation of health particularly in past communities. Factors resulting in a vitamin D deficiency can stem from illnesses that limit mobility and therefore sunlight exposure, as well as underlying episodes of malnutrition or calcium deficiency, or social or cultural practices. These latter causes indicate that this disease may be evident in situations where food supplies are limited, including after natural disasters or bouts of warfare, or where food is deliberately withheld at an individual level of abuse.

APPENDIX: SUMMARY OF PUBLISHED ARCHAEOLOGICAL EVIDENCE FOR VITAMIN D DEFICIENCY

TABLE A1 Summary of Published Archaeological Evidence for Vitamin D Deficiency Rickets in Juvenile Remains

Locations	Date	Total burials	Age/Sex, bones affected	Methods	Attributed cause	Sources
Russia	24,000 BP MUP (LUP)	2	2 children. Long: bending	Mac. Rad.	? rickets. Clothing, lifestyle suited to cold weather and cave shelter, possible limitations of vitamin D	Formicola & Buzhilova (2004); Buzhilova (2005)
Bellanbandi Palassa, Sri Lanka	ca. 125,000–30,000 (P) ?	NS	NS. Femora and Tibiae: Moderate bending	Mac.	Poor-quality nutrition?	Kennedy in Cohen & Armelagos (1984:182)
Mahadaha, Gangetic Plain, South Asia	ca. 125,000–30,000 (P) ?	NS	NS. Femora and Tibiae: Moderate bending	Mac.	Poor-quality nutrition?	Kennedy in Cohen & Armelagos (1984:182)
Chandoli and Nevasa Maharashtrian communities, South Asia	Farming communities (ca. 12,000 BC+?)	NS	NS. Femora and Tibiae: Moderate bending	Mac.	Agricultural economy, population pressure, lower-quality nutrition	Kennedy in Cohen & Armelagos (1984:182)
Yugoslavia	ca. 6300–5300 BC (ME)	NS	Common in children. BNS	NS.	Incipient cultivation; restrictive dietary practices children and females	Y'Edynak (1987) cited in Cohen (1989:214)

Balkans	9000 BC	NS	J. young children. craniotabes, thick legs, characteristic rib cage lesions*	Mac.	Settlement within deep river gorge, dense forest covering prominent shadows and prevention of sunlight exposure*	Zivanovic (1982) cited in Aufderheide & Rodríguez-Martín (1998:310); Barnes (2005:327)*
Denmark	ca. 5000–1800 BC (N)	NS	6. NS.	Mac.	NS	Nielsen (1911) cited in Wells (1975)
South Africa	4910–4730 BP (N)	4	1. J. 3.5–5 months. Bone shaft thickening, metaphyseal swelling, cranial and orbital porosity, growth plate porosity, thickened but porous cortex, angulation distal tibiae, rachitic rosary	Mac. Rad.	Possible inborn error of metabolism, renal malfunction? Unlikely dietary deficiency, tight swaddling suggested for some features	Pfeiffer & Crowder (2004)
Denmark	3000 BC–AD 1000 (R/(IA))	NS	3. NS	Mac.	NS	Nielsen (1911) cited in Wells (1975)
Adittanalur, South Asia	1100–0 BC (IA)	16	5. J. Crania	Mac.	Agricultural economy, population pressure, lower-quality nutrition	Kennedy in Cohen & Armelagos (1984:182)

Continued

TABLE A1 *(Continued)*

Locations	Date	Total burials	Age/Sex, bones affected	Methods	Attributed cause	Sources
Mahujhari, South Asia	1100–0 BC (IA)	NS	J. Long: Diaphyseal bending and flaring	Mac.	Agricultural economy, population pressure, lower-quality nutrition	Kennedy in Cohen & Armelagos (1984:182)
Bahrain, Arabian Gulf	300 BC–AD 250 (IA)	1280 individs over burial period	4 J. 2 × 0–6 months, 1 × 2–3 years, 1 × 6–12 months. Vert. Long: Cupping and flaring of tubular appearance of long bone metaphyses, concave depressions in vertebral endplates	Mac.	Combination of causes: Frequent infections, diarrhoeal disease and inadequate sunlight exposure	Littleton (1998b)
Bahrain, Arabian Gulf	300 BC–AD 250 (IA)	1280 individs over burial period	10. Long bone changes	Mac.	Cultural practices avoiding sunlight	Littleton (1998b)
Italy	ca. 750 BC–AD 550 (R)	139 (total adults and children)	1. Healed rickets	Mac.	NS	Bisel (1991)

England	Late first century AD (R)	11	1. NS. Long: Slight bowing long, severe EH	Mac.	NS	Wakely & Carter (1996)
Fazekasboda, Hungary	NS Roman	NS	2. NS	Mac.	NS	Regöly-Mérei (1962) cited in Wells (1975)
England	AD 410–1050 (AS)	ca. 130	3/64 children under 11 years, 1/6 adolescents (12–17 years). Lower long: Antero-posterior bending, sub-periosteal apposition of new bone	Mac.	NS	Stirland (1985)
England	AD 410–1050 (AS)	NS	1. ca. 6 years. Leg bones. Anterior bowing	Mac.	NS	Ortner & Putschar (1981); Ortner (2003)
England	AD 410–1550 (M)	NS	1 infant. Widening epiphyses, bowing long bones		NS	Manchester pers. com. cited in Stuart-Macadam (1989:212)
Scandinavia (Burgundian tribe)	AD 410–1550 (M)	NS	NS	Mac.	NS	Morel & Demetz (1961) cited in Wells (1975)
Ireland	ca. AD 800–1100 (M)	66	1. ca. 3 years old. NS	Mac.	NS	Power (1994)

Continued

TABLE A1 *(Continued)*

Locations	Date	Total burials	Age/Sex, bones affected	Methods	Attributed cause	Sources
England	AD 850–110 (EM)	142	1. 0.6–2.5 years. Rib and skull: Frayed rib, porotic hyperostosis	Mac.	Diet?	Lewis (2002)
England	AD 950–1550 (M)	200	1. 2.6–6.5 years. Lower long: Bowing and frayed distal surfaces. Dawes (1980:59) refers to 2 classic cases and between 30 and 40 instances of greater than normal femur or tibia bowing	Mac.	Diet? Urban conditions?	Lewis (2002); Dawes (1980)
England	AD 950–1550 (M)	303	6: 3 × New born-0.5 years; 1 × 0.6–2.5 years; 1 × 0.6–6.5 years; 1 × age? Various: Long bones, ribs and cranial; fraying, bending, porosity, beading of ribs	Mac.	Illness, sunshine deficiency	Lewis (2002)

England	AD 950–1550 (M)	687	8 J. 3–18 months. Flared and porous ribs ends, deformed long bones, porous and irregular subchondral bone, porous woven bone deposits cranial, deformed mandibular condyle	Mac.	Ill from other causes and kept inside out of sunlight	Ortner & Mays (1998)
Mixed sites, Hungary	10–13th centuries (M)	NS	Increase in number of cases from 0.7% to 2.5%. NS	Mac.	NS	Nemeskéri & Harsányi (1959) cited in Wells (1975)
Västerhus, Sweden	ca. AD 1100–1300 (M)	364	3 suspected, only 1 definite by Wells' view. NS	Mac.	NS	Gejvall (1960) cited in Wells (1975)
England	ca. AD 1100–1550 (M)	6J/35A	1 J (?) angulation proximal tibia, bowing R tibia, femora, ML flattened fibulae	Mac.	NS. Small, rural hospital	Chundun & Roberts (1995);
England (Jewish burial ground)	ca. AD 1170–1280 (M)	154 (under 20 years)	1 J, less 1 year. Bowing deformity forearm. 3 skulls under 1 year. (Fine pitted bone may be indicative of early rickets)	Mac.	NS	Brothwell & Browne (1994)

Continued

TABLE A1 (*Continued*)

Locations	Date	Total burials	Age/Sex, bones affected	Methods	Attributed cause	Sources
Aebelholt Monastery, Denmark	ca. AD 1250–1550 (M)	800	9. NS	Mac.	NS	Møller-Christensen (1958) cited by Wells (1975)
Lajme and Sellye Hungary	NS Medieval	NS	2. NS	Mac.	NS	Regöly-Mérei (1962) cited in Wells, (1975)
England	AD 1550–1900 (PM)	2/65	1 × 7–8 years; 1 × 11–12 years	NS	N/A	Start & Kirk (1998)
Broadgate, London, UK	AD 1550–1900(PM)	58	7 J. 6 healed rickets 1 × neonate. No correlation with delayed growth	Mac.	Infant feeding practices, low socio-economic status	Pinhasi et al. (2006)
England	AD 1550–1900(PM)	ca. 50	12. Femora, tibiae: Severe bowing, buttressing and splayed epiphyses	Mac.	NS	Wells (1967)
England	1650 (PM)	1	1. Daughter of King Charles I. Cranial changes, scoliosis, arm and leg deformities	Mac.	Sickly constitution, potential retained indoors	Burland (1918)

England	AD 1729–1859(PM)	28/186	6× new born–0.5 years; 19 × 0.6–2.5 years; 1 × 2.6–6.5 years; 1 × 6.6–10.5 years; 1 × age? (20 J)*. Long, ribs and cranial	Mac.	Sunshine deficiency, nutritional practices; climate at birth	Lewis (2002); Molleson & Cox (1993)*
First African Baptist Church, Philadelphia, US	1823–1841 (PM)	10	2 J. 1 × 8 years, 1 × NS. Femora and tibia severe bending.	Mac.		Angel et al. (1987)
Cedar Grove, Arkansas, US	AD 1890–1927(PM/Mo)	NS	J. 1/4 of all children dying aged 3–20 months affected. Cranial rachitic lesions	Mac.	Low protein and calcium diets; socio-cultural status: Free black slaves	Martin et al. (1987); & Rose (1985) cited in Cohen (1989:222)
Denmark	NS	NS	NS	Mac.	NS	Fürst (1920) cited in Wells (1975)
Norway	NS	NS	NS	Mac.	NS	Fürst (1920) cited in Wells (1975)
Sudan and Nubia	NS	NS	NS. Deformed sacrum	Mac.	NS	Ruffer (1914) cited in Wells (1975)
Peru	NS	NS	NS	Mac.	NS	MacCurdy (1923) cited in Wells (1975)

Continued

TABLE A1 (*Continued*)

Locations	Date	Total burials	Age/Sex, bones affected	Methods	Attributed cause	Sources
South America (Patagonia)	Early period	NS	NS	Mac.	NS	Lehmann-Nitsche (1903) cited in Wells (1975)
England	Late R	NS	2 infants (b–9 months). NS	Mac.	NS	Farwell & Molleson (1993)

Notes: *Several cases reported by* Wells (1975) *excluded on basis of* Wells' *subsequent observations. Dates given are approximate and supplied to provide an approximate time line. Date ranges may be subject to regional variations. Time periods given by original authors are indicated in brackets*: LUP, *late upper palaeolithic*; P, *pre-historic*; ME, *Mesolithic*; N, *Neolithic*; IA, *Iron Age*; ER, *Early Roman*; R *Roman*; A-S, *Anglo Saxon*; M, *Medieval*; PM, *post-Medieval*; Mo, *modern*; J, *juveniles*; A, *Adults*; NS, *not specified*; BNS, *bones not specified*; Mac., *macroscopic examination*; Rad., *radiological examination*; Mic., *histological examination. At the time of writing there are a number of cases from various sites not yet been published.*

**Data denoted with asterix are reported in asterixed source.*

TABLE A2 Summary of Published Archaeological Evidence for Adult Cases of Healed Juvenile Vitamin D Deficiency Rickets

Locations	Date	Total burials	Age/Sex, bones affected	Methods	Attributed cause	Sources
Russia	24,000 BP (UP)	1 adult	1 Mature male. Bending forearm, femora, fibulae	Mac. Rad.	? rickets, ? physiological bowing	Formicola & Buzhilova (2004); Buzhilova (2005)
Russia	26,640+/− 110 BP (UP)	3	1 YA. Femoral shortening, *coxa vara*, humerus, radius, ulna bending	Mac.	Congenital? Physiological? Rickets not preferred diagnosis	Formicola et al. (2001)
Yugoslavia	ca. 6300–5300 BC (Me)	NS.	Common in women. NS	NS.	Incipient cultivation; restrictive dietary practices children and females	Y'Edynak (1987) cited in Cohen (1989:214)
US	3250–300 BP (P)	251	133 M, 118 F. Cranium, femur, tibiae measured*	Mac. Quantitative assessment	Potential implications of dietary changes through time	Ivanhoe & Chu (1996)
US	ca. 3000 BC (P)	NS	1. Adult. Femur bowing, *coxa vara*	Mac.	NS	Means (1925) cited by Wells (1975)
US	3000 BC–AD 1850 (15 series) (P+)	359	Cranial vault, cranial base, femora and tibiae measured*	Mac. Quantitative assessment	Dietary calcium deficiency, physiological stress in females, male nutritional advantages	Ivanhoe (1995)

Continued

TABLE A2 *(Continued)*

Locations	Date	Total burials	Age/Sex, bones affected	Methods	Attributed cause	Sources
United Arab Emirates	3000–2000 BC (P)	47	1. A ? Femur bowing	Mac.	Vitamin D deficiency?	Blau (2001)
Egypt	1500–100 BC (NK)	NS	Adult. NS. Long	Mac.	Unlikely lack of sunlight; may be related to ill-health	Nerlich et al. (2000)
England	AD 43–410 (R)	40	1 ?M 18–23 single femur. Marked bowing	Mac. Rad.		Harman et al. (1981)
England	AD 410–1050 (AS)	107	5. Adults. NS. Various tibiae, femora, bending? normal variation	Mac.	NS	Anderson (1993)
England	ca. AD 500 (AS)	NS	NS. Aged 20–55 years, 10 M, 10 F recorded. Cranial, femora and tibiae measured*	Mac. Quantitative assessment	Sunshine and calcium deficits and physiological stress in females	Ivanhoe (1994)
England	AD 43–410 (R)	NS	1 YA F. Bowed femora and tibiae	Mac.	NS	Farwell & Molleson (1993)
England	AD 43–410 (R)	92	2 A M. Healed rickets, bowing tibiae	Mac.	Possible dietary or physiological factors for general lack of vitamin D	Roberts & Cox (2003:143); Boylston & Roberts (2004)

Switzerland	AD 410–1550 (M)	NS	1. Male 50+. Tibiae.	Mac.	NS	Ortner & Putschar (1981); Ortner, 2003
Ireland	ca. AD 800–1100 (M)	200	4. A M. Mainly lower limbs; 1 × case humeri bowing	Mac.	NS	Power (1994)
US	AD 100 (MW)	NS	1 Ad. M. Possible case	Mac.	NS	Neiburger (1989) cited in Ortner (2003:404)
England	ca. AD 1100 (M)	29 (adults only)	1. NS. Lower long. Bowing tibia and femora, angulation tibia, flattening fibulae			Chundun & Roberts (1995)
England	AD 1485–1603 (M)	NS	2 bowed femora, 12 bowed tibiae	Mac.	Poor childhood nutrition	Stirland (2000)
England	AD 1170–1280 (M)	316	2: 1 × A F femora bend, 1× A F, tibiae bend	Mac.	NS	Brothwell & Browne (1994)
US	AD 1790–1820 (PM)	16 adults	A 50% F 75% M affected. Tibiae marked bowing	Mac.	Working conditions	Kelley & Angel, 1987
US	1823–1841 (PM)	NS	30% of A affected. Adults 3 F, 7 M. Tibiae bowing		Socio-economic and nutritional conditions	Angel et al. (1987)
Ireland	AD 1800+ (PM)	17	2. 1 × YA M severe bending deformities legs. 1 × A F legs and arms affected	Mac.	Air pollution and lowered sunlight exposure or conditions affecting calcium and phosphate metabolism	Power (1994)

Continued

TABLE A2 *(Continued)*

Locations	Date	Total burials	Age/Sex, bones affected	Methods	Attributed cause	Sources
England	AD 1718–1845 (PM)	250 recorded	37. A (20–60 years) 18 M, 19 F. Cranial, femora and tibiae measured*	Mac. Quantitative assessment	Sunshine and calcium deficits	Ivanhoe (1994)
England	AD 1550–1900 (PM)	NS	1 A M. Lower legs	Mac.	NS	Manchester (1983) cited in Stuart-Macadam (1989:212); Ortner (2003)
England	AD 1550–1900 (PM)	968	15 adults. NS.	Mac.	Likely dietary and sunshine deficiency, industrialisation	Molleson & Cox (1993)
Egypt	NS	NS	1. M Adult. NS	Rad.	NS	Moodie (1931) cited by Wells (1975)
Australia	NS	NS	1 Adult. Lower long.	Mac.	NS	Ortner (2003:404)

Notes: *Several cases reported by* Wells (1975) *excluded on basis of* Wells' *subsequent observations.* Dates *given are approximate and supplied to provide an approximate time line.* Date *ranges may be subject to regional variations. Time periods given by original authors are indicated in brackets*: P, *Prehistoric*; UP, *upper palaeolithic*; Me = *Mesolithic*; MW, *Middle Woodland*; NK, *New Kingdom*; R, *Roman*; A-S, *Anglo Saxon*; M, *Medieval*; PM, *post-Medieval*; J, *juveniles*; A, *Adults*; NS, *not specified*; BNS, *bones not specified*; Mac., *macroscopic examination*; Mic., *histological examination*; SSEM, *surface* SEM; Rad., *radiological examination*.

*No specific cases RR *referred to.*

TABLE A3 Summary of Archaeological Cases of Adult Vitamin D Deficiency Osteomalacia

Locations	Date	Total burials	Age/Sex, bones affected	Methods	Attributed cause	Sources
Nubia (BMNH 178A)	ca. 3800–2500 BC (PD)	NS	1. A. M Bending arms, legs, deformity pelvis and sacrum, fracture right femur. Hypertrophic bone of femora Paraplegia with possible complication of rickets	Mac.	Sunshine deficiency in adulthood resulting from paraplegia?*	Rowling (1967:275–278); Ortner (2003:403–404)*
Upper Egypt	1991–1802 BC (Twelfth Dynasty)	NS	A. F. Isolated sacrum, small and lightweight, acute 90° angulation	Mac.	NS	Ortner (2003:402)
Austria	1600 BP (R)	38	1. F. 35–45 years Fractures: R scapula ribs, ulnae, pubic ramus, metatarsal Additional disturbed mineralisation lumbar vertebrae	Mac. Rad. Mic. (BSE-SEM)	Roman military hospital	Schamall et al. (2003a)
England	AD 43–410 (R)	NS	2 A. Flattened femora and pelves	Mac.	NS	Farwell & Molleson (1993)
England (Kempston)	AD 43–410 (R)	92	1. YA/MA F. Axial skeleton atrophy, Possible OM, Poor preservation	Mac.	Possible dietary or physiological factors	Roberts & Cox (2003:143); Boylston & Roberts (2004)

Continued

TABLE A3 (*Continued*)

Locations	Date	Total burials	Age/Sex, bones affected	Methods	Attributed cause	Sources
England	AD 1485–1603 (M)	NS	2 A. angled sacrums and 1 curved sternum	Mac.	NS	Stirland (2000)
England	AD 1550–1900 (PM)	143	7 A. 1 MA M, 2 OA F, 1 OA ?, 2 Ad M, 1 Ad F Fractures scapulae ribs clavicles, bone formation ulnae and ribs Additional disturbed mineralisation ribs/femora	Mac. Rad. Mic. (BSE-SEM)	Effects of industrialisation: air pollution, dietary deficiency, socio-economic status and cultural practices	Brickley et al. (2007)
England (3 sites)	(AD 1550–1900) (PM)	831	17 A. Fractures scapulae ribs, clavicles, bone formation ulnae, ribs, femur sub-trochanteric pseudofracture Additional disturbed mineralisation ribs/femora	Mac. Mic. (BSE-SEM)	Industrialisation: air pollution, dietary deficiency, socio-economic status and cultural practices	Ives (2005); Ives (in preparation)
England	AD 1550–1900 (PM)	388	2 Adults, NS	Mac.	NS	White (1985); Roberts & Cox (2003:309)

Italy	1865 (PM)	NS	1.A M. Skull porosity and thinned or low weight of bones and multiple post-cranial fractures	Mac.	NS	Ortner (2003:398)
Strasbourg*	1896 (PM)	NS	1. 18-year old M. Pelvic deformities, angulation lumbar vertebrae, pitted cortical bone skull	Mac.	NS	Ortner (2003:400)
Hamann-Todd Collection, US*	post-1900 (Mo)	1122 black and white males and females	Axial psueofractures, frequent co-occurrence osteopenia with osteomalacia. Different prevalence age, sex, racial groups	Mac. Rad.	Vitamin D deficiency, ancestry, age and sex-related differences	Mensforth, (2002)
Switzerland*	post-1900 (Mo)	NS	13 A. Incomplete adults. Axial, appendicular fractures, femoral bending. Significant rib and pelvic deformities	Mac.	NS	Brickley et al. (2005)
Austria*	post-1900 (Mo)	NS	15 A. Incomplete adults. Axial, appendicular fractures, femoral bending. Significant rib and pelvic deformities	Mac.		Schamall et al. (2003b); Brickley et al. (2005)

Continued

TABLE A3 *(Continued)*

Locations	Date	Total burials	Age/Sex, bones affected	Methods	Attributed cause	Sources
Prague*	NS	NS	NS. Fracture proximal femur, with severe posterior angulation, osteopenia	Mac.		Ortner (2003:399)
England*	NS	NS	1. A F. Axial deformities.	Mac.	NS	Ortner (2003:400)
England	NS	NS	NS. A. Marked pelvic deformities	Mac.	NS	Brailsford (1929) cited by Wells (1975)

Notes: *Dates given are approximate and supplied to provide an approximate time line. Date ranges may be subject to regional variations. Time periods given by original authors are indicated in brackets:* PD, *Predynastic;* R, *Roman;* A-S, *Anglo Saxon;* M, *Medieval;* PM, *post-Medieval;* Mo, *Modern;* J, *juveniles,* A, *adults;* M, *male;* F, *female* ? *Undetermined.* NS, *not specified;* BNS, *bones not specified;* Mac., *macroscopic examination;* Rad., *radiological examination;* Mic., *histological examination;* BSE-SEM, *Back-scattered* SEM *detection.*

*Location *of curated remains, see individual source. * bones affected, asterix denotes additional data provided by asterixed source.*

Chapter 6

Age-Related Bone Loss and Osteoporosis

Osteoporosis is a disease characterized by a decrease in bone amount with consequent increases in skeletal fragility and heightened risk of fractures, which occur with minimal trauma (Riggs et al., 1998, 2001; Rodan et al., 2002; Melton et al., 2003). The decrease in bone tissue can sufficiently alter the microstructural architecture of bone, facilitating increasingly inefficient buffering of load-bearing. These changes contribute to bone fatigue and further increase the risk of fracture (Frost, 2001; Rodan et al., 2002; Frost, 2003; Chapter 3).

DEFINITIONS OF OSTEOPOROSIS

Osteoporosis and its precursor osteopenia (decrease in bone tissue without risk of fracture) can occur in a wide range of circumstances and as such have received designations such as 'primary', 'secondary' and 'post-menopausal' in order to highlight the relevant causative variables. This chapter focuses on the occurrence of age-related osteoporosis in females and males, and post-menopausal osteoporosis in females. Secondary osteoporosis that occurs following trauma, pathology or dietary imbalance is discussed in Chapter 7.

As highlighted in Chapter 3 (and Box Feature 3.1), bone is not a static tissue, but is consistently being maintained through the process of remodelling. With age remodelling becomes unbalanced with too much bone resorption or too little bone formation causing net bone loss (Compston, 1999; Parfitt, 2003:11–15). This imbalance is accepted as a consequence of the normal process of aging and the resulting bone loss itself is not indicative of osteoporosis (Riggs et al., 1998; Parfitt, 2003). The bone that remains following age-related loss is largely structurally normal, in contrast to the defectively mineralised and pathological bone present in other metabolic conditions (e.g. vitamin D deficiency as discussed in Chapter 5, Paget's disease as discussed in Chapter 8).

Clinical osteoporosis is only evident when the reduction in bone amount and consequent structural changes increase the susceptibility to fracture (Riggs and Melton, 1988; Kanis, 1994; Jergas and Genant, 2001; Melton et al., 2003). Prior to the onset of fracture, there are no external indications of osteopenia or osteoporosis. In the developed world, large-scale non-invasive screening of the quantity of bone tissue is routinely undertaken in order to identify older and post-menopausal individuals who have bone amounts with the greatest risk of fracture. Examination of cortical bone thickness, or bone mineral content and density have been utilised to determine bone quantity (see Genant, 1988; Kanis, 2002; and Brickley and Agarwal, 2003 for review of techniques). However, it is difficult to determine at what amount or quality of bone a pathological fracture will occur, and there is considerable variation in fracture prevalence between different population groups, between the sexes and across age groups. Without such screening programmes, it is impossible to know what the true prevalence of age-related osteoporosis really is.

Alterations in bone quality can also occur in osteoporosis, such as localised defects in bone mineralisation (as distinct from systemic vitamin D deficiency, see Chapter 5), or damage (e.g. cortical cracking) at a microscopic level, both of which can increase mechanical weakening and risk of fracture (Hayes and Ruff, 1986; Burr et al., 1997; Grynpas et al., 2000; Marcus and Majumder, 2001; Grynpas, 2003). These skeletal changes increase with age and may be further modulated by lifestyle influences.

Bioarchaeological investigations of age-related osteoporosis can provide an important insight into the evolution of what is considered by many to be a modern disease. There are significant challenges inherent in the examination of bioarchaeological evidence for this condition, and these are discussed later in the chapter (also Box Feature 6.3). What remains a substantial problem evident in many studies of health is the adoption of non-standardised terminology (see Chapter 2). Any detectable bone loss that appears greater than normal in bioarchaeological studies should be considered to be osteopenia. The presence of osteoporosis may be better defined where an osteoporosis-related fracture is evident within the physical remains.

CAUSES OF AGE-RELATED OSTEOPOROSIS

Age-related osteoporosis is a complex condition with a multi-factorial aetiology (Raisz and Seeman, 2001:1949). This metabolic bone disease has been the focus of intense clinical study over many years (e.g. Riggs and Melton, 1988; Kanis, 1994; Marcus et al., 2001). Whilst the broad factors underlying its development are well recognised, many of the specific mechanisms through which it becomes manifest are still incompletely understood. The pace of research into this condition is largely driven by the means for effective treatment for those significantly affected, such as post-menopausal women (Cummings and Melton, 2002). Whilst necessary, this does create a bias in current research focused on this particular demographic group (further discussed in the Anthropological Perspectives Section later). The following section will briefly review the principal factors currently thought likely to contribute to the onset of this condition.

Menopause

Oestrogen can restrain bone turnover and limit the resorptive capability of mature osteoclasts (Riggs et al., 1998; Lindsay and Cosman, 1999; Reid, 1999;

Pacifici, 2001). With loss of oestrogen during the menopause in females, bone remodelling increases (Pacifici, 2001; Raisz and Seeman, 2001). Bone resorption increases by 90% when compared to 45% increase in bone formation and this imbalance results in a 20–30% loss of trabecular bone and a 5–10% loss of cortical bone (Riggs et al., 1998). Bone loss can occur following decreases in ovarian function within the decade prior to the final cessation of menses (e.g. perimenopause phase) (Riggs et al., 1998; Reid, 1999:55). Oestrogen loss may further remove cell sensitivity to strains from physical activity, potentially further contributing to bone loss (see below; Lindsay and Cosman, 1999; Frost, 2001; Lanyon and Skerry, 2001; Raisz and Seeman, 2001; Riggs et al., 2001; Frost et al., 2004).

The greater prevalence of osteoporosis-related fractures in post-menopausal females compared to age-matched males demonstrates the importance of the menopause in impacting on skeletal health (Parfitt, 1986b; Kanis, 1994; Riggs et al., 2001; Pacifici, 2001). Whilst all women are liable to undergo the menopause, not all experience bone loss at a level sufficient to result in osteoporosis or related fractures. Riggs et al. (1998, 2001) have recently reiterated their hypothesis that it is the onset of the menopause coupled with an additional as yet unspecified factor but potentially including poor nutrition or genetic background, which will increase the rate of bone lost and fracture risk. There may be considerable variation within present and past populations in the potential for additional factors to exacerbate the onset of post-menopausal bone loss resulting in osteoporosis. Significant age-related cortical bone loss has been demonstrated in a post-Medieval British population sample, but with a lack of evidence for osteoporosis-related fractures (Mays, 2000) (see later in Section Consequences of Age-Related Osteoporosis: Fractures). Mays (2000) posited whether cultural factors may have mediated the impact of age-related osteoporosis from becoming manifest, highlighting that social and cultural factors inherent across lifestyles in both present and past communities may have the potential to affect the expression of various metabolic bone diseases.

Increased Age

Increased age results in an imbalance in bone remodelling and bone loss (Riggs et al., 1998; Mundy, 1999; Rosen and Kiel, 1999:57; Frost, 2001, 2003). The extent of bone loss with age is actually similar between males and females with estimates of 20–30% losses of both trabecular and cortical bone occurring (Riggs et al., 1998). It is the contribution of post-menopausal bone loss to this age-related bone loss that causes females to become typically more susceptible to risk of osteoporosis-related fracture compared to males.

Increases in osteoclastic resorption depths can perforate trabecular elements contributing to structural weakness and removing the bone surface necessary for bone formation, preventing the means of improving skeletal structural support, as shown in Figure 6.1 (Riggs et al., 2006). A recent study by Banse et al. (2003) has demonstrated that some bone formation does occur to act in

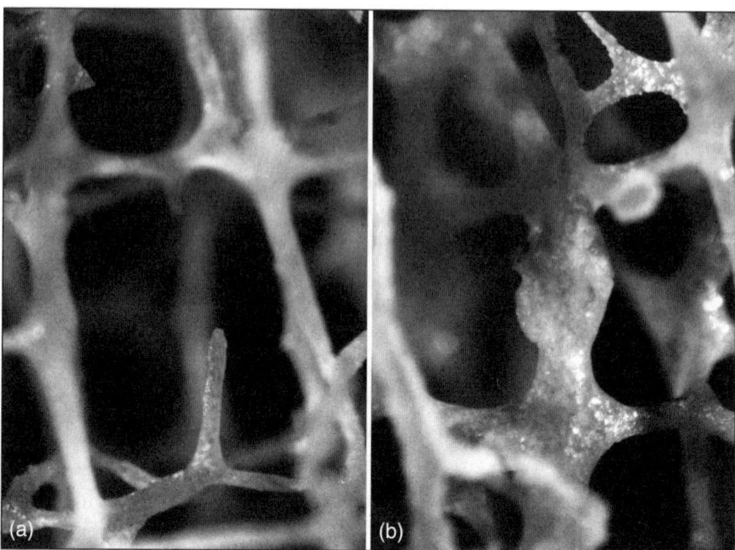

FIGURE 6.1 Light microscopy examination of age-related changes in trabecular bone structure. (a) A 'free-end' or trabecula that has been perforated due to bone resorption. (b) A rounded micro-callus indicating a micro-fracture in a trab-ecula. Archaeological adult human samples.

bridging perforated trabecular struts. While, this has implications in suggesting some potential for structural recovery or adaptation, it is not yet clear how frequently this may occur throughout the skeleton.

Comparatively little is known of the effect of age on bone cell function. Aging may adversely affect osteoprogenitor cells via reducing the number of existing cells as well as limiting the capacity to differentiate new bone cells (see Chapter 3; Pearson and Lieberman, 2004). Aging may also hinder the detection and response to accumulating fatigue damage, adding to the increasing fragility in the skeleton (Schaffler et al., 1995; Burr et al., 1997; Chapter 3). It is not yet clear whether individuals with osteoporosis have a limited function to repair micro-damage, which would also exacerbate the risk of bone weakness and failure (Melton et al., 1988:120; Burr et al., 1997).

Aspects of mineral homeostasis alter with age. Importantly, dietary calcium intake tends to decline with age in many elderly individuals in the developed world (Heaney, 1999:24), as does efficiency of intestinal absorption (Heaney, 1999). Vitamin D concentrations also fall with age (see further Chapter 5). Together these changes can stimulate PTH production increasing bone resorption in order to release calcium from renal metabolism (Chapter 3) and bone into the blood serum (see Chapters 5, 7, 9). A deficiency of calcium is not the cause of age-related and post-menopausal osteoporosis, however alterations in calcium availability have an important role to play in exacerbating adverse aging effects on skeletal health (see Heaney, 1999:25; Chapter 7).

Peak Bone Mass

Peak bone mass refers to the bone tissue that has accumulated by the end of the adolescent growth spurt and prior to the onset of age-related bone loss. This growth phase tends to be achieved between the ages of 18 and 35 years (Orwoll et al., 2001; Bonjour et al., 2003), although some studies have suggested it is achieved between the range of 15 and 20 years (Raisz and Seeman, 2001). The development of peak bone mass is a complex process regulated by environmental factors such as health during childhood growth, nutritional quality, degree of physical activity and physiological factors including concentration of sex steroids, as well as genetic determination (Lindsay and Cosman, 1999; Lloyd and Cardamone Cusatis, 1999; van Der Meulen and Carter, 1999; Orwoll et al., 2001; Recker and Barger-Lux, 2001; Van der Sluis and de Muinck Keizer-Schrama, 2001; Bonjour et al., 2003).

Failure to attain adequate peak bone mass means that with the onset of normal remodelling balances with age, a quantity of bone will be reached at which fractures may occur at an earlier age compared to individuals who have attained greater amounts of bone during growth (Recker and Barger-Lux, 2001). The later completion of puberty in males compared to females enables larger bones and muscles to be developed. Males therefore tend to have greater bone mass from larger bone size and stronger bones, which are more resistant to fracture than females (Ruff et al., 1994; Ruff, 2000; Orwoll and Klein, 2001; Bonjour et al., 2003; Parfitt, 2004). This is an additional contributing factor to increased female prevalence of age-related osteoporosis-related fractures compared to males (Orwoll and Klein, 2001). The potential to attain adequate peak bone mass may be an important variable in the risk of osteoporosis determinable in past populations (see later in this chapter).

Mechanical Loading

Physical activity is important in building bone strength, and is a significant component in mediating the risk of osteoporosis-related fractures (Friedlander et al., 1995; Nikander et al., 2005; Riggs et al., 2006). Limited mechanical loading or disuse increases bone resorption and is a significant consequence of paralysis and of zero-gravity situations such as space flights (see further Chapter 7; and Bickle et al., 1997; Vico et al., 2000; Giangregorio and Blimkie, 2002). It is possible that the influence of age-related changes in bone strength on osteoporosis-risk has been underestimated (Nikander et al., 2005). In particular, assessments of bone mineral density may not be able to accurately detect age-related changes in bone structure and quality (Huuskonen et al., 2001; Riggs et al., 2006; see also discussion in Ruff et al., 2006:487). Increased sedentism in the developed world, potentially stemming from the onset of agriculture with reduced physical activity may be related to increased levels of osteopenia (Ruff, 2000), although interaction with changes in dietary quality, reproductive stresses and pathology also need careful determination (see Section Anthropological Perspectives later in this chapter, as well as Bridges, 1989; Larsen, 2003).

There are age- and sex-related trends in the efficiency of skeletal tissue response to loading. In adolescence and young adulthood, bone is particularly adaptive to mechanical stimuli, frequently mediated by osteoprogenitor cells within the periosteal membrane (Ruff et al., 1994; Larsen, 1997; van Der Meulen and Carter, 1999; Lloyd and Cardamone Cusatis, 1999; Ruff, 2000; Parfitt, 2004; Pearson and Lieberman, 2004; Ruff et al., 2006). Intense loading during growth increases sub-periosteal bone expansion in males, widening the cross-sectional shape of long bones, increasing bone strength and reducing fracture risk. Bone periosteal expansion can occur in females, but bone tends to be deposited more on the endosteal surface, which is not as efficient in loading as sub-periosteal apposition (Ruff et al., 1994; Petit et al., 2004; Nikander et al., 2005; Riggs et al., 2006).

Individuals undertaking increased physical activity may reduce the rate of endosteal resorption and bone loss in weight-bearing bones (Pearson and Lieberman, 2004:76; Peck and Stout, 2007). Activity during later adulthood may result in small gains of bone at the periosteal margin (Eisman et al., 1991; Ruff et al., 2006), but these take longer to accumulate and may require a greater intensity of activity in order to become manifest. Ruff et al. (2006) have argued that many short-term or cross-sectional investigations could have limited potential in detecting the consequences of loading-related remodelling in adults. This may be an important factor in bioarchaeological reconstructions of the causes of osteoporosis-risk in past populations that are inherently cross-sectional.

Osteoblasts appear increasingly unable to respond to mechanical stimuli with age (Lanyon and Skerry, 2001; Pearson and Lieberman, 2004:74, 79), although modern populations tend to demonstrate decreasing physical activity with age, complicating interpretations of purely age-related changes. It is also important to highlight that strength is not a lone factor affecting bone health. Decreases in muscle strength with age may exacerbate the risk of falling and fractures (Nguyen et al., 2000; Szulc et al., 2005).

Extremes of Exercise

Extreme exercise can result in hypothalamic suppression and hypo-oestrogenism causing amenorrhoea (cessation of menstruation). Athletes who combine intense physical activity with low dietary calorie intakes (e.g. gymnasts, ballet dancers, endurance runners, figure skaters) may experience amenorrhoea (New, 2001; Cobb et al., 2003; Rome and Ammerman, 2003:422; Castelo-Branco et al., 2006). This physiological state increases bone resorption resulting in lowered bone density, which is significant for the attainment of peak bone mass and risk of osteopenia and fractures (Kanis, 1994:166; Cobb et al., 2003; Rome and Ammerman, 2003:422; Goodman and Warren, 2005; Punpilai et al., 2005).

Amenorrhoea can be further exacerbated by starvation and eating disorders, with risk for the onset of osteopenia (Chapter 7 and Box Feature 9.2).

Continuing Sub-Periosteal Apposition

During adult life continuing sub-periosteal bone apposition partially offsets endosteal resorption strengthening the load-bearing axis particularly of long bones and providing a means of greater resistance to compressive and tensile strains (Lazenby, 1990; Garn et al., 1992; Ruff, 2000; Orwoll and Klein, 2001; Pearson and Lieberman, 2004). The rate of age-related endosteal and trabecular bone loss in males and females in the vertebrae and iliac crest is similar (see Duan et al., 2001a). However, females are affected to a greater extent by age-related bone fragility partly due to oestrogen deficiency, but also because they frequently fail to gain as much sub-periosteal bone as males appear to (Garn et al., 1992; Duan et al., 2001a; Seeman, 2001). Changes in bone tissue with age are not a single simplistic process. The variability of alterations in rates of bone loss and gain between the sexes over age, and also across different skeletal locations, require continued investigation (Duan et al., 2001a) and careful interpretation regarding possible mechanisms of age-related osteoporosis in bioarchaeological research.

Genetics and Population Groups

To date, no single, straightforward genetic contribution to age-related osteoporosis has been identified (Livshits et al., 1998; Rosen et al., 2002; Seeman, 1999a; Stewart and Ralston, 2000). Stewart and Ralston (2000) and Eisman (1999) have reviewed how genetic defects in the various principle regulators of bone metabolism may influence age-related osteoporosis risk. These factors include the vitamin D receptor, the oestrogen receptor, genes responsible for the Type I collagen protein, as well as TGF-β, osteocalcin, PTH gene and interleukin-6 (IL-6) (see also Zmuda et al., 1999). These data indicate that the genetic influence on age-related bone loss may stem from actions on specific cells involved in bone remodelling (see Chapter 3).

Importantly in modern contexts, age-related osteoporosis may have an inherited component. For example, mothers with osteoporosis-related fractures appear to have daughters with relevant skeletal sites exhibiting low bone density suggesting inherited defects in bone mass (Eisman, 1999:790), although it is difficult to rule out diet, lifestyle and physiological similarities which may be expected within familial units (see further Cooper, 2005:36). The genetic influence on age-related osteoporosis-risk is unlikely to be static throughout the life course. Genes may act variably on bone metabolism, including development of bone morphology, the attainment of peak bone mass and with influence during phases of the menopause (e.g. early or late), with resulting implications for bone health during life (e.g. Slemenda et al., 1996; Eisman, 1999; Brown et al., 2004; Huang and Wai Chee Kung, 2006; Demissie et al., 2007).

Bone remodelling rates may differ between groups of different ancestry. For example, young adult African-Americans have higher concentrations of parathyroid hormone and circulating vitamin D, and can utilise and conserve dietary calcium more efficiently than white Americans (Slemenda et al., 1997; Heaney, 1999:23; Bilezikian and Silverberg, 2001). These changes may derive from a secondary hyperparathyroidism caused by variation in vitamin D synthesis in response to increased skin pigmentation (see Chapters 5 and 9). Increases in active vitamin D decrease the urinary excretion of calcium and may aid increases in bone mass, although the mechanisms and manifestations of population differences with age require further documentation (Heaney, 1999; Bilezikian and Silverberg, 2001). Consideration of differences in bone mass, patterns of bone loss with age and risk of age-related osteoporosis-related fractures between population groups is expanding in both clinical and anthropological research (Solomon, 1986; Riggs et al., 1998; Heaney, 1999; Riggs et al., 2001; Cho and Stout, 2003; Robling and Stout, 2003; Pearson and Lieberman, 2004; Pfeiffer et al., 2006; Peck and Stout, 2007).

Nutrition and Lifestyle

A range of dietary factors may limit the potential development of peak bone mass and/or exacerbate age-related bone loss (e.g. Martin et al., 1985), and this relationship is considered in detail in Chapter 7. Lifestyles typical of the developed world also include high caffeine and alcohol intake and cigarette smoking (van der Voort et al., 2001). Whilst these variables have been linked to reduced bone mineral density in cross-sectional studies, the long-term effects on skeletal health have been less well investigated (Jergas and Genant, 2001:421).

SKELETAL FEATURES OF AGE-RELATED OSTEOPOROSIS

The macroscopic, radiological and histological features of age-related osteoporosis are presented later in this chapter. However, Table 6.1 outlines the bone changes observed in this condition and changes in trabecular bone are illustrated in Figure 6.1 and in cortical bone in Figure 6.2. There are sex-specific differences in the degree of bone loss and structural responses to the risk of fracture. These changes are further manifest in the differences in prevalence of age-related osteoporosis-related fractures in the developed world.

CONSEQUENCES OF AGE-RELATED OSTEOPOROSIS: FRACTURES

Reduction of bone quantity and quality from age-related remodelling imbalance increases bone fragility and consequently require only minimal trauma

TABLE 6.1 Bone Structural Changes in Age-Related Osteoporosis and Implications for Fracture Risk

Bone tissue type	Bone cell actions	Skeletal changes	Sources
Trabecular bone	Increased osteoclastic resorption depth Osteoblastic failure to re-fill resorption cavities	Perforation trabeculae Trabecular thinning Trabecular micro-fractures Increasing trabecular strut length Reduced trabecular connectivity Removal of bone surface for reducing future remodelling Compensatory thickening of remaining trabeculae	Melton et al. (1988), Roberts & Wakely (1992), Brickley & Howell (1999), Compston (1999:259), Marcus & Majumder (2001), Orwoll & Klein (2001), Banse et al. (2003), Frost et al. (2004)
Cortical bone	Increased osteoclastic resorption depth Osteoblastic failure to re-fill resorption cavities	Increased intra-cortical porosity Thinned cortical bone Increased endosteal resorption and bone loss Coalescence of resorption cavities Increased formation of trabecular structure Increased accumulation of fatigue damage Increased medullary space	Parfitt (1986b), Kanis (1994), Parfitt (1994), Compston (1999), Recker & Barger-Lux (2001), Frost et al. (2004), Cho et al. (2006)

in order to result in fracture (Parfitt, 1988:77; Kanis, 1994:3). Fractures do however, derive from many inter-related factors, including the risk of falls (e.g. environmental surround, neurological capability), the energy released by the impact of the injury or fall, as well as the ability of soft tissues and muscles to dissipate such energy (see Harkess et al., 1984; Melton et al., 1988:125; Parfitt, 1988:76; Cummings and Melton, 2002).

Three sites in the skeleton are predilected to osteoporosis-related fracture: the distal radius, vertebral fractures and fractures of the femoral neck (Melton

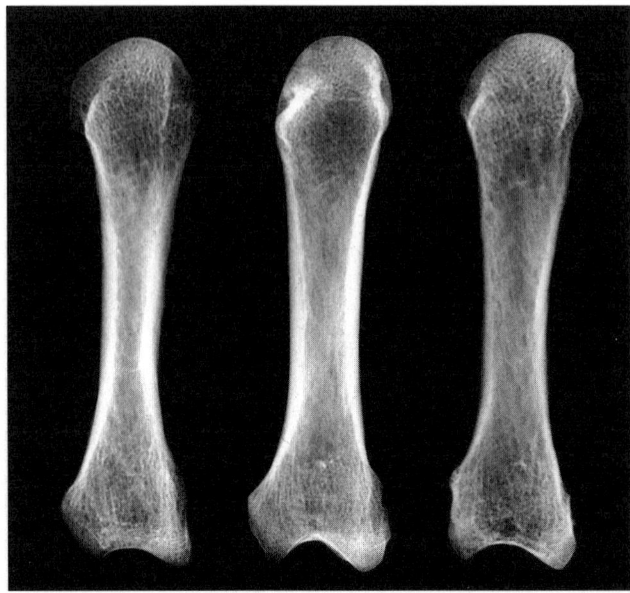

FIGURE 6.2 Thinning of cortical bone and loss of trabeculae in the adult second metacarpal with increased age. Macroscopic observations of bone loss are improved with quantification, for example using metacarpal radiogrammetry to determine age-related changes in cortical thickness (see text). Post-Medieval archaeological adult human samples from St. Bride's Lower Churchyard, London. From left to right: FAO2073 young adult female with percentage cortical thickness (CI index) as 58.61%, FAO2122 middle adult female CI 48.3% and FAO1350 old adult female with CI 34.94%. Radiograph by Rachel Ives, courtesy of the Museum of London.

et al., 1997; Arden et al., 1999; Melton et al., 2003). However, age-related osteoporosis can increase the susceptibility to fracture throughout the skeleton (Cummings and Melton, 2002:1761). Post-menopausal females who have suffered one osteoporosis-related fracture are at greater risk of sustaining further fractures (Riggs et al., 1998; Cummings and Melton, 2002; Briançon et al., 2004; Melton and Kallmes, 2006).

Distal Radius Fractures (Colles' Fractures)

The distal radius is largely composed of cortical bone, although the proportion of trabecular bone increases towards the epiphysis in order to protect the joint surface by dissipating loading through the networked structure (Melton et al., 1988). Age-related cortical thinning, intra-cortical porosity and trabecular bone loss have significant implications for the strength of the distal radius. Limited, if any, periosteal apposition appears to occur at the distal radius (see Melton et al., 1988:118; Lazenby, 1990) potentially due to lack of weight-bearing mechanical stimuli. Therefore, age-related bone loss does not appear to be compensated for, further limiting the resistance to loading during fracture impacts.

TABLE 6.2 Summary of the Features and Complications of Distal Radius Colles' Fractures

Features of Colles' fracture	Pathological changes	Sources
Fracture	Metaphyseal fracture distal radius Dorsal displacement of distal fragment Trabecular compression Healed: fracture line on anterior surface Impaction of posterior cortex Shortening of shaft Loss of slight medial curvature, increased posterior angulation	Melton et al. (1988), Gartland & Werley (1951), Mays (2006b)
Additional bone involvement	Intra-articular fracture of distal radius joint surface Lack of ulna involvement Avulsion or subluxation of fibro-cartilage at the distal radio-ulnar joint Ulnocarpal abutment (abnormal contact) Injury to scapholunate interosseous ligament	Melton et al. (1988), Gartland & Werley (1951), Dai & Jiang (2004), Mays (2006b)
Complications	Severe fractures: reduced function of wrist Secondary osteopenia, impaired local blood supply, necrosis of bone, bone resorption Reflex sympathetic dystrophy (Sudeck's dystrophy): local pain, tenderness, swelling, instability, loss of joint motion stiff fingers, reduced grip strength, limited supination and pronation, osteoarthritis in wrist	Melton et al. (1988), Gartland & Werley (1951), Dai & Jiang (2004), Mays (2006b; Chapter 7)

Colles' fractures typically derive from falls onto an outstretched hand, resulting in posterior displacement of the distal part of the radial shaft. The manifestations of a Colles' fracture together with potential complications are summarised in Table 6.2 and illustrated in Figures 6.3 and 6.4. The appearance and severity of deformity of Colles' fractures can vary, as discussed by Gartland and Werley (1951) as well as in detail recently by Mays (2006b). However, the consistent posterior angulation of the distal fracture segment is a vital component in enabling differentiation from other traumatic episodes that can affect the radius (e.g. Smith's fractures).

In the developed world there is an increased prevalence of Colles' fractures in post-menopausal females, occurring between 50 and 60 years. The loss of

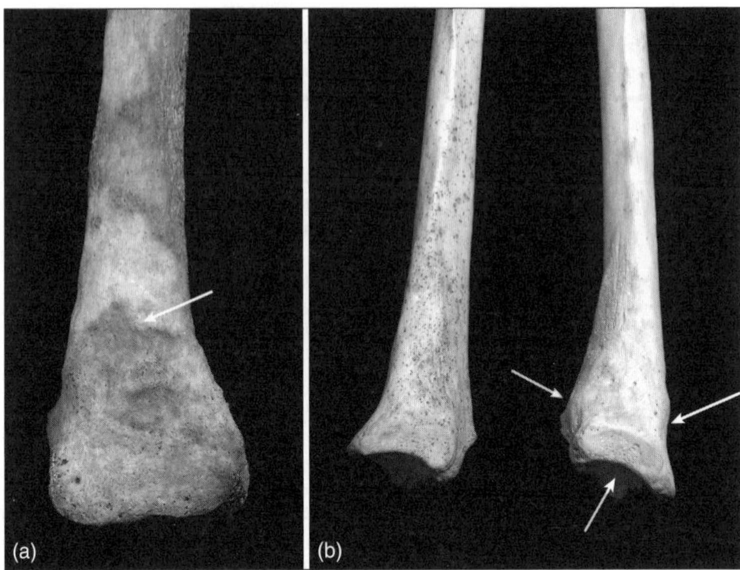

FIGURE 6.3 Colles' fractures of the distal radius. (a) The anterior surface of the radius shows the line of the fracture (arrow) and irregularities in the surface below linked to the posterior displacement of the fractured part. Post-Medieval example from St. Bride's Lower Churchyard, London, old adult female (FAO1547). (b) A medial view of a normal distal radius is compared to a Colles' fracture, with arrows highlighting the posterior displacement marked on the anterior aspect, impaction of the shaft on the posterior and change in direction of the joint surface. Post-medieval example from St. Bride's Lower Churchyard, London, middle adult female (FAO2383). Photographs by Rachel Ives, courtesy of the Museum of London.

bone with the menopause in females together with differences in bone size, mean that fractures can occur at this location more readily following a fall in females as smaller forces are required compared to those in males (Melton et al., 1988; Cummings and Melton, 2002). Colles' fractures present in post-menopausal females are considered as an indicator of age-related osteoporosis and a predictor of future osteoporosis-related fractures. This type of fracture tends to become less frequent in the elderly, even though the risk of falling increases with age due to deterioration in cognitive function and balance, as well as muscle atrophy (Cummings and Melton, 2002).

Colles' fractures can result from falls without any underlying bone loss, and may reflect the nature of the surrounding physical environment. For example, in the developed world, Colles' fractures can also occur in young adult males, resulting from falls and various accidents. Evidence of osteopenia may help differentiate those fractures that are purely traumatic from those that are pathological in origin. Clinical investigations demonstrate a consistent pattern of low bone amount in individuals affected by Colles' fracture (Dai and Jiang, 2004). Treatment of a Colles' fracture may however induce bone loss through immobilisation, often resulting in time-limited and localised muscle and bone atrophy

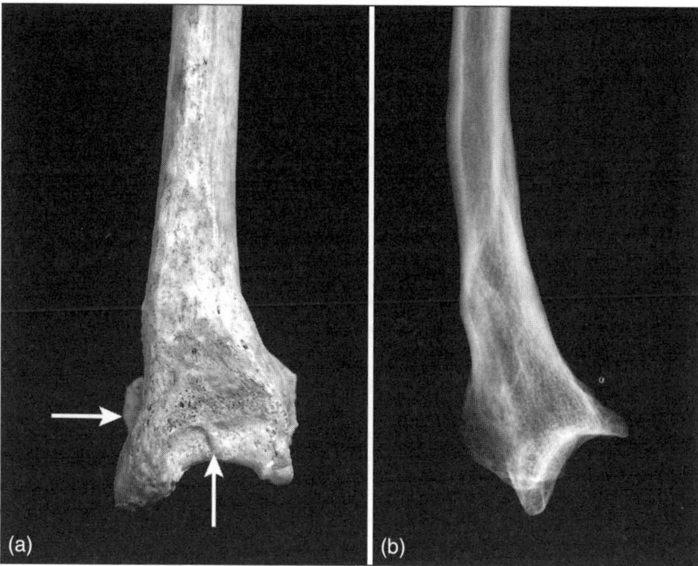

FIGURE 6.4 (a) Archaeological example of a Colles' fracture with arrows illustrating posterior angulation of fractured segment and related fracture of the intra-articular surface. Post-Medieval example from Chelsea Old Church, old adult female (OCU587). Photograph by Rachel Ives, courtesy of the Museum of London. (b) A medial radiograph of a healed Colles' fracture with posterior displacement and marked change in the direction of the distal joint surface. Post-Medieval example from St. Martin's Churchyard, Birmingham.

(Dai and Jiang, 2004). These factors can complicate the identification of osteoporosis-related Colles' fracture particularly in bioarchaeological research.

Vertebral Fractures

Vertebral fractures are an important manifestation of osteoporosis in both post-menopausal females and aging men (Jergas and Genant, 2001; Melton and Kallmes, 2006). Trabeculae in the vertebral body provide most of the bone strength. Decreased bone mass and structural changes significantly reduce bone strength (see Table 6.3). A variety of daily activities can cause compressive forces needed to keep the spine upright and to carry the forces of the weight of each vertebra (Melton et al., 1988:119). Activities such as lifting, coughing, laughing and jumping can increase vertebral compressive loads and in osteoporotic individuals can result in vertebral fractures, as can falling (Melton et al., 1988:119; Cummings and Melton, 2002; Melton and Kallmes, 2006). Load bearing may vary according to body position.

Both sexes lose significant amounts of bone with age (see Duan et al., 2001a, 2001b; Seeman et al., 2001; Melton and Kallmes, 2006). Endosteal

TABLE 6.3 Summary of Age-Related Changes in the Vertebrae and Implications for Fracture Risk

Age-related vertebral changes	Consequences	Sources
Trabecular resorption and perforation	Reduced bone strength particularly in women Smaller loads required for vertebral body collapse	Kneissel et al. (1997), Brickley & Howell (1999), Duan et al. (2001a, 2001b)
Loss of horizontal trabeculae Increasing length of vertical trabeculae	Inefficient dissemination of stresses Increased susceptibility to fracture	Melton et al. (1988), Resch et al. (1995), Kneissel et al. (1997), Brickley & Howell (1999)
Loss of vertical trabeculae Compensatory thickening of remaining trabeculae	Inefficient buffering of loads Increased compressive loads on remaining trabeculae	Melton et al. (1988), Resch et al. (1995), Kneissel et al. (1997), Brickley & Howell (1999)
Thickening may increase bone brittleness Thinning of cortical bone walls	Exacerbated development of micro-fractures Increased risk of collapse	Melton et al. (1988), Duan et al. (2001a, 2001b), Orwoll & Klein (2001), Seeman et al. (2001), Melton & Kallmes (2006)
Age-related degeneration intervertebral disc	Reduced elasticity Reduced cushioning of compression Increased risk of fracture	Melton et al. (1988), Resch et al. (1995), Melton & Kallmes (2006)
Age-related periosteal apposition	Increased bone cross-sectional area Increased buffering of compressive loads Reduced fracture risk Greater in males than females	Duan et al. (2001a), Seeman et al. (2001), Melton & Kallmes (2006)

resorption is an important cause of bone loss, but reduced rates of periosteal apposition in individuals with age-related osteoporosis may also be a significant contributory factor in reducing bone strength (Duan et al., 2001a). Clinical research indicates that both males and females with osteoporosis-related vertebral fractures tend to have smaller vertebrae with lower bone densities and cortical thicknesses than those unaffected (Ritzel et al., 1997; Duan et al., 2001a; Orwoll and Klein, 2001; Seeman et al., 2001). Whether the size differences result from age-related changes or from inefficient bone development needs further investigation.

Vertebral fractures in osteoporosis are classified into three broad types: wedge, compression (crush) and concave (codfish) fractures (Table 6.4). However, there is significant inconsistency in identifying a fracture. It is likely that only severe compression fractures are identified in bioarchaeological contexts. Wedge fractures may be underestimated or not considered in relation to potential underlying osteoporosis. The range of vertebral deformity as illustrated in Figure 6.5 together with grades for assessing the extent of compression or change, would provide useful data if routinely applied and reported on in archaeological analyses. Examples of vertebral fracture from archaeological contexts are also shown in Figure 6.5.

TABLE 6.4 Summary of Osteoporosis-Related Vertebral Fractures Types

Type of vertebral fracture	Features
Wedge	Anterior collapse, maintained posterior height Lateral collapse, maintained body height on opposing aspect Frequent in the lower thoracic and lumbar Mid-thoracic vertebrae may show slight anterior wedge as normal May result in increased curvature in remainder of spine (kyphosis) Creates additional compression on surrounding vertebrae
Compression (collapse/ crush) fracture	Reduction in all aspects of body height Quite severe compression of entire body Typically affects single vertebra
Concave ballooning ('codfish shape')	Collapse/compression of central portion of the superior, inferior, or both surfaces Relative maintenance of anterior and posterior body heights May occur in multiple vertebrae with bone softening (mineralisation defects) in vitamin D deficiency (Chapter 5)

Source: Melton *et al*. (1988:119).

Observational changes in body deformity and measures of the range of decrease in vertebral body height in fracture cases compared to normal, have been suggested as potentially useful means to help determine fracture present in clinical investigations (Genant et al., 1988; Cummings and Melton, 2002; Melton and Kallmes, 2006). There is potential for such a method to be applied to osteological remains. This type of recording may help in distinguishing fractures, if suitable standard deviations and normal ranges are routinely utilised alongside measures of the affected vertebra.

Recognition of the extent of normal variation in vertebral morphology in bioarchaeological examinations is vital (Ericksen, 1978a, 1978b; Kneissel

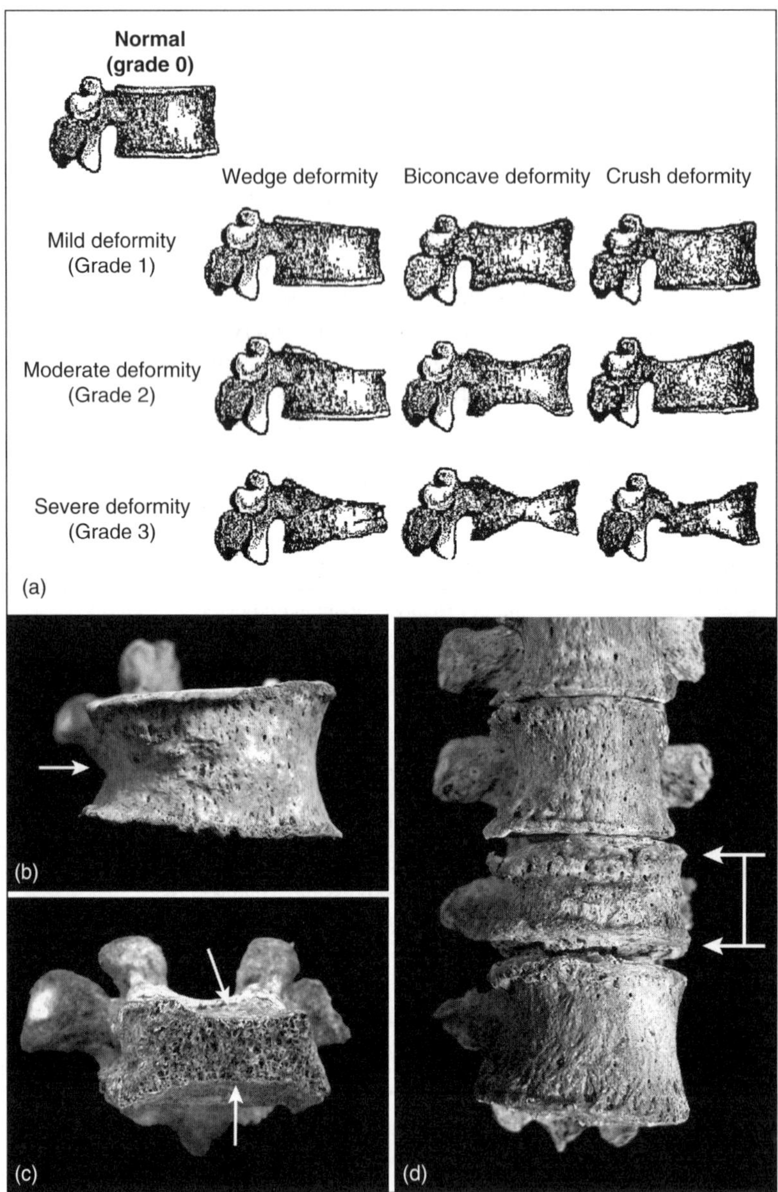

FIGURE 6.5 (a) The range of vertebral deformities possible in age-related osteoporosis together with grades for assessing the severity of change. Reproduced with permission from Genant et al. (1993). Vertebral fracture assessment using a semi-quantitative technique. (b-d) Range of vertebral fractures identified in archaeological bone including (b) lateral wedge fracture in a post-Medieval example from St. Bride's Lower churchyard, London, old adult male (FAO1408); (c) compression of the body particularly marked towards the anterior aspect, post-Medieval example from St. Martin's churchyard, Birmingham, old adult male (SMD705), (d) compression with whole body loss of height, post-Medieval example from the Cross Bones burial ground, London. Photographs (b) and (d) by Rachel Ives, courtesy of the Museum of London..

et al., 1997; Rühli et al., 2005), as is understanding the range of pathological conditions that can manifest in osteopenia and vertebral collapse, including for example myeloma and tuberculosis (Melton and Kallmes, 2006; Chapter 7). Maat and Mastwijk (2000) have discussed the range of pathological changes that can be identified in skeletal vertebral remains. Age-related osteoporosis fractures of vertebrae occur frequently from age 50 years and onwards (Cummings and Melton, 2002), but trauma and vertebral fracture can be quite common in young adult males in the developed world (Melton and Kallmes, 2006). Trauma and occupational injuries of the spine may have affected males and females to a greater extent in the past than is seen today and need careful differentiation from pathological fractures. Analysis of bone quantity and quality may be required to help achieve this in bioarchaeological research.

Consequences of vertebral fracture vary depending on severity and the number of vertebral bodies affected, but can include reduction in stature and kyphosis, which can lead to respiratory complications (Melton and Kallmes, 2006). Severe fractures may impair urinary function and lead to neural cord compression (Orwoll and Klein, 2001; Watts, 2001).

Femoral Fractures

Osteoporosis-related femoral neck fractures have serious consequences with increased mortality following blood loss, shock and generalised declines in overall health reported (Kanis, 1994:10; Cummings and Melton, 2002). Clinical estimates suggest that 90% of these fractures occur from moderate trauma in falls from standing height and frequently in backwards or sideways motion onto the hip (Kanis, 1994:10; Cummings and Melton, 2002; Mayhew et al., 2005). Fractures of the femur occur in two locations affecting both cortical and trabecular bone: intra-capsular fractures affecting the femoral neck and extra-capsular fractures affecting the trochanteric region (Sevitt, 1981; Melton et al., 1988). Age-related changes in cortical and trabecular bone as outlined in Table 6.1 and progressive loss of bone tissue result in failure to support loads applied to bone in the femoral neck in falls, causing fractures.

The manifestations of femoral neck fractures and subsequent complications will vary depending on the severity of injury, attributable largely to the extent of associated soft tissue and blood supply damage (Phemister, 1934; Sevitt, 1981: Chapter 13). For example, severe fractures of the femoral neck have been associated with the loss of between 1 and 2.5 litres of blood (Harkess et al., 1984:26).

The range of fractures that can occur at the femoral neck have been previously discussed (see Phemister, 1934; Sevitt, 1981; Mayhew et al., 2005; Damany and Parker, 2005). Fractures may result in the impaction of bone in the neck causing shortening, and angulation of the femoral head. These fractures tend to heal relatively well (Sevitt, 1981) and are likely to be recognisable in bioarchaeological remains (see Figure 6.6b). Small fissure fractures within the cortex can result from fracture compression, where fracture energy is not efficiently disseminated, such as where trabecular connectivity is reduced (see Kanis, 1994:10; Mayhew et al., 2005; Damany and Parker, 2005). This partial fracture has significant effects on reducing bone resistance to bending. Tensile stresses are increased in the area unaffected by the fracture and if the remaining bone is not strong enough to facilitate the redistribution of such stress, it may also succumb to further fracturing (Mayhew et al., 2005). A recent archaeological example of an incomplete fracture at the femoral neck is discussed by Mays (2006a). Fissure or incomplete fractures at the femoral neck can also occur in vitamin D deficiency (see Chapter 5) and Paget's disease (Chapter 8). However, it is likely that other pathological changes across the skeleton would enable these conditions to be identified.

Femoral neck fractures can be subject to malunion with angulation of the neck and head, or non-union and avascular necrosis (Sevitt, 1981; Kanis, 1994; Center et al., 1999; Cummings and Melton, 2002). It is likely that some individuals affected by severe femoral neck fracture in the past may not have survived for long following the injury, and these cases would show little definite evidence of ante-mortem bone changes. However, physical evidence for survival after more severe femoral neck fracture in older individuals does exist from past communities (e.g. Dequeker et al., 1997; Sheldrick, 2007; see Figure 6.6a),

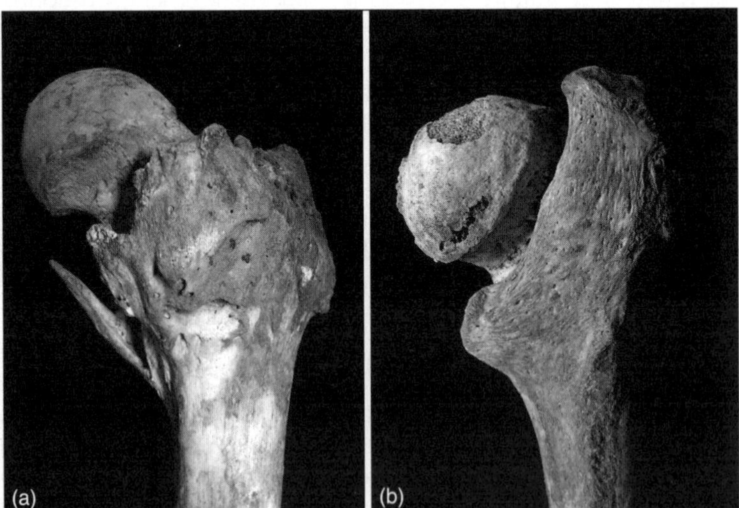

FIGURE 6.6 Femoral neck fractures in archaeological human bone. (a) An inter-trochanteric fracture with significant ossification of soft tissue (anterior view). These fractures can occur with age-related osteoporosis but may be difficult to differentiate from traumatic rather than pathological causes. Post-Medieval example from St. Bride's Lower Churchyard, London, old adult male (FAO1853). (b) Fracture of the femoral neck with impaction and consequent shortening of the neck (posterior view). Example from post-Medieval St. Bride's Lower Churchyard, London, middle adult male (FAO1881). Photographs by Rachel Ives, courtesy of the Museum of London.

demonstrating that there is potential for the effects of this condition to be recognised in the past. These perspectives are discussed further in the chapter.

Changes in trabecular bone structure with age in the femoral head and neck have been well documented (e.g. Melton et al., 1988; Genant et al., 1988). Based largely on the work undertaken by Singh and co-workers (1970), an age-progressive resorption of load-bearing trabeculae is known to occur, which impairs the means of withstanding loading and buffering of forces contributing to fracture risk (see Figure 6.7; Table 6.5).

Significant changes also occur in cortical bone at the femoral neck with implications for age-related osteoporosis fracture risk. Cortical bone thinning occurs by endosteal resorption with age, together with increased intra-cortical porosity. These changes diminish the elasticity of bone, reducing bone strength and increasing fragility. It is significant that cortical thinning occurs at different rates throughout the femoral neck (Ruff et al., 1991:404; Bell et al., 1999a; Bell et al., 1999b; Crabtree et al., 2001; Allen et al., 2004:1008; Loveridge et al., 2004; Mayhew et al., 2005). Recent studies have demonstrated that whilst continued loading with mobility can have beneficial effects on cortical bone in the inferior cortex at the femoral neck, maximum compression tends to occur in the superolateral aspect during a fall (Bell et al., 1999a; Mayhew et al., 2005). It is not yet clear to how uniform cortical bone loss at the femoral neck is between the sexes, or how it responds under a variety of loading situations. Continued development of this understanding has implications for determining how physical activity can benefit bone structure and diminish osteoporosis-related fracture risk.

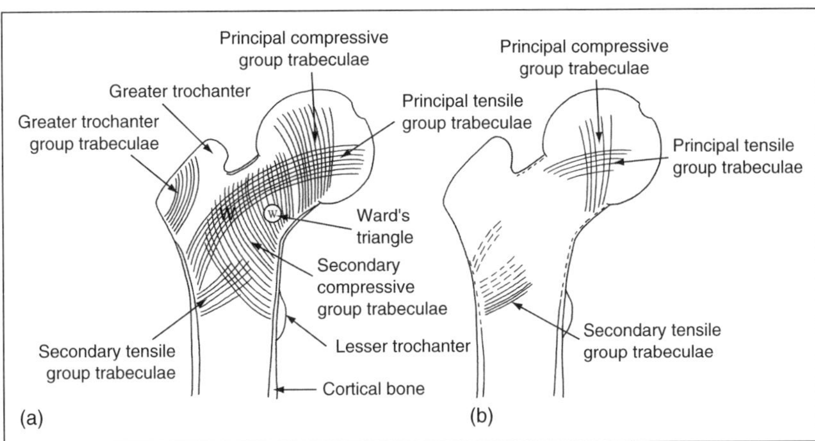

FIGURE 6.7 Schematic line drawing illustrating changes in trabecular structure with age and osteoporosis in the femoral head and neck. (a) Outline of the principal structural components of the femur prior to age-related change. (b) Loss of trabeculae with age and in osteoporosis occurs in a progressive fashion as documented by Singh et al. (1970; see Table 6.5) and leads to considerable loss of support in the femoral neck and resistance to fracture in the event of a fall. Trabeculae that remain are thinned and there is additional thinning of cortical bone in the femoral neck.

TABLE 6.5 Age-Related and Osteoporotic Trabecular Bone Changes in the Femoral Neck

Cause of structural change	Features of changes in trabecular structure	Sources
Age-related	Resorption of trabeculae in Ward's triangle Thinning and resorption of the greater trochanter trabeculae Resorption of the secondary tensile group Resorption of the secondary compressive group	Singh et al. (1970), Genant et al. (1988:194), Melton et al. (1988: Chapter 4)
Osteoporosis	Initial resorption of the principal tensile group towards the greater trochanter Complete resorption of the principal tensile group Resorption of the principal compressive group	Singh et al. (1970), Genant et al. (1988:194), Melton et al. (1988: Chapter 4)

The potential impairment of bone formation may also have a role in increasing bone fragility in individuals with age-related osteoporosis (Power et al., 2003). In particular, cortical bone in fracture samples appears less well mineralised than in healthy age-matched comparisons, indicating a decrease in bone stiffness and resistance to fracture (Allen et al., 2004; Allen and Burr, 2005). There may be further co-existing effects from vitamin D deficiency or collagen defects in individuals with age-related osteoporosis-related fractures that require further investigation (Loveridge et al., 2004; see also Chapter 5).

Anatomical investigations by authors early in the twentieth century determined that there was a lack of osteogenic periosteum at the femoral neck (Phemister, 1934; Sevitt, 1981:206). More recent evidence has demonstrated periosteal membrane covering at least 20% of the femoral neck, capable of producing osteoblasts although likely at a reduced capacity when compared to that at the femoral diaphysis (see Allen et al., 2004; Allen and Burr, 2005). These findings suggest that periosteal apposition can occur at the femoral neck, although how apposition may vary with age, between the sexes and in response to mechanical stimuli at this location are incompletely understood at present.

OSTEOPOROSIS IN THE MODERN PERSPECTIVE

Age-related osteoporosis is currently of significant concern in the modern world, given the projected increase in the aging population and the health

TABLE 6.6 Estimates of the Number of Age-Related Osteoporosis-Related Fractures Occurring Annually in the United States

Fracture type	Estimated occurrence
Hip	300,000
Vertebrae	700,000
Wrist	250,000
Other	300,000
Total annually	Over 1.5 million

Source: *National Osteoporosis Foundation* (2007).

impacts that this will bring. Recent estimates of the number of fractures resulting from age-related fractures in the United States are shown in Table 6.6. The economic cost related to treatment of these fractures was estimated to be 18 billion U.S. dollars in 2002 (National Osteoporosis Foundation; Malabanan and Holick, 2003). Estimates of the potential future occurrence of osteoporosis-related hip fracture predict an expansion from 1.7 million experienced worldwide in 1990 to 6.3 million expected in 2050 (Cummings and Melton, 2002:1762).

ANTHROPOLOGICAL PERSPECTIVES

The investigation of the experience of aging in the past is developing considerably, generating possibilities of correlating the physical evidence of health with additional evidence of cultural and social experiences from past lives (see Box Feature 6.1).

Age-related osteoporosis has been the focus of previous investigations into past health despite some varied conceptions of survival with age in the past (see Box Feature 6.1 and reviews by Martin et al., 1985; Stini, 1990; Agarwal and Grynpas, 1996; Mays, 1999; Agarwal and Stout, 2003). However, evidence of age-related osteoporosis can derive from a broader range of contexts than that typically exist in the developed world. As with many of the metabolic bone diseases, humans are not exclusively affected by age-related osteoporosis. As discussed in Box Feature 6.2 below, many non-human primates also exhibit skeletal changes in line with this condition.

Box Feature 6.1. Historical and Anthropological Perspectives of Aging

The generally widespread belief that it was rare to grow old in the past stems from erroneous interpretations of life expectancy when estimated from birth as evidence of lifelong survival, as recently argued by Thane (2005). In many cultures, the critical years of infancy and childhood are those which pose the most serious threat to survival. Once overcome, many individuals in the past had a reasonable chance of living into old age (Thane, 2000). Evidence of aged individuals in the past does exist. Examination of prehistoric remains has demonstrated the presence of old adults as classified in the 50+ osteological age category (e.g. Rogers, 1990). A recent study has determined that prolonged life is an evolved trait of *Homo sapiens* and has not derived from factors influencing mortality such as the transition to agriculture or industry (Blurton Jones et al., 2002; see also Alvarez, 2000; Peccei, 2001). Elderly individuals are also present within contemporary hunter–gatherer groups (see Cohen, 1989:200; Shanley and Kirkwood, 2001).

Recent developments in historical research are showing a growing interest in the experiences of old age in the past (e.g. Kertzer and Laslett, 1995; Thane, 2005). Analysis of historical evidence has identified the presence of aged individuals in the past. Estimates of historical and demographic records suggest that at least 10% of the populations in England, France and Spain during the eighteenth century were aged over 60 years (Thane, 2005:9; Troyansky, 2005:175). Examination of burial registers and aspects of funerary regalia such as coffin plates also demonstrate the presence of elderly individuals in historical groups, particularly in post-Medieval cemeteries such as St. Martin's, Birmingham (Brickley and Buteux, 2006), Christ Church, Spitalfields, London (Molleson and Cox, 1993) and St. Thomas' Anglican Church, Belleville (Saunders et al., 2002:132).

Anthropological and historical evidence may be selective in the representation of certain demographic groups leading to distortion in the reconstructions of the lives of aged individuals. For example, increased marginalisation of the elderly was evident in ancient Rome, with a loss of empowerment and public authority, increasing dependence and vulnerability notably experienced by aging males. Even more apparent is that limited evidence attesting to the experiences of elderly Roman females to compare with those of males (see discussion in Harlow and Laurence, 2002:118–131). Investigation of the physical evidence in combination with historical and anthropological perspectives may provide a valuable means of investigating health at different stages of the life course for many societies.

Age-Related Osteoporosis in Men

Recent investigations have demonstrated that almost 20% of males aged over 50 years have osteoporosis in either the spine, hip or wrist (Melton, 2001:179; see also Duan et al., 2001b; Orwoll and Klein, 2001). Despite protection by continuing periosteal apposition, the rate of male cortical bone loss at 5–10% per decade is considerably faster than previously estimated from cross-sectional studies (see Orwoll and Klein, 2001:108–109; Seeman, 2001). The female pattern of age-related increase in osteoporosis-related fractures is also evident in men (Orwoll and Klein, 2001).

The specific causes of age-related bone loss in males are not yet clear. Age-related declines in androgens such as testosterone may stimulate remodelling, as may decreases in vitamin D, potentially involving secondary hyperparathyroidism (Boonen et al., 1997). Declines in oestrogen in males may also have more significant implications on bone health than previously realised (Riggs et al., 1998; Bilezikian, 2006). In the developed world, males appear to suffer greater mortality associated with age-related osteoporosis-related fractures (Orwoll and Klein, 2001). It would be of interest to determine if this trend is also apparent in developing countries as well as in past contexts.

The extent of relative bone fragility and age-related osteoporosis risk of males in past populations has received little study in comparison to females, but osteological evidence has demonstrated that males were also affected (see Sambrook et al., 1987; Roberts and Wakely, 1992; Mays, 2001; Brickley, 2002). Whilst age may play a significant role in such skeletal changes, differences between the sexes regarding the impact of challenges in early environments on

Box Feature 6.2. Animal Studies in Osteoporosis I: Age-Related Bone Loss

Recent research has investigated the nature of age-related osteopenia occurring in animal species other than humans and has demonstrated degenerative changes in a range of non-human primates, including macaques, baboons, gorillas, Gombe chimpanzees, cynomolgus monkeys, rhesus monkeys and Cayo Santiago macaques. For recent reviews of the relevant literature see Cerroni et al. (2000) and Grynpas et al. (2000) as well as studies by Sumner et al. (1989), Colman et al. (1999a, 1999b), Black et al. (2001) and Havill et al. (2003).

In particular, combined analysis of age-related bone loss (bone mineral density) and osteoporosis-related vertebral fractures have been recognised in free-ranging non-human primate colonies, such as those from Cayo Santiago studied by Cerroni et al. (2000:405) (Figure 6.8). Reviews of investigations of fracture prevalence in various non-human primates have demonstrated marked increases of fractures in older primates (see review in Lovejoy and Heiple, 1981). However, these trends are likely to be attributable to the cumulative nature of fractures acquired throughout the individual lifespan. Future investigations utilising the methodology adopted by Cerroni et al. (2000) may yield more specific evidence for pathological fractures and osteoporosis presence affecting non-human primates.

Some non-human primate species also undergo a natural menopause. For example, data from captive rhesus macaques indicates that reductions in bone mass and higher bone turnover appear similar to those changes observed in post-menopausal humans (Colman et al., 1999a:4147; see also Walker, 1995a; Havill et al., 2003). Grynpas et al. (2000) have importantly argued that the timing of menopause may not be consistent within and between species, and may occur much later than in equivalent human models. Therefore, detectable patterns of bone loss could be exacerbated by age as well as hormonal status. What is perhaps less well understood in non-human primate investigations are the potential effects of variations in dietary intake, as well as physical activity, to alter bone amount and trends of age-related bone loss in ways that may differ to those observed in humans.

Continued

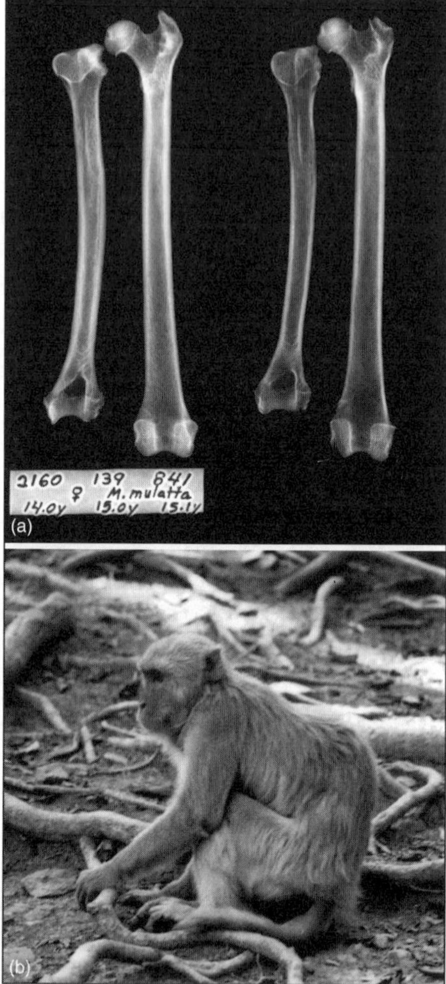

FIGURE 6.8 (a) Osteoporotic female rhesus monkey from Cayo Santiago, Puerto Rico. Radiograph of left humerus and left femur of no. 841 (15.1 years old; on the right) with age/sex-matched control specimens on the left (b) elderly female rhesus monkey from the colony on the island of Cayo Santiago, Puerto Rico, showing dorsal kyphosis of the spine (Dowager's hump). Courtesy of Dr. A.M. Cerroni, University of Toronto.

health and bone development are of interest (Stinson, 1985; Seeman, 1999b; Seeman, 2001). Such sex-specific differences may result in loss of sexual dimorphism and have potential effects on the adequate attainment of peak bone mass and osteoporosis-risk.

In order for patterns of age-related bone loss to be meaningfully interpreted in bioarchaeological studies, the age and sex of the individuals under study need to be accurately assessed. As discussed in Chapter 2, the analysis of archaeological skeletons is fraught with many complicating factors. The inherent challenges in identification of age and sex are highlighted in Box Feature 6.3. Further difficulties in the identification of age-related osteoporosis are discussed below.

Box Feature 6.3. Problems in the Determination of Age-Related Bone Changes in Biological Anthropology

Osteological determination of age and sex

At present, osteological analyses cannot consistently and accurately identify individuals into narrow age bands once over approximately 50 years. This creates substantial problems for investigations of many age-related diseases (Mays, 2006a). Methods are being developed to improve the accuracy of osteological aging (e.g. Buckberry and Chamberlain, 2002; Mulhern and Jones, 2005; Falys et al., 2006; Storey, 2007). However, the large osteological grouping of 50+ years means that it is difficult to extrapolate females affected by menopause-related bone loss from those affected by age-related bone loss. The age structure of a sample may further confound evidence of age-related bone loss (Mensforth and Latimer, 1989; Mays et al., 2006b; see Chapter 2).

Furthermore, sex determination is problematic (e.g. Walrath et al., 2004), and confounded by morphological developments inherent with age. Walker has highlighted the increasingly masculine development of the supraorbital ridge in older females (Walker, 1995b:36), as well as alterations in greater sciatic notch morphology with increased age and between different population groups (Walker, 2005), complicating the means of determining sex. Continued investigation into the range of skeletal development throughout the life course and between the sexes needs to be undertaken together with research into the nature of age-related pathologies in past populations.

Pathology and sex determination

Pathological changes may further influence the various skeletal regions relied upon for age and sex determination. For example, Walker's (2005:390) analysis of individuals affected by vitamin D deficiency (see Chapter 5) demonstrated significantly altered pelvic morphology including the greater sciatic notch required for sex determination. Gross cranial changes in Paget's disease (Chapter 8) can distort many of the skeletal traits necessary for sex determination. Pathological changes may place increasing emphasis on unaffected skeletal areas for determination with potential bias in the accuracy of aging and sexing methods used.

As discussed in Chapter 2, bone preservation may be influenced by aspects of the burial environment or by pathological conditions. Bone loss with age can increase skeletal fragility, potentially resulting in greater fragmentation of affected bones following burial. This may influence the recognition of this condition in past populations. The composition of human burials are culturally construed, and there is an inherent assumption in the investigation of this condition in the past, that all demographic groups of a society are treated in a similar manner or are potentially recognisable within a burial group.

PALAEOPATHOLOGICAL CASES OF AGE-RELATED OSTEOPOROSIS

Palaeopathological analyses require radiological or histological quantification in order to demonstrate age-related and post-menopausal loss of bone (e.g. Carlson et al., 1976; Ericksen, 1976; Martin and Armelagos, 1979; Martin et al., 1985; Pfeiffer and Lazenby, 1994; Mays, 1996b, 1999, 2006a; Lynnerup and Von Wowern, 1997; Drusini et al., 2000; Holck, 2006). It is difficult to differentiate on the basis of bone quantity normal reductions indicative of aging, from pathological changes indicative of heightened risk of osteoporosis and related fracture. In contrast to clinical studies, there are no standards designated for this purpose in paleopathology and inherent variation within populations may make the adoption of such applications difficult (see Melton, 2001). Comparative rather than single-site analyses are likely to provide the best means for improving the interpretation of normal and abnormal patterns of age-related bone loss in past populations (e.g. Cook, 1984; Rewekant, 2001; Mays, 2006a; Mays et al., 2006b).

Interpretations of comparative analyses are currently hampered by the use of a wide variety of techniques in order to assess age-related bone loss (Agarwal and Grynpas, 1996; Mays, 1999; Brickley and Agarwal, 2003). More fundamental may be the differences in remodelling rates apparent throughout the skeleton and the implications these have for reconstructing normal or abnormal age-related bone loss (e.g. Lees et al., 1993; Ekenman et al., 1995; Mulhern and Van Gerven, 1997; Mays, 2000; Robling and Stout, 2003; Pearson and Lieberman, 2004; Pfeiffer et al., 2006; Peck and Stout, 2007). This lack of consistency presents significant difficulties in attempts to gain a perspective on patterns of age-related osteoporosis risk from across past populations beyond descriptions of normal age-related bone loss.

The only direct evidence that can indicate the health impact of age-related osteoporosis is fracture, particularly at the distal radius (Colles' fracture), vertebrae and femoral neck (Brickley, 2002; Mays, 1999, 2006b). Fundamental problems currently exist in the reporting of potential cases of age-related osteoporosis-related fractures. Often in pathology reports, age and sex are not attributed to a specific description of pathological change, preventing interpretation of fracture prevalence in relation to potential underlying disease. We recognise that there are inherent challenges in determining at what age an individual sustained a fracture, unless evidence of healing and recent bone remodelling is evident. Even though osteoporosis-related fractures tend to occur in an age-progressive pattern in modern populations, evidence of a well-healed Colles' fracture in an old adult female may not necessarily or causatively equate with evidence of age-related osteoporosis (see Roberts and Manchester, 2005; Mays, 2006a). However, without data attempting to define trauma by affected age and sex groups (and by bone elements to enable accurate prevalence rates to be determined), there remains a significant barrier to meaningful interpretation of this condition from trauma analyses.

Specific descriptions of fractures (and pathology in general) need to be consistently adopted in paleopathology (e.g. Ortner, 2003; Roberts and Manchester, 2005). Vague descriptions of 'fracture of the proximal femur' cannot be meaningfully interpreted as to whether this fracture is of the femoral neck, trochanteric, or sub-trochanteric/proximal third of the shaft. Equally, the state of the fracture, healed, healing, unhealed, its extent (e.g. complete or fissure), affecting which aspects and additional complications (e.g. angulation, impaction, necrosis, ossification of soft tissue, dislocation) would provide the necessary data in order to better interpret this aspect of past health. Age-related osteoporosis is not a modern disease, and there is clear evidence of the condition in past populations (see in particular Lovejoy and Heiple, 1981; Sambrook et al., 1987; Mensforth and Latimer, 1989; Mays, 1996b; Dequeker et al., 1997; Mays, 2000; Mays, 2006a, 2006b; Sheldrick, 2007). In order for us to understand the extent of this condition in the past, improvements in the manner of disease recording are needed.

A small number of studies have correlated measures of bone loss individuals with age-related osteoporosis-related fractures (Foldes et al., 1995; Frigo and Lang, 1995; Kilgore et al., 1997; Mays, 2000; Strouhal et al., 2003; Mays, 2006a, Mays et al., 2006b; Domett and Tayles, 2006). Individuals with fractures typical of age-related osteoporosis tend to exhibit low bone amounts compared to others within the population sample. Whilst this is strongly suggestive of evidence for osteoporosis, there is potential for bone loss to have occurred during treatment of the fracture or subsequently from it. Despite the range of difficulties in accurate identification, these insights do demonstrate plausible evidence for age-related osteoporosis in the past.

Despite the challenges outlined above and in Box Feature 6.3, paleopathological evidence of osteopenia can prove insightful regarding the efficiency of subsistence strategies in past populations (see Pfeiffer and King, 1983; Cook, 1984; Martin et al., 1984; Bridges, 1989; Larsen, 2003). The mechanisms of dietary composition in contributing to osteopenia are discussed in detail in Chapter 7.

Reproductive stress may impact on skeletal health in a number of ways (see Chapter 5). In particular, pregnancy necessitates large amounts of calcium for the growing foetus, which can derive from increased intestinal absorption of calcium rather than from the maternal skeleton (Eisman, 1998; Laskey et al., 1998; Ritchie et al., 1998; Prentice, 2003). A recent anthropological study has stated that it would be evolutionary maladaptive for pregnancy to result in detrimental maternal health, given the essential role that it plays in these terms (Agarwal and Stuart-Macadam, 2003). However, little is highlighted clinically regarding the potential skeletal effects if dietary calcium supply is poor. This situation could result in imbalance in serum calcium levels with consequent secondary hyperparathyroidism and bone resorption if marked and prolonged.

Metabolic impacts are likely to have been significant where socio-cultural practices further influenced the availability of dietary resources. For example, differential diets between the sexes, where males acquire more animal protein and females more cereals, may affect health in a number of ways (Larsen, 1997:72–76). The latter diet may inhibit calcium availability through phytate binding and vitamin D metabolism (see Chapter 5), which when combined with the calcium requirements in pregnancy could manifest in osteopenia. Similar trends could be apparent if subsistence strategies are influenced by status rather than by sex (see Box Feature 7.3; Larsen, 1997:76).

Significant losses of maternal bone tissue have been demonstrated between three and six months of lactation even where dietary calcium is high (Laskey et al., 1998; Ritchie et al., 1998) although there is some inconsistency (e.g. Sowers et al., 1991), as recently reviewed by Agarwal and Stuart-Macadam (2003) and Karlsson et al. (2005). Increased bone turnover has been reported during pregnancy and lactation in some studies, with loses of 5–7% bone mineral density in the vertebrae and hip (Eisman, 1998; Karlsson et al., 2005). In contrast to pregnancy, high calcium demands in lactation do not derive from the maternal diet but from skeletal stores. Decreases in oestrogen following parturition may remove the inhibitory mechanisms of resorption, although the causes are still incompletely understood (Eisman, 1998; Laskey et al., 1998; Karlsson et al., 2005).

The identification of certain reproductive stresses such as premature osteopenia may be variable depending on social position within a society or cultural practices within the population under study. For example, many high status post-medieval British females during the seventeenth and early eighteenth centuries favoured wet-nursing rather than breastfeeding (Fildes, 1986). This may suggest a means of protection against lactation-related osteopenia for the mothers, but with greater challenges for skeletal health on those employed as wet-nurses.

Whilst skeletal losses during pregnancy and lactation are usually recoverable and would rarely impact on lifetime risk of age-related osteoporosis, additional poor health and dietary change may affect the skeleton during reproduction in different manner to that typically experienced in the developed world. Evidence of detrimental impacts on maternal health and early onset of osteopenia have been identified in females from various past contexts (e.g. Ericksen, 1976; Martin and Armelagos, 1979; Martin et al., 1984; White and Armelagos, 1997; Agarwal, 2001; Poulsen et al., 2001; Turner-Walker et al., 2001; Cho and Stout, 2003; Mays et al., 2006b). The role in which differences in remodelling rates between population groups may further contribute towards premature osteopenia need further investigation (see above in this chapter).

Differences in bone amount between population groups may reflect on challenges to health and development sustained during earlier periods of life (see Box Feature 3.1). For example, in modern populations, dietary deficiencies during childhood can contribute to reduced peak bone mass and increased risk of age-related osteoporosis and fractures (see also Box Feature 9.2). Similar processes may have impacted on health risks in the past. However, there may be inherent difficulties in such reconstructions from osteological analyses, as selective mortality and the demographic composition of many archaeological samples may pose significant biases preventing recognition of such factors. The risk of age-related osteopenia as manifest in old adults from a sample may not be reflective of the health experience as indicated by the young adults, who represent peak bone mass attainment. These interpretative challenges should not discourage future investigations into this metabolic condition, but as argued in Chapter 2, should instead promote cautionary and meaningfully constructed bioarchaeological analyses.

CO-MORBIDITIES

There is often co-existence between osteoporosis and adult vitamin D deficiency particularly in older adults (see Chapter 5), although this group tends to show an increased number of independent conditions that may co-occur with bone loss simply as a function of age. Osteopenia can be a significant factor in many pathologies or following trauma and immobilisation (Chapter 7) as well as in dietary conditions including vitamin C deficiency (scurvy) (Chapter 4) amongst other nutritional disorders (Chapters 5 and 9).

DIAGNOSIS OF AGE-RELATED BONE LOSS AND OSTEOPOROSIS IN ARCHAEOLOGICAL BONE

Macroscopic Features of Osteoporosis

Bone structural changes that occur in age-related osteoporosis will not be visible macroscopically until they result in fracture (Table 6.7). Generalised skeletal changes may occur, for example light-weight bones, but there is potential for post-mortem damage or pathological conditions to have exacerbated bone loss and give an erroneous impression of osteoporosis. Osteoporosis can occur in a range of pathologies and dietary conditions (discussed further in Chapter 7).

Radiological Features of Osteoporosis

Radiological osteopenia may be generalised or localised particularly affecting the regions most susceptible to related fractures (Table 6.8). Simple visual observations of bone radiographs are unlikely to be sensitive to clear evidence of bone loss until a significant amount of bone has been resorbed. Clinical research has suggested that between 20% and 40% of bone mass is lost before clear radiological changes will become apparent (Jergas and Genant, 2001:414). However, quantitative recording of bone tissue amount may prove useful in identifying broad trends of age-related bone loss and osteoporosis risk. Measures of age-related bone loss have been applied to archaeological bone, in particular using the second metacarpal (metacarpal radiogrammetry), which tends to survive the burial environment well (see Mays, 1996b, 2000, 2001, 2006a; Ives and Brickley, 2004, 2005). Analyses of bone mineral density

TABLE 6.7 The Macroscopic Features of Osteopenia

Bone affected	Features	Code	Differential diagnosis	Sources
Cranium	Fine pitting/ diffuse porosity of ectocranium	G	Paget's disease, Vitamin D deficiency	Ortner (2003)
	Significant thinning and depression of parietal bone Usually bilateral	G	Anaemia, Aging	
Dentition and mandible	AMTL Resorption of alveolar ridges	G	Dental disease, Age, Infection, Various metabolic	August & Kaban (1999), Jeffcoat et al. (2001)
Vertebrae	Fracture: wedge, compression, biconcave deformities Usually single site Upper thoracics or lumbar	D	Vitamin D deficiency, Infection (e.g. tuberculosis), Neoplastic (e.g. myeloma)	Melton et al. (1988), Brickley (2002), Ortner (2003), Melton & Kallmes (2006)
	Kyphosis if severe	G	Congenital, Developmental (e.g. Scheuermann's disease), Trauma, Vitamin D deficiency	
Ribs and sternum	Potential thinning of rib cortex	G	Osteopenia, Trauma, Vitamin D deficiency	Ortner (2003), Brickley (2006)
	Rib fractures	G		
Pelvis and sacrum	Cortical thinning/ translucency ilia	G	Osteopenia, Vitamin D deficiency, Trauma	Ortner (2003)
	Insufficiency fracture pubic rami	G		
Long bones	Fracture distal radius (Colles' fracture)	D	Trauma, Vitamin D deficiency, Paget's disease (see Chapter 8)	Gartland & Werley (1951), Melton et al. (1988), Brickley (2002), Ortner (2003), Dai & Jiang (2004), Mays (2006b)
	Fracture of femoral neck (complete or fissure) or fracture through trochanter	D		

Note: The features listed demonstrate the skeletal changes useful for pathological identification. Diagnosis can be determined on the strength of each feature: S – strongly diagnostic feature, S features are normally required for a diagnosis. D – diagnostic feature, the presence of multiple D features are required for a diagnosis. G – general changes, can occur in many of the metabolic bone diseases, as well as in other conditions. G features can aid a diagnosis together with S or D features but cannot be used alone to suggest a diagnosis. Commonly occurring conditions that display these pathological features are listed to aid differential diagnosis, although this list is not exhaustive. AMTL, ante-mortem tooth loss. See additional discussions of pathology in Chapter 7.

TABLE 6.8 The Radiological Features of Osteopenia

Bone affected	Features	Code	Differential diagnosis	Sources
Cranium	Thinning of cortical bone, particularly parietal ectocranium	G	Paget's disease, Vitamin D deficiency, Anaemia, Aging	Ortner (2003)
Dentition and mandible	Generalised osteopenia Cortical and trabecular bone loss in mandible May be difficult to differentiate from normal	G	Dental disease, Age, Infection, General metabolic	August & Kaban (1999), Jeffcoat et al. (2001)
Vertebrae	Generalised osteopenia Thinning of cortical bone Residual cortical outline maybe enhanced 'picture-frame effect' Loss of trabeculae (primarily vertical) Coarsening of remaining trabeculae, striated appearance	D	Vitamin D deficiency, Paget's disease	Genant et al. (1988), Melton et al. (1988), Jergas & Genant 1999, 2001), Brickley (2002; Chapter 7)
Ribs and sternum	Potential thinning of rib cortex – may be difficult to discern	G	Secondary osteopenia	Chapter 7
Pelvis and sacrum	Cortical thinning/ translucency ilia	G	Age, may be inconsistent throughout skeleton, other metabolic e.g. Vitamin D deficiency	Ortner (2003), Brickley et al. (2005)
Long bones	Thinning of cortical bone* Increasing width of medullary cavity Striated cortex if severe intra-cortical porosity Loss of trabecular bone Coarsening of remaining trabecular bone Loss of trabeculae in femoral neck: e.g. Singh's index. Juxta-articular osteopenia	D	Vitamin D deficiency, Paget's disease, Hyperparathyroidism, Renal dysfunction, Joint diseases	Singh et al. (1970), Genant et al. (1988), Melton et al. (1988), Jergas & Genant (1999, 2001), Brickley (2002), Ortner (2003; Chapter 7)

Note: See Table 6.7 for definition of codes used in diagnosis.

**Quantitative measures recommended (e.g. metacarpal radiogrammetry), see text.*

and content have also been undertaken using archaeological human bone (see Lees et al., 1993; Mays et al., 1998), but there are significant problems with the application of this technique, including the potential for diagenetic alteration to distort bone density estimates, as recently reviewed by Brickley and Agarwal (2003).

Histological Changes of Osteoporosis

Bone structural changes are significant in underlying osteoporosis-related fractures (Table 6.1). Microscopic analyses can aid the recognition of age-related bone loss and osteopenia, and have been applied to bioarchaeological samples (Roberts and Wakely, 1992; Brickley and Howell, 1999; Agarwal et al., 2004) (see Table 6.9). However, these analyses require destructive sampling and rely on good bone preservation. Comparisons with healthy individuals and across age and sex groups are necessary for meaningful interpretations. Age-related osteopenia needs to be determined from pathological causes, as discussed throughout in Chapter 7.

TABLE 6.9 The Histological Features of Osteopenia

Bone affected	Features	Code	Differential diagnosis	Sources
Cortical bone	Increased number of resorption sites	D	Many metabolic conditions (see Chapters 5, 7 and 9), especially vitamin D deficiency	Martin & Armelagos (1979), Thompson & Gunness-Hey (1981), Burr et al. (1997), Peck & Stout (2007)
	Incomplete filling of osteons	D		
	Increased resorption depths	D		
	Coalescence of resorbing spaces	D		
	Potential increase in micro-cracks/ fatigue damage*	G		
	May be localised mineralisation defects	G		
Trabecular bone	Trabecular resorption	D	Various metabolic conditions (see Chapter 7)	Roberts & Wakely (1992), Kneissel et al. (1997), Brickley & Howell (1999), Agarwal et al. (2004)
	Loss of connectivity	D		
	Increased trabecular perforation	D		
	Micro-fractures	D		

Note: See Table 6.7 for definition of codes used in diagnosis.

**Micro-damage in cortical bone may be difficult to determine from post-mortem damage in archaeological samples.*

CONCLUSIONS

Remodelling imbalances derived from the menopause and increased age are vital factors in the pathogenesis of osteoporosis and related fractures. Broader examination of bioarchaeological evidence may demonstrate that this condition is not simplistically related only to the domain of the post-menopausal female. While males may be less affected by this disease than females, the range of potential factors that could impact on both sexes particularly in past communities should not be overlooked. In particular, factors that may influence the development of peak bone mass prior to the onset of age-related bone loss may be significant in many past contexts. The inter-relationship between the diverse factors that may mediate behaviour, cultural practices, reproduction, dietary quality and health at various stages across the life course is of particular concern when investigating osteoporosis. Past perspectives on this condition may illustrate a different experience of this condition, which is extensively prevalent in the developed world.

Chapter 7

Secondary Osteopenia and Osteoporosis

Secondary osteopenia and osteoporosis can be underlying complications of a large range of diseases. As discussed in Chapter 6, abnormal loss of bone can have severe consequences for increasing bone fragility. The recognition and implications of secondary pathological changes are frequently overlooked in relation to identifying the primary changes in bioarchaeological analyses. However, the presence of conditions such as osteopenia can enable a meaningful interpretation of the impact or consequences of disease or injury on an individual's life. As future research develops to investigate further the various co-morbidities potentially identifiable, there is increasing scope for recognising the factors that can contribute to secondary osteopenia in both present and past populations.

CAUSES OF SECONDARY OSTEOPENIA AND OSTEOPOROSIS

There are three broad mechanisms that can result in the skeletal manifestations of secondary osteopenia. Injury or pathological change may affect a specific limb or the spinal cord, subsequently limiting mobility and leading to disuse atrophy. Osteopenia may occur as a specific mechanism of a disease. Finally, dietary inadequacy can disrupt mineral metabolism with skeletal effects including osteopenia. This range of factors relating to secondary bone loss will be reviewed in this chapter.

OSTEOPENIA AND MOBILITY

Injury affecting a limb or the spinal cord can result in limited movement or paralysis, and osteopenia. Relatively little attention is awarded to the interpretation of consequences of injuries and pathology in bioarchaeology (see Roberts, 2000:340–341, 350; Box Feature 7.2). Enhanced recognition of potential complications of a range of causes may improve the understanding of how individuals or communities coped with various challenges, and may be valuable in comparisons of past and present communities.

Effects of Immobilisation

The mechanisms of immobilisation that cause increased bone loss are incompletely understood. Immobilisation can result in increased calcium excretion, which stimulates the PTH to initiate remodelling and osteoclastic resorption to release skeletal calcium (Kiratli, 2001; Epstein et al., 2003). The increase in the number of active remodelling sites can result in a temporary bone loss as resorption occurs faster than bone mineralisation. The removal of load-bearing forces such as muscle pulls during immobilisation, are significant factors

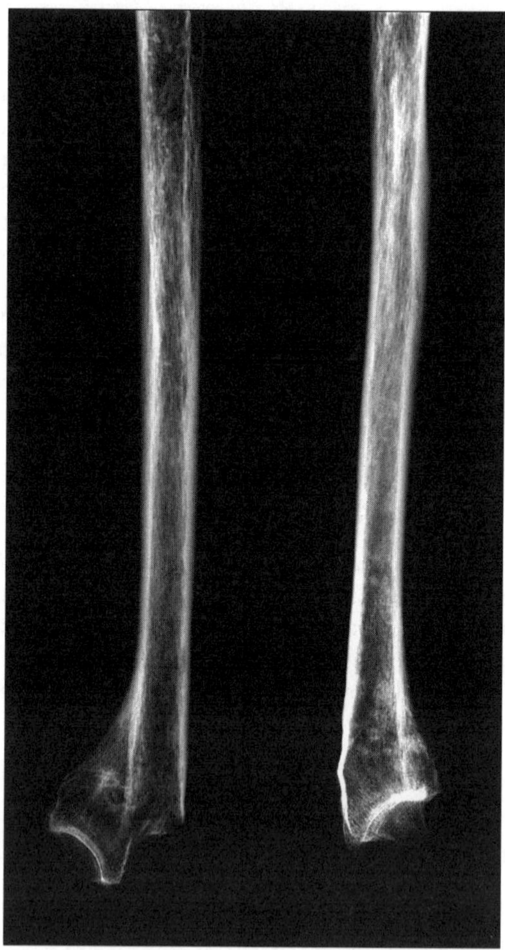

FIGURE 7.1 Medial radiograph of bilateral Colles' fractures in distal adult radii from an archaeological example. Variation in severity of fracture is evident with slight posterior angulation on right of image and marked changes on left. Modern treatment of severe fractures can include immobilisation with risk of temporary osteopenia. Bilateral fractures may also reduce normal wrist function prior to recovery (see chapter text and also Chapter 6). Post-Medieval archaeological adult from St. Martin's Churchyard, Birmingham. Radiograph courtesy of M. Brickley.

contributing to muscle and skeletal atrophy (Epstein et al., 2003; Frost, 2003; Rittweger et al., 2005; see also Chapters 3 and 6). This mechanism is currently of particular concern in situations of micro-gravity, with osteopenia frequently recognised in astronauts (Bickle et al., 1997; Vico et al., 2000; Giangregorio and Blimkie, 2002).

The occurrence of osteopenia following immobilisation can be rapid, with clinical studies demonstrating rapid and significant loss of bone before eventually reaching a plateau (Lazo et al., 2001; Epstein et al., 2003). Depending on the precise cause of the immobilisation, manifestations of secondary osteopenia may be localised or generalised. The treatment of an injury may paradoxically require immobilisation and contribute to temporary periods of osteopenia. Such changes have been observed in the modern treatment of long bone fractures such as Colles' fractures (see Dai and Jiang, 2004). It is of interest to determine whether this practice was likely to have affected individuals in the past, or whether different lifestyles cannot support such periods of immobility or use different treatments (see Figure 7.1). Of further interest is to what extent the skeletal effects of disuse are similar between humans and animals (see Box Feature 7.1).

Box Feature 7.1. Animal Studies in Osteoporosis II: Immobilisation-Related Osteopenia

Animal species may be at risk of osteopenia from secondary mechanisms in a similar manner to humans, although the potential of this to occur has been much less studied than age-related osteopenia. Immobility is a significant cause of secondary osteopenia in humans. Some animal models have also demonstrated disuse osteopenia occurring, although the mechanisms behind these alterations are not yet completely understood (Donahue et al., 2003). Animals that undertake seasonal hibernation provide a good resource for attempting to better understand the mechanisms of disuse osteopenia. The long periods of immobility adopted by these animals frequently occur with periods of reduced food intake or fasting.

Various species of hibernating animals including squirrels, hamsters and bats have been shown to experience disuse osteopenia, although variation exists between species in the degree of bone loss or location of defects (Jee and Ma, 1999). There is also variation in the extent of bone loss experienced with increased age (Perrien et al., 2007). Theoretical models have predicted various implications of such bone changes. For example, Donahue et al. (1999:1481) have postulated the skeletal consequences of hibernation on black bears. These bears undertake hibernation for prolonged periods of up to five or six months. In consequence, once awake the animals should not have sufficient time in which to replenish the bone that is lost during immobility, unless there are significant differences in rates of bone loss and other potential physiological adaptations under conditions of hibernation between species (see Donahue et al., 1999:1481; Donahue et al., 2003). In contrast, research shows age-related increases in bone strength, bone

Continued

mineral content and decreases in cortical porosity in black bears suggesting that bone formation is not decreased during immobilisation in black bears (Donahue et al., 1999:1484). It is probable that hormone levels may help to regulate bone turnover in the absence of calcium excretion during hibernation (see Donahue et al., 1999:1485). Adaptable mechanisms utilised by different species to overcome potential challenges to skeletal health may provide additional insights into the mechanisms of human skeletal remodelling. Importantly, such research may help determine methods of treatment or management of related pathological conditions in humans in the future.

Trauma and Causes of Immobility

The sequelae of trauma may include complicating factors such as infection, which may prevent the use of an injured limb, as outlined in Table 7.1 (Sevitt, 1981:177–178; Salter, 1999:473). Bioarchaeological evidence indicates that trauma is not always successfully managed. In particular, fracture non-unions in archaeological skeletal remains provide an insight into the extent to which limited function may occur following such trauma (e.g. Nystrom et al., 2005). Figures 7.2 and 7.3 illustrate various archaeological examples of trauma in which normal limb function may have been impaired during recovery.

Various injuries may require removal of bone and affected soft tissue, including modern industrial accidents (e.g. Stanbury et al., 2003 in the United States and Liang et al., 2004 in Taiwan, see Ponce et al., 2005 for past perspective), transport accidents even in the relatively recent past (see Rundle, 1886; see also Treves, 1887a) or in relation to various vascular pathologies, malignant diseases or infections (see Mitchell, 1999; Weaver et al., 2000) (see Figure 7.4). Intentional violence can lead to limb removal, but may occur at varying stages prior to death. Cases of foot removal undertaken by the Moche of ancient Peru discussed by Verano et al. (2000), were of sufficient long-standing to enable successful adaptation preventing atrophy. Integrated analyses based on historical evidence and skeletal observations may provide an insight into the various mechanisms which existed for coping with such trauma in the past (see Mitchell, 1999).

Non-Long Bone Trauma and Additional Causes of Disuse Osteoporosis

The majority of bioarchaeological evidence of trauma concerns healed long bone fractures. Determining additional causes of injury potentially through ethno-historical research may broaden the understanding of factors capable of limiting mobility or function. For example, injuries among the Shiwiar forager-horticulturalists of Ecuadorian Amazonia (Sugiyama, 2004:382) that caused prolonged disability included: knee injury, tooth abscesses with subsequent

TABLE 7.1 Summary of Factors that can Lead to Immobility and Possible Secondary Osteopenia

Trauma	Features	Skeletal effects	Sources
Trauma/fracture complications: soft tissue damage	Damage: nerve vessels, blood supply, muscles, periosteal and endosteal membranes (see Chapter 3)	Time limited Severe blood vessel damage may cause bone necrosis Pathology during growth may result in atrophy/osteopenia	Sevitt (1981: Chapter 8), Stirland (1985:56), Lovell (1997a), Salter (1999: Chapter 15, 473), Ortner (2003:128–136)
Fracture healing	Extensive callus formation may span joint surfaces Malunion of fracture and residual deformity of bone shape and function Influence of treatment practices Subsequent re-injury/dislocation risks further damage	Time limited Recovery of normal function Continued deformity – may result in immobility and localised osteopenia	Churchill & Formicola (1997:37), Kilgore et al. (1997), Lovell (1997a), Schultz (2003:177)
Sequalae of trauma: infection	Open fractures and exposure of blood vessels to bacterial infection Pain and tissue swelling Prevent mobility Severe infections (e.g. osteomyelitis) may exacerbate avascular necrosis	May be time limited May result in localised osteopenia If severe may result in bone destruction and immobility	Sevitt (1981:177–178), Lovell (1997a), Salter (1999:473), Özbek (2005)
Fracture non-unions	Fibro-cartilaginous bridging of fracture Development of a false joint (pseudo-arthrosis)	May result in significant limb disfunction If prolonged disuse, localised osteopenia	Sevitt (1981: Chapter 11), Nystrom et al. (2005)

Continued

TABLE 7.1 *(Continued)*

Trauma	Features	Skeletal effects	Sources
Amputation	Soft tissue and bone removal	May occur at varying stages prior to death	Kuhns & Wilson (1928), Wingate Todd & Barber (1934), Brothwell & Møller-Christensen (1963a, 1963b), Ladegaard Jakobsen (1978), Sevitt (1981: Chapter 12), Lazenby & Pfeiffer (1993), Mays (1996a), Mitchell (1999), Weaver et al. (2000), Ponce et al. (2005)
	Range of causative injuries/ intentional violence, various vascular pathologies, malignant diseases or infections	Disuse atrophy and marked osteopenia if long-standing loss of mechanical force	
Non-long-bone trauma: location of fracture	May hinder joint movement, e.g. fractures of scapula or dislocations Secondary soft tissue pathology may affect function, e.g. trauma to spinal cord and subsequent paralysis	Localised osteopenia or generalised osteopenia	Schutkowski et al. (1996), Wakely (1996), Charlton et al. (2003)
Additional causes of disuse	Infection, tumours, pain from soft tissue afflictions	May result in localised osteopenia	Milgram (1990)

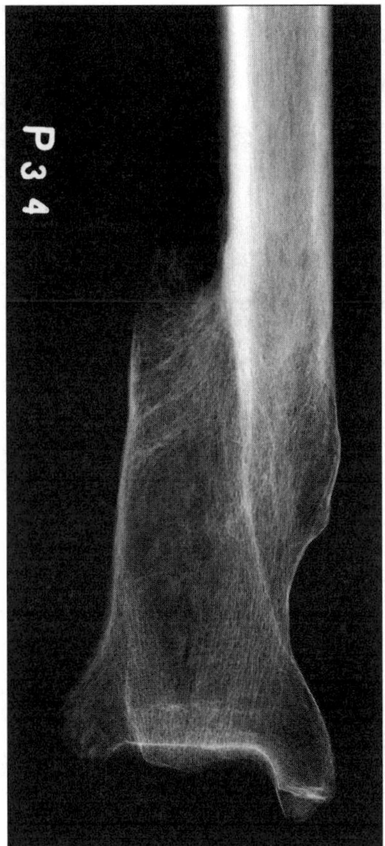

FIGURE 7.2 Antero-posterior radiograph of a distal tibia fracture with severe medial displacement of fractured end in an archaeological example. Temporary osteopenia is likely to occur during healing due to limited weight-bearing. Pain and soft tissue damage may further exacerbate temporary immobility, although there is no skeletal evidence for any complicating secondary infection in this example. Post-Medieval archaeological adult from St. Martin's Churchyard, Birmingham. Radiograph courtesy of M. Brickley.

systemic infection, stroke, malaria, whooping cough, postpartum infection, respiratory illnesses and severe foot fungus. In particular, an untreated case of snakebite in one young adult male caused severe infection and permanent disability and disuse of the foot (Sugiyama, 2004:382–383).

Pain from soft tissue injury or nerve damage can be a significant cause of secondary osteopenia (see Figure 7.5), as can haemarthrosis due to pathology or trauma, with subsequent prolonged limb immobility (Joffe, 1961; see also Chapter 4). Paleopathological examples of long bone atrophy and poorly developed muscle attachment sites indicating limited limb use have been identified in adolescents affected by conditions such as syphilis (Erdal, 2006) and Langerhans cell histocytosis (Barnes and Ortner, 1997; see also Churchill and

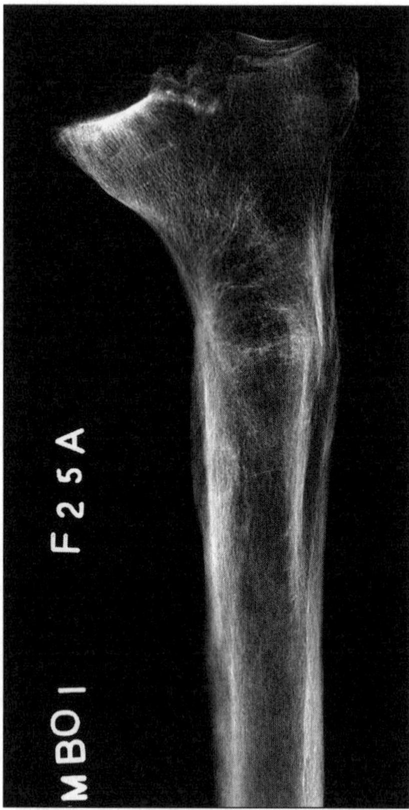

FIGURE 7.3 Antero-posterior radiograph of a fracture in a proximal tibia with impaction of the shaft and resulting angulation of the proximal joint surface. In severe cases, loss of continuity of the joint surfaces could limit normal limb function or lead to alterations in the manner of gait. Post-Medieval archaeological adult from St.Martin's Churchyard, Birmingham. Radiograph courtesy of M. Brickley.

Formicola, 1997). Prevention of pain or general illness associated with these conditions may have limited individual mobility, although potential complicating factors such as dietary quality or affects on the spinal cord need consideration (see later in this chapter). In paleopathology, long-standing conditions are those most recognisable (see Figure 7.6). Potential identification of localised bone loss surrounding pathologies may be possible in such cases, together with more systemic expressions of osteopenia.

Bone Loss in Infectious Diseases

Secondary osteopenia can be a consequence of infectious conditions. Trauma may be complicated by infection significantly causing immobility in a manner similar that discussed above. Whilst many non-specific infections result in bone apposition, in severe cases affected limbs are likely to be painful and

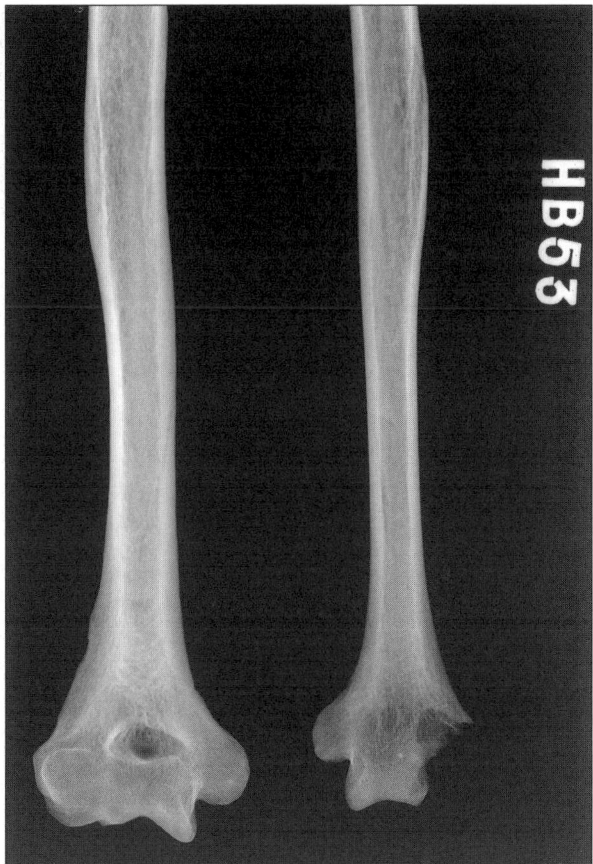

FIGURE 7.4 Radiograph of distal humeri from a mature adult female who experienced an amputation of the left forearm, which was well healed prior to death. The distal humeri show a reduction in cortical thickness and marked asymmetry in bone size resulting from reduced arm function compared to the normal right arm. Post-Medieval archaeological example from St. Peter's Churchyard, Wolverhampton, courtesy of BARC, Archaeological Sciences, University of Bradford.

swollen, potentially limiting function (see Figure 7.7). The consequences of long-standing changes may have an effect on skeletal tissues. Specific infections, such as syphilis and leprosy, can result in neurotrophic arthropathies with implications for normal mobility (see Illarvamendi et al., 2002:46). Loss of motor and autonomic nerve systems can result in loss of limb function and may prevent the detection of trauma. Repeated trauma may cause significant bone and joint destruction, and subsequent ankylosis, resulting in a charcot joint and functional complications (see Aufderheide and Rodríguez-Martín, 1998; Salter, 1999:287; Nawaz Khan et al., 2003). A recent review by Buckley and Tayles (2003) has considered further the impact on mobility and function of joint and bone destruction particularly in treponemal infections.

FIGURE 7.5 Fracture resulting in impaction of an adult proximal humerus. Fractures can vary in severity at this location. Whilst there may be minimal bone disruption to normal joint function, as in this example, the extent of possible pain and soft tissue injury leading to nerve damage, impinagement or disruption of the rotator cuff may be significant limiting limb use. Post-Medieval archaeological example from Chelsea Old Church, London, old adult male (OCU948). Photograph by Rachel Ives, courtesy of the Museum of London.

Severe bone destruction in the spine with vertebral body collapse and kyphosis (Pott's disease) are well-recognised manifestations of tuberculosis (e.g. El-Najjar, 1979; Powell, 1988:155–157, 2000:19–28; Ortner, 2003:255–261). The sequelae of these pathological changes has been relatively overlooked to date, yet secondary osteopenia can develop via several mechanisms.

Abscess formation with bone necrosis can cause bone debris to protrude into the spinal canal, risking compression and paraplegia (Luk, 1999:339). Subsequent vertebral collapse and kyphosis increases the risk of compression of the spinal cord (Duggeli and Trendezenberg, 1961:14), as can irregular bone formation aimed at stabilising collapsed vertebrae (Luk, 1999:339). A recent

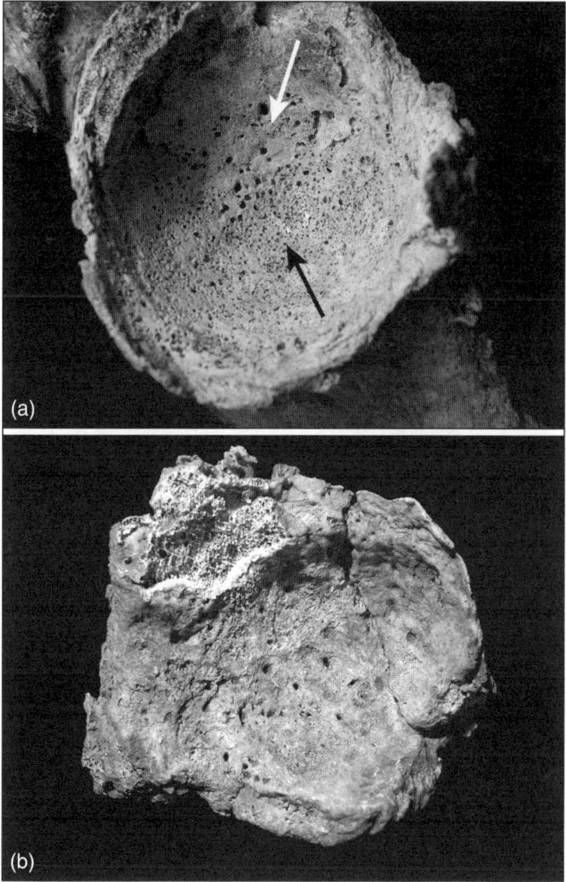

FIGURE 7.6 Pathological changes affecting two pelves from archaeological examples. The acetabulum shown in (a) displays a partially eburnated bone surface, indicative of osteoarthritis and some degree of joint function (black arrow), yet together with a portion of sclerotic bone (white arrow), indicating a lack of the full range of normal movement in the hip joint. The acetabulum in (b) shows a fractured portion of the joint together with a sclerotic reaction over the entire joint surface following trauma. This change indicates no normal function or movement of the hip was occurring prior to the time of death. (a) Post-Medieval archaeological example from St. Bride's Lower Churchyard, London (FAO1549). (b) Figure from Medieval Bermondsey Abbey, London, (BA843274). Photographs by Rachel Ives, courtesy of the Museum of London.

clinical study estimated that 41% of cases of spinal tuberculosis subsequently developed paraplegia (Luk, 1999:343; see also Duggeli and Trendezenberg, 1961). Severe Pott's disease together with infection of the sacrum and pelvis occurred with marked atrophy of the legs in a young adult female from seventh to eighth century Hungary (Marcsik et al., 1999:336; see also Knick, 1981;

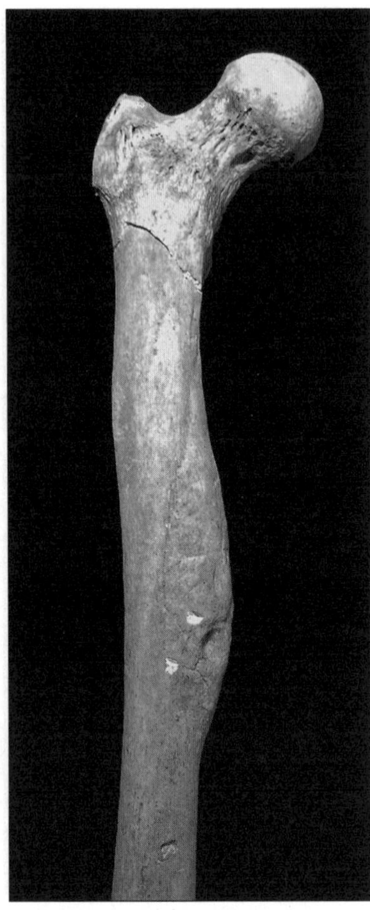

FIGURE 7.7 Healed, non-specific infection of the femoral midshaft from an archaeological example. Pain and soft-tissue inflammation and swelling following infection, or as a secondary consequence of trauma, may result in temporary disuse of a limb. The identification of any resulting osteopenia may however be complicated by skeletal response to the infection. Medieval archaeological example from East Smithfield Black Death cemetery, London (MIN11480). Photograph by Rachel Ives, courtesy of the Museum of London.

Mann et al., 1991). These changes indicate immobility due to paraplegia secondary to tuberculosis.

Tuberculosis of a joint may result in severe bone destruction and soft tissue alterations that can limit normal function (Treves, 1887b:18–19; Kelley and El-Najjar, 1980; Cook, 1984; Ortner, 2003:262). Localised osteopenia surrounding the joint and the limbs affected by limited mobility may develop (e.g. Sissons, 1952). These secondary characteristics are not frequently reported on (see also Roberts and Buikstra, 2003:97, 108).

Specific aspects of the tuberculous condition may result in localised osteopenia distinct from that caused by immobility. Tuberculous arthritis can cause juxta-articular osteopenia (Panuel et al., 1999:233). Transient osteopenia has also been reported in the vertebral bodies prior to any destructive foci, suggesting an additional cause of underlying bone loss (Duggeli and Trendezenberg, 1961:15). Greater consideration of the range of pathological expressions and consequences of tuberculosis may enhance reconstructions of disease impact on individual health.

Immobility in Viral Conditions

Paralysis and osteopenia can result from viruses that target the central nervous system, such as poliomyelitis (Aufderheide and Rodríguez-Martín, 1998). Recent estimates suggest that only a small number of susceptible individuals will develop paralysis (0.5%) (Duintjer Tebbens et al., 2005), but increasing outbreaks of new cases are reported in endemically affected countries (Lahariya and Pradhan, 2007:61) despite world-wide programmes attempting to eradicate this disease. Poliomyelitis is evident in past communities, ran-ging from late Saxon England through to ancient Egypt, although accurate diagnosis of this condition can be problematic (Ortner, 2003; Roberts and Manchester, 2005:181).

Congenital and Developmental Conditions

Various congenital or developmental skeletal alterations may affect mobility leading to secondary pathology. Disturbances of epiphyseal development may particularly affect limb function (see further Salter, 1999:33). Dislocation following congenital hip dysplasia can affect normal motion with consequent effects on the surrounding bone structure. Gracile long bone shafts with limited muscle markings and lack of secondary joint degeneration may indicate that normal movement or gait was dysfunctional. Whilst this condition is not widespread in the past, secondary pathological defects may be apparent in communities that have limited access to medical treatments to correct developmental defects (see Murphy, 2000:63–72; Mitchell and Redfern, 2008).

Osteopenia in Spinal Cord or Neuromuscular System Afflictions

Trauma or pathology of the spinal cord can have serious consequences and contribute to the expression of secondary osteopenia. Birth trauma may damage the spinal cord resulting in disuse atrophy in post-paralytic skeletal deformities (Boylston and Roberts, 2004:343). The severe condition spina bifida cystica causes the meninges and nerve roots to protrude through unfused vertebral arches leading to compression and paralysis (Aufderheide and Rodríguez-Martín, 1998; Torwalt et al., 2002; Roberts and Manchester, 2005).

Few cases of this condition have been described from past communities, probably owing to its serious consequences. However, one reported case is an adolescent from an Early Archaic site in Florida with severe disuse atrophy of the legs (Dickel and Doran, 1989). Defective skeletal development can impact on skeletal function, such as severe congenital scoliosis which may be associated with partial paraplegia and secondary osteopenia (see Murphy, 2000). Neuromuscular conditions can inhibit movement and Stirland (1997) has discussed a neuromuscular or dystrophic condition with severe effect on an adult male from Medieval England, with marked atrophy and severe generalised osteopenia likely attributable to long-standing disuse. The inherited condition osteogenesis imperfecta can also result in generalised osteopenia with extreme bone fragility and the onset of multiple fractures (see Chapter 9).

Spinal trauma is a significant cause for concern in many modern contexts, particularly for instance in severe car accidents. The relationship between spinal cord injury and paralysis resulting in secondary osteopenia has been established in clinical studies (e.g. De Bruin et al., 1999; Lazo et al., 2001). Bone loss can occur rapidly following disuse and there can be increased risk of spontaneous osteoporotic fracture, despite the risk of fracture from falls being reduced (De Bruin et al., 1999). Inadequacy of the diet in disabled persons can further exacerbate osteopenia (Bertoli et al., 2006; dietary effects on osteopenia are discussed below). Prolonged survival of many individuals affected by limited mobility may have a bearing on the interpretation of social aspects of individual care (see Box Feature 7.2).

Box Feature 7.2. Implications of Immobility and Inferences of Disability

Mobility is likely to be a crucial factor in many communities. Seasonal migrations between temporary settlements, hunting and foraging subsistence strategies and maintenance of herding animals, are all factors that require levels of mobility and are likely to have been important components of life in the past. Modern analyses are investigating methods to reconstruct patterns of mobility in relation to subsistence and settlement in past societies (e.g. Holt, 2003). In the present, individuals within many communities are required to be mobile in order to obtain water supplies from neighbouring community wells, or to obtain medical or educational resources.

Mobility can be limited due to a diverse range of pathological causes as well as from increased age (Chapter 6). The implication of identifying an individual that has been affected by immobility conjures a range of responses but primarily it is inferred that there is an evident disability with the requirement of needing care. In some cases it remains difficult not to envisage some means of care and provision required those struck by near complete paralysis (e.g. Goodman et al., 1988:178; Hawkey, 1998). Roberts (2000) has recently discussed the nature of these connotations with regard to the reconstruction of life and health in the past. There is potential for ethno-historical research to contribute to the understanding of mobility in social groups (e.g. Blurton Jones et al., 2002).

Continued

Research has also discussed the nature of disability as evident in various historical sources (e.g. Haffter, 1968). Recent investigations into the study of disability have demonstrated how cultural concepts and interpretations of disability have tremendous diversity (Barnes et al., 1999:14). It is important to recognise that such variation may also have been implicit in past communities with inherent implications for anthropological reconstructions of health and life. Barnes et al. (1999) have argued that much more research is required in order to develop our understanding how and why attitudes and practices towards disability have varied throughout history. It may be possible to significantly improve knowledge of disability in the past, from both its causes and consequences, through combined investigation and interpretation of archaeological, anthropological and historical evidence.

OSTEOPENIA IN PATHOLOGICAL CONDITIONS

Joint Disease

In many joint diseases osteopenia can occur via immobilisation and as part of the pathological process (Rogers and Waldron, 1995). Bone loss and degenerative joint changes both accumulate with age leading to difficulties in differentiating between age-related and secondary osteopenia. However, the expression of skeletal changes related to the specific joint conditions may help in extrapolating the cause of any bone loss. Methods for identifying osteopenia are discussed both in Chapter 6 and later in this chapter.

Progressive bone fusion affecting the sacroiliac joint, vertebrae and ribs in ankylosing spondylitis can lead to immobility-related secondary osteopenia if severe (see striking case discussed by Hawkey, 1998). Studies have however demonstrated vertebral osteopenia at early stages of the disease, indicating that limited mobility is not the only cause of bone loss in this condition (Will et al., 1989; El Maghraoui, 2004; see also Aufderheide and Rodríguez-Martín, 1998). The mechanism of osteoclastic reaction to inflammatory stimuli under many conditions needs better understanding (Will et al., 1989; El Maghraoui, 2004). There is a widespread range of geographic evidence of this condition in past populations (see Aufderheide and Rodríguez-Martín, 1998; Rogers, 2000; Ortner, 2003; Powers, 2005; Roberts and Manchester, 2005). Further consideration of the extent of secondary pathological changes may enhance the interpretation of the impact on functionality of this disease.

In rheumatoid arthritis (RA), a systemic inflammatory disorder, inflamed synovial fluid ('pannus') can destroy articular cartilage and bone following increased osteoclastic resorption (Harper and Weber, 1998; Goldring, 2001). The specific cause of the condition is unknown, but may involve both environmental and genetic components (Rogers and Waldron, 1995; Goldring, 2001). Both periarticular and generalised osteopenia are consistent features of this condition. Limited joint movement due to pain may contribute to localised osteopenia, but does not explain the occurrence of generalised osteopenia, which has been associated with increased risk of osteoporosis-related fracture (Harper and Weber, 1998; Zak et al., 1999; Goldring, 2001). A number of studies have identified RA in various past populations (e.g. Ortner and Utermohle, 1981; Rogers et al., 1981; Thould and Thould, 1983; Blondiaux et al., 1997; Hacking et al., 1994; Waldron et al., 1994; Inoue et al., 1999). The radiographic identification of osteopenia has been incorporated into the diagnostic criteria of RA in paleopathology and is a consistently recognised feature (see Blondiaux et al., 1997).

Haematopoietic Conditions

Various haematopoietic conditions that alter blood marrow productivity (haemopoiesis) can result in a secondary loss of bone tissue via expansion of marrow forming sites such as the medullary cavity of long bones and diploic space of the cranium (Tyler et al., 2006). Abnormal rates of osteoclastic resorption have been documented in specific types of inherited anaemic conditions such as thalassaemia and sickle cell anaemia (Voskaridou et al., 2001; Eren and Yilmaz, 2005). Osteopenic skeletal changes have been demonstrated in both children and adults under these conditions (see Hershkovitz et al., 1997; Faerman et al., 2000; Kosaryan et al., 2004; Miller et al., 2006; Tyler et al., 2006; Lagia et al., 2007). Marrow expansion can affect the vertebral bodies causing an expansion in body height together with trabecular thinning. Continued weight-bearing may sufficiently weaken pathologically altered vertebrae resulting in collapse and fracture at multiple sites throughout the spine (Faerman et al., 2000; Tyler et al., 2006). To date, there is limited skeletal evidence of osteopenia in investigations of these inherited anaemias in past populations (see reviews by Tayles, 1996; Lovell, 1997b).

Neoplastic and Malignant Conditions

Various neoplastic conditions have been associated with secondary osteopenia, although the specific causative mechanisms in many cases are unknown (see Mundy, 1999:32). Several mechanisms of immune defence can trigger increases in reactionary osteoclastic resorption to abnormal accumulations of plasma cells, as well as activated T- and B-cells, and immune reactions and bone cell responses are also stimulated where there is sufficient vitamin D (Holick, 2003; Ortner, 2003; Clowes et al., 2005). The implications of osteopenia may vary across conditions, with slow, localised or generalised bone loss having little skeletal impact. Rapid bone loss can result in sufficient risk of pathological fracture, including vertebral compression fractures as occur in multiple myeloma. Table 7.2 highlights several of the conditions that have been linked with secondary osteopenia and that may be recognisable in past populations. For detailed reviews see Resnick (1988) and Ortner (2003).

TABLE 7.2 Factors Within Various Neoplastic Conditions With Risks for the Development of Secondary Osteopenia

Condition	Features	Sources
Multiple myeloma	Vascular spaces between trabeculae aid spread of metatastic disease Increased bone resorption. Increased risk of vertebral fracture Modern treatments can increase secondary bone loss	Joyce & Keats (1986), Genant et al. (1988a), Resnick (1988), Coleman (2001), Atoyebi et al. (2002), Ortner (2003), Uetani et al. (2004), Melton et al. (2005)
Leukaemia	Malignant white blood cells infiltrate soft tissues Replacement of normal bone marrow with tumour cells. Resorption of trabeculae Increased endosteal and intra-cortical bone resorption Defective bone formation may exist Generalised bone loss in vertebral bodies causing fracture	Resnick (1988), Rothschild et al. (1997), Aufderheide & Rodríguez-Martín (1998), van der Sluis et al. (2002), Ortner (2003), Goldbloom et al. (2005), Roberts & Manchester (2005)
Neuro-fibromatosis	Hereditary condition Affects the central and peripheral nervous system with numerous tumours that develop following nerve pathways Destruction of cortical bone adjacent to soft tissue tumour masses Tumours in spinal canal compress the cord causing paralysis and osteopenia Impaired remodelling	Swann (1954), Knüsel & Bowman (1996), Aufderheide & Rodríguez-Martín (1998:421), Illés et al. (2001), Kuorilehto et al. (2005)
Meningioma	Benign growths causing tissue pressure (spinal cord, brain stem) Damage blood vessels and nerves in cranial base Early stages lytic focus Destruction of sella turcica, pathological disruption of the pituitary gland Osteopenia: paralysis by compression of skull base; excess amounts of corticosteroid hormones secreted from pituitary gland (relation to osteopenia poorly understood) For example adult male, Alaska, destruction of sella turcica, post-crania unusually light, suggestive of ante-mortem bone loss	Jónsdóttir et al. (2003), Ortner (2003:513–514)

THE INFLUENCE OF DIET ON OSTEOPOROSIS RISK

A delicate balance is required between multiple components of the diet in order to maintain an adequate mineral homoeostasis (see Chapter 3). Where a nutritional balance is not achieved, bone turnover may be altered. Prolonged nutritional inadequacy can result in various metabolic bone diseases (see Chapters 4, 5 and 9). Osteopenia in particular can occur following dietary insufficiency (see in particular Box Feature 9.2) (New, 2003; Putnam et al., 2007). It is likely that dietary quality constituted a significant risk for osteopenia in many past communities (see Martin et al., 1985; Ortner et al., 2006, 2007).

Dietary Acid Load and Proposed Mechanisms of Bone Loss

All foods are determined to have either a net acidifying or alkalizing effect on the extracellular fluid (ECF) pH balance (see Cordain et al., 2000; Putnam et al., 2007; Cordain, 2007). The ECF is typically maintained between pH 7.35 and 7.45 (Putnam et al., 2007) and a reduction in the pH level (increased acidity) can stimulate osteoclast activity and bone resorption releasing calcium from the skeleton to rebalance the ECF (Anderson, 1999; Arnett, 2003; New, 2003). Fish, meat, poultry, eggs, shellfish, cheese, milk and cereal grains are acid producing foods, while fruit, vegetables, tubers, roots and nuts are alkaline based (Cordain et al., 2000:349–350; Cordain, 2007:375). A balanced diet could theoretically negate detrimental effects of acid foodstuffs by the consumption of foods with high alkaline content.

Diets with continual acid intake (metabolic acidosis) may suffer skeletal effects (Eaton, 2007:389) and have been postulated as one of the potential contributory causes for increased levels of disease, including osteoporosis (see Cordain et al., 2000:350; Sebastian et al., 2002; Cordain, 2007:375). This risk may be exacerbated in old age, where kidney (renal) function becomes less efficient in excreting acid (Arnett, 2003; Putnam et al., 2007). Knowledge surrounding these skeletal consequences is incomplete (New, 2003:892), with implications for reconstructions of health based on ancient hominid diets (see Section Anthropological Perspectives later).

Calcium

Calcium forms 99% of the mineral in the skeleton and is also present in the ECF (Heaney, 1997a; Broadus, 1999). The importance of calcium in the body is reviewed in Chapter 3 (Civitelli et al., 1998; Heaney, 2002). The calcium contents of naturally occurring foodstuffs are shown in Table 7.3. Adequate dietary calcium intake needs to be balanced by good intestinal calcium absorption as well as between renal (kidney) and faecal excretion and renal re-absorption of calcium (Hoenderop et al., 2000; Houillier et al., 2003). Calcium homoeostasis is controlled by the PTH and by mechanisms of vitamin D

TABLE 7.3 Calcium Content of Various Food Sources.

Food source	Raw calcium content per 100 g (solids), 100 ml (liquids)	Food source	Raw calcium content per 100 g (solids), 100 ml (liquids)
Apple with skin	6	Mustard greens	103
Apple without skin	5	Mustard seed, yellow	521
Avocado	12	Nuts, acorn	41
Banana	5	Nuts, almond	248
Blackberries	29	Nuts, brazil	160
Broccoli	47 (cooked 40)	Nuts, cashew	37
Brussel sprouts	42 (cooked 26)	Nuts, chestnuts (European)	27
Cabbage	40 (cooked 48)	Nuts, coconut flesh	14
Carrots	33 (cooked 30)	Nuts, hazel	114
Corn bran crude	42	Nuts, macadamia	85
Corn white	7	Nuts, pecan	70
Corn yellow (maize)	7	Nuts, pistachio	107
Cranberries	8	Nuts, walnuts (English)	98
Egg	53 (fried 59)	Onions	23
Figs	35	Onions, spring	72
Fish, cod	7 (cooked 9)	Orange, juice natural	11
Fish, wild salmon	12 (cooked 45)	Orange, whole	40
Fish, mackerel	12 (cooked 29)	Oysters	(N = 12) 230
Honey	6	Parsley	138
Green beans	37	Parsnip	36 (boiled 37)
Liver, beef	5 (cooked 6)	Peas, green	25 (cooked 27)
Liver, lamb	7 (cooked 9)	Peppers, sweet green	10

Continued

TABLE 7.3 (*Continued*)

Food source	Raw calcium content per 100g (solids), 100ml (liquids)	Food source	Raw calcium content per 100g (solids), 100ml (liquids)
Meat, beef	3 (broiled 4)	Potatoes	9 (baked with skin 5) (boiled, peeled 8)
Meat, half lamb shank	9 (roasted 10)	Pumpkin	21 (cooked 15)
Meat, chicken, light meat	11 (cooked 13)	Raspberries	25
Meat, caribou hind quarter	5	Rhubarb	86 (cooked with sugar 145)
Meat, seal	4	Scallops	(N = 6) 120
Milk, cow, calcium fortified	204	Spinach	99 (cooked 136)
Milk, cow, whole milk	113	Strawberries	16
Milk, cow, non-fat milk	123	Tomatoes	10 (cooked 11)
Milk, goat	134	Tomatoes, sun-dried	110
Milk, sheep	193	Turnips	30
Milk, human	32	Turnips, greens	190
Milk, canned, condensed, sweetened	284	Watercress	120

Sources: USDA *National Nutrient Database for Standard Reference*, 2006 (*Release* 19) *and see also* Prince (1999:483).

production which balance skeletal and ECF reserves of calcium (see Chapter 3; and Heaney, 1997a:487; Heaney, 1999). Calcium functions as a threshold or plateau nutrient with excess mineral excreted and further losses can occur during sweating and shedding skin, nails and hair as highlighted by Heaney (1997a:487).

Table 7.4 lists the recommended daily amounts of calcium for various demographic groups (see Civitelli et al., 1998:167). Recommended amounts

TABLE 7.4 Recommended Daily Amounts of Dietary Calcium Intake Based on Survey of US Whites

Age group		Optimum calcium intake (mg/day)
Infant	Birth to 6 months	400
	6 months to 1 year	600
Children	1–5 years	800
	6–10 years	800–1200
Adolescents/young adults	11–24 years	1200–1500
Adult men	25–65 years	1000
	65+ years	1500
Adult women	Pregnant and nursing	1200–1500
	25–50 years	1000
	50+ years	1500

Sources: *National Institutes of Health Consensus Conference*, US, 1994 (Civitelli *et al.*, 1998:167).
Note: *Recommendations can vary between population groups as well as through the life course, see chapter text.*

will vary depending on age, ancestry and the range and quantities of other components in the diet (see further New, 2001:269). A good intake of calcium during adolescence and young adulthood can help optimise the peak bone mass and is recommended for adult health throughout later life (Miller et al., 2001:179s). Decreased calcium intake or reduced intestinal absorption of calcium will result in bone resorption via secondary hyperparathyroidism (see Chapters 3 and 9; and Heaney, 1997a:487).

A range of foodstuffs can interfere with the bioavailability of calcium, as well as other minerals such as magnesium, zinc and iron, and it is important to consider such dietary interactions in reconstructing dietary adequacy (Miller et al., 2001:176s). Phytates contained in cereals, seeds and nuts and oxalates in spinach, rhubarb, sweet potatoes and walnuts can prevent the binding and utilisation of calcium once consumed (Miller et al., 2001:176s). Phytates can also affect the utility of vitamin D, with implications for the efficiency of calcium absorption (Chapter 5; Anderson, 1999).

Protein

Dietary protein intake can be attained from a variety of foodstuffs, including animals, fish and plants. Prior to the agricultural revolution, protein would have been obtained from animal and fish sources and potentially some seeds and beans (Peters, 2007:238). Following the adoption of agriculture, cereals are likely to have provided varying sources of protein as summarised in Table 7.5.

TABLE 7.5 Components of Various Cereal Sources of Dietary Protein

Cereal	Geographic region	Protein content	Complicating for nutritional utility
Wheat	Europe, North America, Australia, New Zealand, parts of Russia, high altitude regions of North India	High	Milling to produce white flour reduces protein contained in outer layers Milling reduces vitamin B1 (thiamin) by 80%
Maize	Central and South America, Mexico, South Africa	Less protein than wheat. Contains more carbohydrate than wheat	Yellow maize contains vitamin A Vitamin B complex (nicotinic acid) is contained in maize but in an unavailable (bound) form. The amino acid tryptophan required to synthesise nictotinic acid is also limited Treatment with an alkali such as limes can convert bound nicotinic acid to usable form (see Chapter 9)
Rice	Eastern and Southern Asia	Less protein than wheat and maize More carbohydrate than wheat	Hand-husked rice contains vitamin B1 (thiamin). Machine-refined and polished rice contains little vitamin B1
Millet	Africa, Asia and South America	Low protein Less carbohydrate than rice	See Chapter 9

Source: *Widdowson* (1991).

Current understanding of the potential implications of protein consumption on bone health are summarised in Table 7.6 (see reviews Kerstetter et al., 2003a; Rizzoli and Bonjour, 2004). The high acid-load base of animal protein has been postulated to negatively affect the skeletal reservoir (e.g. New, 2003:892). Increased oxidation of sulfur-containing amino acids from protein may increase acid production and renal excretion, which is balanced with urinary calcium excretion. This may result in increased bone resorption to restore the calcium balance (Anderson, 1999:229; Rizzoli and Bonjour, 2004:527). However, Kerstetter et al. (2007) have suggested that the excess calcium excreted in high protein diets may derive from saturation of calcium absorption threshold levels rather than derives from skeletal reserves (see also Anderson, 1999:229; Wengreen et al., 2004:543). A recent clinical review indicates the largely consistent beneficial effect of increased protein intake on the skeleton (see Rizzoli and Bonjour, 2004:527).

Low dietary protein intake is currently of greater concern as a risk factor for osteoporosis (Ammann et al., 2000; Hannan et al., 2000 Rizzoli and Bonjour, 2004; Kerstetter et al., 2003b). Low protein intakes decrease intestinal calcium absorption and increased bone remodelling via secondary hyperparathyroidism to balance serum calcium (see Rizzoli and Bonjour, 2004:527 and Chapter 3). Low dietary protein may relate to poor physical fitness, muscle weakness, impaired co-ordination and increased risk of falls (Ammann et al., 2000:683).

Protein-calorie malnutrition can have a severe effect on health, particularly children who have been weaned onto inappropriate adult foodstuffs leading to stunted growth, osteopenia, severe malnourishment (Adams and Berridge, 1969; Garn et al., 1969; Widdowson, 1991:293). The relationship between dietary protein intake and skeletal health may be modified by additional components in the diet, particularly calcium, and further research is needed to investigate such inter-relationships (see Massey, 2003:864s).

Fatty Acids

Recent research has questioned the role that fats and in particular fatty acids have in skeletal health (see Table 7.6). Whilst excess fats appear harmful for calcium metabolism (Corwin, 2003; Corwin et al., 2006), deficiencies can reduce intestinal calcium absorption. Recent studies indicate that the consumption of polyunsaturated fatty acids (omega-3) found in lean meat, wild plants, eggs, fish, nuts and berries can enhance calcium absorption and reduce calcium excretion. They may have beneficial effects on bone mineral density (BMD), although more research is required in this area (Moyad, 2005:42; Simopoulos, 2006:503).

Fruit and Vegetables

A positive association may exist between the quantity of fruit and vegetables consumed and skeletal health (New, 2001; New, 2003:892; Putnam et al.,

TABLE 7.6 Summary of the Proposed Impacts of Protein and Fatty Acids on Skeletal Health

Dietary component	Quantity	Skeletal effects	Sources
Protein	High intake	Dietary acid may stimulate bone resorption Potential suppression of osteoblast function High intake associated with increased urinary calcium excretion (hypercalciuria) Unclear whether resorption contributes to hypercalciuria High intake may increase intestinal calcium absorption Excess calcium excretion may result from increased calcium absorption Long-term effects on BMD are unproven. Clinical consensus no detrimental skeletal effects. Implications of high protein with low calcium and vitamin D are unclear.	Anderson (1999), Hannan et al. (2000), Kerstetter et al. (2003a), New (2003), Rizzoli & Bonjour (2004), Putnam et al. (2007)
	Low intake	Impairs intestinal calcium absorption Development of secondary hyperparathyroidism Increased bone loss Suggestive of significant risk for osteopenia Effects with increased or decreased levels of dietary calcium unclear	Hannan et al. (2000), Ammann et al. (2000), Kerstetter et al. (2003b), Rizzoli & Bonjour (2004)
	Phytate, e.g. cereals, legumes and some seeds and nuts	Binds with calcium Calcium becomes insoluble and is excreted Inhibits iron and zinc absorption Enzyme activity can limit detrimental effects	Miller et al. (2001), Sandberg (2002)
Fatty acid	High fat	Rich saturated fatty acids decrease BMD Increase risk of fracture May exacerbate the uncoupling remodelling. May inhibit formation of mature osteoblasts. May increase loss of urinary and faecal calcium.	Corwin (2003)
	Low fat	Deficiency dietary fats decrease intestinal absorption of calcium and increase renal excretion.	Corwin (2003)

Note: *Explanations of the various bone cells and actions are provided in Chapter 3.* BMD, *bone mineral density.*

2007; Zalloua et al., 2007). Table 7.7 summarises the potential role that various fruits and vegetables may have in influencing bone health at the cellular level. Increased consumption of fruit and vegetables may result in a generalised lowering of the acid-load base of the diet, potentially limiting levels of bone resorption (New, 2003:894). It is possible that such actions could potentially complicate specific effects of individual dietary components.

ANTHROPOLOGICAL PERSPECTIVES

The composition and health implications of the human diet is an exceptionally complex area of anthropological research (Nestle, 2000; Lee-Thorp and Sponheimer, 2006; Ungar, 2007). For example, a recent synthesis of reconstruction of ancient hominin diets has demonstrated the vast range of resources and methodologies available (see Ungar, 2007). However, it may not be feasible to make specific inferences regarding the implications for modern health that dietary evolution may have yielded (see Box Feature 7.3). Subsistence regimes may vary throughout a given region if environmental conditions and climate differ considerably. Dietary reconstructions, particularly of past populations, need awareness of the diversity that inter-site variation may demonstrate (see e.g. Ubelaker and Newson, 2002:347).

Calcium in the Evolutionary Perspective

The consumption of calcium is likely to have varied throughout the evolution of past diets. Ancestral hominids and primates may have obtained calcium from leafy greens, fruits and nuts as well as small vertebrates, as suggested by Eaton and Nelson (1991), Eaton et al. (1997), Heaney (1997b), and see also Peters (2007). Comparative investigations of traditional hunter–forager diets together with estimates of ancient dietary composition have enabled some researchers to suggest that pre-historic diets would have been substantially in excess of the average calcium dietary intakes in the developed world (e.g. Eaton and Nelson, 1991; Heaney, 1997b). However, this does not necessarily mean that the calcium requirements have changed over time (see Heaney, 1997b). The utility of calcium obtained in early diets was likely enhanced by plentiful sunlight exposure and vitamin D synthesis, the consumption of lots of fruits (vitamin C) and probable high levels of physical activity.

However, there are significant difficulties in attempts to determine the quantity of components of past diets (Eaton et al., 1997; Nestle, 2000). We face great challenges in attempting to determine any subsequent health implications, which include determining how patterns of selective eating, or effects of climate seasonality or variability in habitat productivity and changing population pressure may have altered the composition of the diet (see Cook, 1979; Lambert, 2007; Reed and Rector, 2007). We also know little regarding what interactions existed between dietary variables within ancient diets, which may

TABLE 7.7 Summary of the Interaction Between Various Fruits and Vegetables and Bone Cell Actions

Food source		Skeletal effects
Fruits *containing flavonoids, phenols*	Citrus fruits	May inhibit bone resorption May affect osteoclast numbers High content of vitamin C essential for bone collagen and osteoid formation (Chapter 4)
	Prunes *containing boron*	May inhibit bone resorption May help restore BMD and improve trabecular bone micro-architecture
	Apples	May prevent bone loss only in inflammatory conditions
	Grape seeds in a high calcium diet	May increase BMD trabecular and cortical bone May inhibit proteolytic enzyme role in bone resorption
	Pomegranate extract	Beneficial effects on BMD and trabecular bone formation
	Olives	Potential protective effect on bone mass only in inflammatory conditions
	General consumption fruits and vegetables	High quantity during childhood may increase femoral neck BMD and whole body and radial bone area
Vegetables	Mushrooms *containing polysaccharides*	Promote mobilisation of haematopoietic stem cells Act on bone marrow cells to enhance growth and differentiation May reduce bone resorption May increase osteoblast activity May slow decreases in BMD
	Garlic, garlic supplements, onions and leeks *containing sulfur compounds*	May slow decreases in BMD May slow urinary excretion of calcium and phosphate. Promote intestinal transfer of calcium May aid osteoprogenitor differentiation to osteoblasts

Continued

TABLE 7.7 (*Continued*)

Food source		Skeletal effects
Vegetables (*continued*)	Fennel, celeriac, French beans, arugula, broccoli, cabbage, lettuce, tomatoes Wild yam	May inhibit bone resorption Stimulates osteoblast proliferation Inhibits osteoclast formation May slow reductions of BMD
	Vegetables with β-carotene: carrots, sweet potatoes, oranges, cantaloupes, squashes, tomatoes and green leafy vegetables (spinach, lettuce)	Contains vitamin A, essential for bone growth Required in bone cell differentiation and remodelling
	Green leafy vegetables: spinach, lettuce, cabbage, kale	Contain vitamin K for protein production, including osteocalcin in bone matrix formation
Herbs	Sage, rosemary, thyme, dill, parsley	May inhibit bone resorption

Sources: *Anderson (1999) and Putnam et al. (2007).* **Note**: *Many studies have been conducted on animal models, particularly of rats and mice and data is not currently available to determine if human effects will be similar.* BMD, *bone mineral density. Bone cell actions described in detail in Chapter 3.*

have interfered with calcium bioavailability. Ulijaszek (1991) has also argued that modelling of the adequacy of components of past diets requires considerations of body size and mass as well as the likely levels of energy expenditure and physical activity undertaken. We do not yet fully understand how increased levels of activity would benefit or exacerbate any skeletal changes caused by any dietary imbalances.

The Effect of Meat Eating on Calcium Adequacy

The role of meat in the human diet has been discussed in various studies (Aiello and Wheeler, 1995; Wrangham et al., 1999; Bunn, 2007) and Cordain (2007:366) has cautiously estimated that it was likely that over 50% of the average daily energy intake for most Paleolithic hominin species was obtained from animal foods.

According to the acid base theory outlined above, the displacement of fruits, vegetables, lean meats and seafood by milks and grains would have detrimental effects on health (Cordain, 2007:374). Cordain (2007) has suggested

Box Feature 7.3. The Health of Adaptive and Transitional Diets: Integrated Approaches?

Dietary composition and subsistence strategies may have helped shape the adaptation and evolution of the human species. It is possible that settlement and social organisation developed with subsistence changes, such as meat-eating and cooking strategies (Cachel, 1997; Wrangham et al., 1999; Sealy, 2006; Ungar, 2007). Researchers have also investigated how diets may adapt during times of seasonal subsistence stress (e.g. Lambert, 2007; Reed and Rector, 2007) and how adaptations may have influenced population expansion (see Cachel, 1997). Complex and conflicting dietary reconstructions frequently develop to explain the same evidence, particularly concerning meat-eating (see Bunn, 2007).

Archaeological evidence has been used to determine dietary components through stable isotope analysis (Richards et al., 2003; Lee-Thorp and Sponheimer, 2006; Sealy, 2006), and floral and faunal composition of excavated sites (e.g. Reinhard et al., 2007). However, there remains little utility of the assessment of health in contributing towards interpretations of the quality of past diets. As discussed in Chapter 2, it is possible that there may be problems linked to skeletal fragmentation, or small sample numbers from archaeological sites of considerable antiquity in the assessment of diet and health in the past (see also Katzenberg and Lovell, 1999). However, anthropological resources from periods spanning dietary transitions do now exist. These include the Natufian culture in the Levant (Eshed et al., 2004), Mesolithic and Neolithic sites in Western Europe (Meiklejohn et al., 1984), and coastal and inland Late Holocene hunter–gatherer sites in South Africa (Sealy, 2006). Comparative analyses in these areas could yield important insights into past diets and health.

Dietary composition may be influenced by the nature of the demographic samples analysed. Variation in dietary quality according to age or sex is evident in many contemporary cultures, for example in sex-biased breastfeeding practices in Caribbean communities (see Quinlan et al., 2005). Age, sex and status-related dietary variation are likely to be evident in past communities, as has been demonstrated recently in ancient Rome by Prowse et al. (2005) as well as Cohen and Armelagos (1984:599), Larsen (1997:72–76) and Walker and Thornton (2002:513).

Dietary analyses could potentially provide a means of insights into social organisation within a community. Combining multiple anthropological perspectives (nutritional, biological and physical) in attempts to determine the manner and effects of dietary composition and quality on health is a necessary goal of future research into both modern and past diets.

that an ancestral diet high in meat intake may have been relatively beneficial by including lean tissue and muscle as a good sources of protein. However, Cordain (2007:375) has also determined that 'virtually all pre-agricultural diets were net-base yielding because of the absence of cereals and energy-dense nutrient-poor foods' (see also Sebastian et al., 2002). Fish and meat have been recently listed as the dietary factors creating the highest net acid load, followed by egg, shellfish, cheese and milk and with cereals as the lowest acid producers (Cordain, 2007:375). These data would imply that habitual or relatively persistent consumption of meat, may have resulted in diets relatively high in acid load (acidosis), which is suggested to be one of the harmful effects within the modern diet (Arnett, 2003; New, 2003:891; Eaton, 2007:389) and which could result in increased bone resorption.

High protein consumption appears better for skeletal health than low consumption (see Table 7.6 above), only where there is good calcium supply (Mazess et al., 1985:145; Massey, 2003:864s). Various researchers have suggested that as meat eating became increasingly adopted in ancestral diets, there may have been parallel decline of the consumption of plant matter, fruits, nuts and seeds (Ulijaszek, 1991:276; Wrangham et al., 1999). If so, this may have resulted in a limited availability of some sources of calcium within the diet, altering the beneficial effects of protein consumption.

Anthropological research has questioned the relationship between protein and calcium intakes as responsible for marked bone loss represented by reduced bone formation, evident in Inuit Alaskan Native American samples when compared to Arikara Plains Native Americans and Pueblo Native American from Southwest United States (see Richman et al., 1979; also Lazenby, 1997; New, 2003:892). However, continuing histological research has questioned the role of physiological stresses caused by pregnancy (Iwaniec, 1997 in Cho and Stout, 2003:214; see also Martin et al., 1984:211 for discussion of evidence from Nubia), as well as physical activity on bone remodelling parameters (Mulhern and Van Gerven, 1997; Cho and Stout, 2003; Pfeiffer et al., 2006). Furthermore, the effects of variation in remodelling with age, between the sexes and between groups of different ancestry are also at present not well understood.

At present it is difficult to extrapolate the effects of diet on bone health from histological analysis. It is also difficult to determine disease etiologies from specific dietary origins. It is likely to become increasingly necessary to conduct large-scale investigations with due consideration of the dietary base, range of across-skeletal effects of physical activity together with age and sex for multi-period comparisons in order to gain a better understanding of how osteopenia may have been modified by diet and additional factors in past communities (see discussions in Bridges, 1989; Pfeiffer and Lazenby, 1994). Keesing and Strathern (1998:111–123) provide further insight into the potential pitfalls of inter-linking dietary inadequacy with anthropological determinants of behavioural reconstructions.

Calcium Availability with the Onset of Domestication

The transition to the domestication of plants and animals is likely to have had a range of impacts on dietary quality (see Box Feature 4.2), including dietary calcium availability. The domestication of animals is thought likely to have occurred at approximately 10,000 BC although varying throughout world (see discussion in Zeder, 2006; and in relation to health Roberts and Manchester, 2005:184–185). The herding of cows in particular increases supplies of calcium through the production of milk, cheese and butter (see Keesing and Strathern, 1998:94–103), although consumption may remain limited for some social groups who instead rely on income generated from dairy production (Jelliffe, 1955:29). The consumption of raw milk (pasteurisation was not accepted as safe treatment until 1925; Bell and Palmer, 1983:592), which is untreated for various pathogens and bacteria, can result in severe intestinal infections (Pratt, 1984; Jones, 2004). If infections or diarrhoeal diseases were frequent and/or prolonged, then this source of calcium would be limited under malabsorption conditions and may well further restrict the intestinal absorption of other dietary micronutrients (see also Barnes, 2005). Furthermore, milk availability is by no means a measure of milk quality. In the more recent past, historical evidence clearly indicates the extent to which supplies were adulterated deliberately (Picard, 2000:65; see also Roberts and Cox, 2003:310) and accidentally (see Higgins et al., 2002:167).

Ill-health and disease vectors affecting animals may further limit milk production. The tuberculosis bacteria in cattle (*Mycobacterium tuberculosis bovis*) is an important means of human transmission of this disease (see further Roberts and Manchester, 2005:184–185), but can be associated with decreased milk supply (Hernandez and Baca, 1998; Johnson et al., 2001) limiting calcium availability. Parasitic infections can also be associated with decreased milk production in cattle (Vercruysse and Claerebout, 2001) and are likely to have been severe in communities without access to modern treatments. For example, the tropical bont tick (*Amblyomma variegatum*) is currently of significant concern affecting the West Indies requiring persistent treatment of cattle, sheep and goats (Pegram et al., 2004). The effect of this tick on cattle is to cause severe skin infections and prevent milk production (Food and Agricultural Organisation, 2004). We do not yet have adequate knowledge regarding the health of animals in the past, nor the manner in which disease-vectors may have been spread (see Box Feature 8.1).

Research suggests that there is an increasing tendency for osteopenia to have become more apparent after the onset of agriculture (Cohen and Armelagos, 1984:587; Cohen, 1989:119, 124) in several regions including the United States, Mediterranean and Near East and Africa (Perzigian, 1973; O'Connor, 1977 cited in Cook, 1984; Martin et al., 1984; see reviews Bridges, 1989; Ulijaszek, 1991; particularly Pfeiffer and Lazenby, 1994; Larsen, 2003). Cereals contain no vitamin C (see Box Feature 4.2), vitamin A (except yellow maize), or vitamin D, and are poor sources of calcium, sodium, B complex vitamins (Table 7.5) and iron (Widdowson, 1991).

Increased consumption of cereals containing phytates, which interfere with calcium absorption, as occurred during Neolithic times, may have contributed to inefficient absorption of calcium (Ulijaszek, 1991). The range of causative factors may also include secondary hyperparathyroidism caused by dietary imbalances of calcium and phosphorus exacerbated by diets high in maize (Stout, 1978 reviewed by Cook, 1984:244) or millet (Martin et al., 1984:208). Protein-calorie

TABLE 7.8 Summary of Potential Identifying Features for Secondary Osteopenia from Skeletal Remains

Macroscopic features	Complicating factors
Bilateral asymmetry bone size, under-development of muscle attachments Osteoporosis-related fractures, e.g. vertebrae, distal radius, femoral neck	Difficult to differentiate from other pathologies Over-compensation of limb use may exaggerate atrophy and muscle attachment development between limbs
Spontaneous fractures wide range of bones, e.g. tibiae and fibulae	Atrophy may not be severe to result in osteopenia Immobility may reduce risk of falls and resulting fractures Age, sex, ancestry, nutritional health of affected individuals need consideration The pattern of changes throughout the skeleton may enable differentiation from age-related changes
Radiological features	**Complicating factors**
Thinning cortical bone Increased intra-cortical porosity Thinning/removal of trabeculae Compensatory thickening with maintained load-bearing When immobilized, all trabeculae may be thinned Sub-periosteal bone loss if immobilisation preventing apposition Periarticular osteopenia, small, spotty or speckled areas of bone loss	Osteopenia may be localised or generalised Compensatory biomechanical use may alter bone structure complicating radiographic comparisons between limbs Osteopenia may be part of pathological conditions More apparent in juxta-articular locations in various joint pathologies
Histological features	**Complicating factors**
Increased bone resorption Enlargement of Haversian canals Increased coalescence resorption spaces Multiple sites Howship's lacunae Greater osteon size Limited bone formation	Normal number of active remodelling sites or 'normal' size of Haversian canals may not be identifiable Complicated by age, sex, ancestry, health, physical activity Primary pathology may affect histological manifestations

Note: *See also description of skeletal changes in age-related osteoporosis in Chapter 6.*

malnutrition, low dietary calcium, high iron intake and high disease loads, could also be responsible for differences in bone remodelling rates and risk of osteopenia in various skeletal samples (see reviews above). Reconstructions of the skeletal consequences dietary alterations are however complicated by the potential effects of additional factors, probably including decreased physical activity (see Larsen, 2003), although the nature of this effect being site specific within the skeleton needs further consideration (see Bridges, 1989; Chapter 6).

Continued analyses of skeletal collections using better techniques for assessing bone amount and loss with age (Chapter 6 and Box Feature 6.3), together with increased analysis of past diets using techniques such as stable isotope analysis would significantly develop our knowledge of the impact of diet on health particularly across transitional periods (Box Feature 7.3). Such research may clarify the influence of additional life-style-related factors in contributing to skeletal changes as well as dietary change, and the demographic nature of those previously analysed also need consideration in reconstructions of the impact of health.

DIAGNOSIS OF SECONDARY OSTEOPENIA IN ARCHAEOLOGICAL BONE

Many of the identifying features of secondary osteopenia will match those described for age-related bone loss (Chapter 6) and these changes are summarised in Table 7.8. However, the typical pattern of skeletal regions affected will vary according to the location and nature of the primary pathologies in causing secondary bone loss.

CONCLUSIONS

Significant understanding of the impact of disease on life could be made through consideration of secondary pathologies and adaptations to illness. In addition, a well-balanced and nutritional diet is important in preventing osteopenia, but unfortunately there are many circumstances in which this balance is not maintained. Determining the composition of past diets is becoming increasingly viable through utilising a wide array of techniques and approaches and will remain an important focus of future bioarchaeological research into metabolic bone disease.

Chapter 8

Paget's Disease of Bone

Unlike many of the conditions covered by this book, the exact causes of Paget's disease of bone (PDB) are still unknown. This position is both frustrating and exciting. The lack of an agreed aetiology for the condition makes interpreting cases of the disease in past populations, and using information on PDB in wider anthropological debates, difficult. However, as is discussed below, the symptoms experienced by individuals with the condition are known, so diagnosis of PDB in past populations will provide additional perspectives on experiences of individuals with the condition. The lack of a known aetiology means that this is one condition where diagnosis of cases from archaeological skeletal assemblages really does have the potential to provide significant new evidence that can help with understanding its present distribution.

PDB is a chronic disease that results in disruption of bone remodelling in affected bones characterised by gross deformity and enlargement of parts of the skeleton. It is named after Sir James Paget, who made a key description of the condition in 1876 in a lecture given to the Royal Medical and Chirurgical Society of London. However, cases that were almost certainly Paget's disease were described by Wilkes in 1869 (Dalinka et al., 1983) and in 1873 by Czerny (Haibach et al., 1985). James Paget originally believed that the disease was an inflammatory condition of bone and so termed it osteitis deformans (Paget, 1882). Although this is now known to be inaccurate, PDB is still occasionally referred to as osteitis deformans (see discussion by Rhodes and Jawad, 2005).

POSSIBLE CAUSES OF PAGET'S DISEASE

The causes of Paget's disease have long been debated, and there is currently no consensus. As discussed previously in this book, the link between infectious diseases and a number of metabolic diseases have been investigated recently (e.g. Chapter 4). Possible links to infectious diseases have also been investigated for PDB including bacterial infection, particularly of oral bacteria (Dickinson, 2000). A number of viral conditions have also been linked to Paget's disease, and Rousière et al. (2003) provide a review of recent evidence for viral causes of PBD. These researchers suggest that increased vaccination against measles in humans and canine distemper in dogs may have had a role in influencing the occurrence of PDB in humans (Rousière et al., 2003).

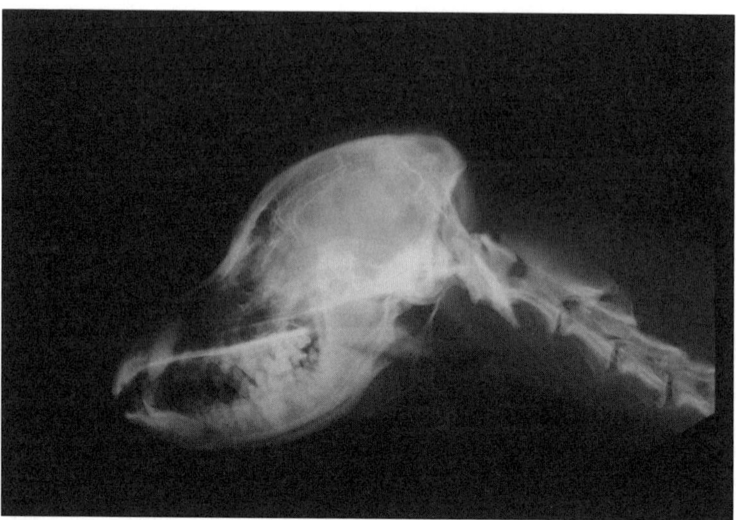

FIGURE 8.1 Lateral radiograph of a dog skull showing a markedly thickened calvaria. Courtesy of Dr Yamada (Yamada et al., 1999:1056), reproduced with permission from the Japanese Society of Veterinary Science.

Osteoclasts from pagetic bone have been shown to have viral inclusions linked to measles and canine distemper and it has been suggested that these may be a causative factor in the onset of the disease (Roodman, 1994, 1999). Paget's disease has been linked to dog ownership, as well as ownership of birds, cats and cattle (Hadjipavlou et al., 2002). Zoonoses, infectious conditions of animals that can be passed to humans, have the potential to provide important information on a range of social and cultural practices in both past and present societies. Yamada et al. (1999) reported a case of a dog with a number of pathological conditions, including thickening of the skull (as shown in Figure 8.1) that resemble PDB. Cases of this condition in animals could be more widespread, but without the use of radiography or MRI they may be undetected. Dog skeletons are frequently reported from archaeological sites (e.g. Brickley and Thomas, 2004), and this is one area that zooarchaeologists might fruitfully investigate (see Box Feature 8.1).

Familial occurrence of PDB has indicated that genetic defects could play an important part in its development. Daroszewska and Ralston (2005) have identified four possible causal genes, but there is still much that is not understood. Roodman (1999) in particular has questioned the existence of genetic links in familial occurrence of the condition.

Epidemiological similarities between patterns of PDB and childhood vitamin D deficiency (discussed in Chapter 5) have been noted (Barker and Gardner, 1974). However, no conclusive results on a link have been produced, and it seems unlikely that childhood vitamin D deficiency will prove to be a

Box Feature 8.1. Animal Paleopathology

Animal pathology has the potential to contribute significant information to those engaged in the study of past societies. As the examples included in this chapter and Chapters 4, 5 and 7 illustrate, many of the metabolic bone diseases affect both animals and humans. In addition to providing information on the history and development of diseases, such conditions can also contribute to an understanding of past husbandry practices and economic activities.

The first significant publication on animal paleopathology was *Animal Diseases in Archaeology* (Baker and Brothwell, 1980), but over the next 20 years little work was done on the subject. Recently, there has been a renewal of interest in this field of enquiry with the publication of Davies et al. (2005b) and Miklíková and Thomas (in press). Although there are many veterinary texts that cover metabolic bone disease in animals (e.g. Vrzgula, 1991), to date investigation by zooarchaeologists there has been limited. Studies by zooarchaeologists have mainly dealt with arthropathies and pathological change thought to be due to infection (Thomas personal communication).

One of the problems faced by zooarchaeologists is that many of the animal skeletons that they deal with are disarticulated, and this can make suggesting a differential diagnosis difficult, as the pattern of pathological change across the skeleton cannot be viewed. However, future research such as that being undertaken by Thomas, and the greater development of histological and aDNA investigation will undoubtedly facilitate valuable insights into past animal health.

causative factor in the development of PDB. A number of toxic substances that could trigger the occurrence of PDB have also been considered, and with clear geographical differences in the occurrence of the condition, links to toxins that occur in certain regions, or are used in particular industries are an obvious area of investigation (e.g. Spencer et al., 1992).

From considering the available evidence it seems unlikely that PDB has a single cause. Investigations have been extensive, but have yet to produce evidence of a clear link between PDB and any of the factors considered to date.

CONSEQUENCES OF PAGET'S DISEASE

One unusual feature of PDB is that the condition is often only active for a finite period of time (Milgram, 1990:869). In cases of PDB only selected portions of the skeleton are affected by the changes, and additional bones may become involved at any stage of the process (Mirra et al., 1995a). As a result the condition does not fit with the definition of metabolic bone diseases used by Albright and Reifenstein (1948) (see Chapter 1). However, a number of authors include the condition as a metabolic bone disorder (e.g. Cooper et al., 1999) and as affected bones do exhibit changes due to disruption of remodelling it is included here. Cases involving a single bone are termed 'monostotic', and involving between two and four bones are described as 'oligostotic'. Where at least five bones are involved, the condition has been described as 'polyostotic' (Renier et al., 1996). Monostotic lesions have been found more often in females than males (Renier et al., 1996). Any bone can be affected, but the most commonly involved bones are those of the pelvis, lower spine, cranium, and long bones of the legs (Walsh, 2004).

Three phases have commonly been identified in PDB, but there is some variation in the clinical terminology used to describe each phase. In the first phase, bony changes are characterised by an increase in resorption caused by a massive increase in osteoclastic activity (Mirra et al., 1995a). As a result hypervascularisation is apparent histologically due to increased resorption (Gruber et al., 1981). This stage has been referred to as the 'osteolytic phase' (Roodman and Windle, 2005).

The second phase of the condition is characterised by marked bone formation and this has been referred to as the mixed (Gruber et al., 1981) or intermediate phase. During this phase, a very characteristic pattern of cement lines develops within bone tissue, often referred to as a 'mosaic appearance' (see Figure 8.2). Repeated episodes of resorption, which create scalloped borders of resorption typical of Howship's lacunae, are in-filled with newly formed bone, creating a series of cement lines around irregular areas of newly formed woven bone (Milgram, 1990). Unless there is further disruption of the remodelling process, due to a fracture or prolonged immobilisation, bone formation

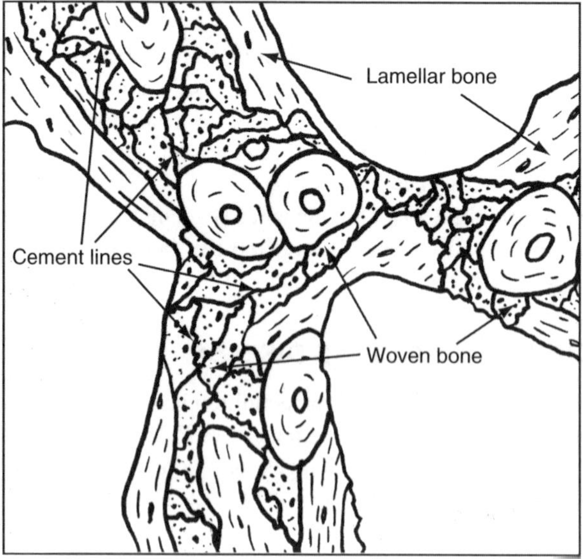

FIGURE 8.2 Diagram illustrating some of the histological features seen from the end of the first phase of the condition when the 'mosaic' pattern starts to develop. Repeated episodes of resorption followed by new bone formation result in a series of cement lines separating small and irregular areas of bone tissue.

and resorption will be relatively balanced but proceed at a faster rate than normal (Moore et al., 1994). Bone turnover during the second phase has been measured to be up to five times the rate expected in normal bone (Krane and Simon, 1987) (the different bone types are discussed more fully in Chapter 3). However, due to the disorganised way in which new bone is produced, the bone is of poor quality, and much weaker than normally formed bone (Roodman, 1999; Roodman and Windle, 2005). Pagetic bone is predominantly lamellar bone, although woven bone can be present in some areas (Krane and Simon, 1987). The disturbance in remodelling seen in PDB is primarily caused by osteoclasts, and in individuals with the condition osteoclasts are increased in size and number (Roodman, 1994, 1999). Towards the end of the mixed phase, bone formation can proceed at a faster rate than bone resorption, resulting in thickening of bone structures (Mirra et al., 1995a).

The third phase of PDB is marked by the development of sclerotic bone and a decrease in vascularity (Gruber et al., 1981), and so has been referred to as the late or sclerotic phase (Wittenberg, 2001) or inactive phase (Milgram, 1990:871). At the very last stage of the disease process there is often an almost complete cessation of osteoclastic activity and only minimal osteoblastic activity (Mirra et al., 1995a). Jaffe (1972) referred to this as the burned out phase. The reasons why changes apparent in each of the phases of the disease occur are still poorly understood.

Pelvic Changes

The pelvis is an early skeletal site for the development of pagetic lesions (Renier et al., 1996), with initial development of lesions occurring most frequently in the ilium, close to the sacroiliac joint. Severe involvement of the pelvis can lead to *protrusio acetabuli*, and possibly *coxa vara* (Mirra et al., 1995a). Involvement of the pelvis is not bilateral, and the right side is more frequently involved than the left (Dalinka et al., 1983).

Cranial Changes

Any bone of the skull can be involved, but the areas most commonly affected are the cranial vault and bones of the skull base. However, no site in the skull has emerged as a preferential location of lesion development (Renier et al., 1996). Basilar invagination can occur in the third phase of PDB due to softening of the bone (Greenfield, 1990:117). Changes in the skull can affect hearing (Milgram, 1990:869), possibly leading to deafness where the petrous temporal bone is involved (Walsh, 2004). Involvement of the maxilla is about six times more common than the mandible, but when both are involved leonine features can develop (these can be confused with leontiasis ossea, a rare condition involving excessive bone formation in the skull, Ortner, 2003:416). Mirra et al. (1995a) suggest that such changes could also be confused with fibrous dysplasia.

Long Bone Changes

Deformities of the long bones follow characteristic patterns; in the femur bowing occurs laterally and in the tibia it occurs anteriorly (Mirra et al., 1995b). Bowing of the lower extremities can lead to shortening of the limbs, and where shortening is uneven will result in the development of a limp or other gait disturbance (Milgram, 1990:869; Fumio et al., 2004). Once pagetic bone has accounted for two-thirds or more of a bone it is difficult to identify the initial site of the lesion, and this is almost impossible in bones that were completely altered (Renier et al., 1996). Wedged or V-shaped areas (or V-shaped cutting cone, Mirra et al., 1995a) of radiolucency can appear in long bones affected by PDB, as shown in Figure 8.3. These features, which appear early in the course

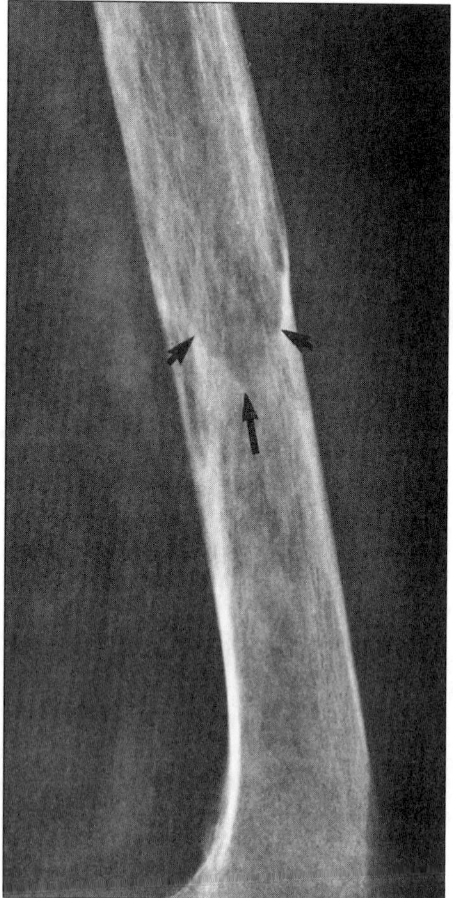

FIGURE 8.3 Lateral view of distal femur with a V-shaped area of radiolucency indicated by arrows. This feature has been referred to as the blade of grass or flame sign. The bone above this region is slightly expanded and has a coarsened trabecular pattern. Courtesy of Wittenberg 2001. Radiology 221:199–200. The Radiology Society of North America.

of the disease, have also been referred to as the blade of grass or flame sign (Wittenberg, 2001). This lytic process can spread at a rate of about 1 cm a year (Mirra et al., 1995a), but there may be variation in the rate at which normal bone is converted to pagetic bone (Resnick, 1988).

Other Bones that Can be Affected

Although the skeletal areas listed above are the most common sites of pagetic changes, the condition can affect most parts of the skeleton. For example, the spine can be involved and changes are recorded in 50% of patients with polyostotic PDB (Saifuddin and Hassan, 2003). The lumbar and sacral areas the most frequently involved regions of the spine (Mirra et al., 1995b), with the thoracic the next most frequently affected, and the cervical is least commonly affected (Saifuddin and Hassan, 2003). Changes include enlargement of the vertebral bodies, with coarsening of trabeculae (Mirra et al., 1995b). When the spine or cranial areas are affected nerve compression can occur (Wittenberg, 2001). Dalinka et al. (1983) provide a review of possible consequences of spinal involvement. A 'picture frame' appearance in vertebrae (seen in Figure 8.4),

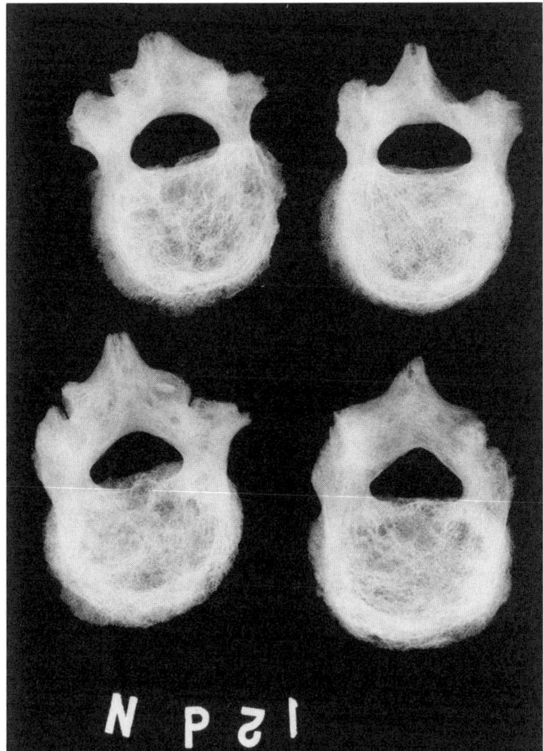

FIGURE 8.4 Vertebrae of one of the individuals from Norton Priory, these bones have characteristic 'picture frame' appearance. Courtesy of Anthea Boylston, Alan Ogden and BARC, Archaeological Sciences, University of Bradford.

where there is coarsening of the trabecular bone, surrounded by more sclerotic bone can develop (Wittenberg, 2001). This feature is caused by increased sclerosis of the vertebral margins, in particular the superior and inferior margins (Mirra et al., 1995b). Sclerotic changes across vertebrae can result in an increase in density that has been referred to as 'ivory vertebra' due to the radiological appearance (Saifuddin and Hassan, 2003). At a macroscopic level such vertebrae may be enlarged compared to neighbouring bones (Dalinka et al., 1983).

Pagetic lesions can also affect the scapula, and in a study of 200 patients with PDB undertaken by Renier et al. (1996), 9.8% were found to have lesions. The glenoid fossa was the site at which lesions developed most frequently. The bones of the hands and feet are occasionally involved, and an accentuated trabecular pattern can develop in tubular bones in these areas (Mirra et al., 1995b). Gruber et al. (1981) report that in some cases the clavicles and ribs can also be affected.

CO-MORBIDITIES

Paget's disease can co-exist with a wide range of conditions, and in older individuals there are many degenerative diseases that can co-exist with PDB (Milgram, 1990:870). Features of PDB such as gait disturbance will contribute to the development of conditions such as osteoarthritis. Osteoarthritic changes are particularly common around the hips and may be secondary to the replacement of cartilage with pagetic bone (Dalinka et al., 1983).

Paget's disease in vertebral end plates can resemble changes caused by infections (Moore et al., 1994). However, as Moore et al. point out the exact sequence of events in such cases is complex, and it is possible that PDB may co-exist with infectious conditions. Another factor that should be considered is that the distribution of infectious diseases caused by blood borne infection, like osteitis and osteomyelitis, is similar to the distribution of pagetic lesions in the limbs (Renier et al., 1996). The similarity in distribution between these conditions may be due to the carrying of pagetic osteoclasts by the bloodstream (Renier et al., 1996). Care should therefore be taken that these conditions are not confused.

Neoplasms have often been reported to arise in individuals with PDB (Haibach et al., 1985). Up to 10% of individuals with PDB will develop associated malignant changes (Mirra et al., 1995b), but reported incidence of such changes varies considerably. However, Moore et al. (1994) found that it was not possible to state whether metastases occurred more frequently in pagetic than normal bone. The combination of PDB and neoplastic conditions can make it difficult to identify blastic or lytic changes due to neoplastic conditions (Moore et al., 1994).

In particular osteosarcomas have been widely reported as a relatively common complication, and between 5.5% and 14% of cases reported in the modern population are associated with PDB (Cheng et al., 2002). Individuals with both conditions have a poorer survival rate than those that only have an osteosarcoma. Mortality rates from osteosarcomas associated with Paget's disease are reported to be 86.6% within three years of diagnosis (Cheng et al., 2002). Possible suggestions about why this might happen include factors such as an increased blood flow in Pagetic bone facilitating more aggressive growth of tumours (Haibach et al., 1985). However, individuals who develop osteosarcomas do tend to be older, and it is possible that increased age in these individuals may play a role. Fibrosarcomas are also reported to be quite common in individuals with PDB as are cases of giant cell tumour (Dalinka et al., 1983). Other sarcomas are occasionally reported (Hadjipavlou et al., 2002). In the current population, even where there is access to good levels of health care, the prognosis for individuals that develop malignant conditions in addition to PDB is not good.

In many individuals PDB is asymptomatic, certainly in the first phase of the disease, but in approximately 30% of cases there may be morbidity such as pain and deafness. Bone deformity can also contribute to the development of osteoarthritis (Daroszewska and Ralston, 2005).

PAGET'S DISEASE IN THE MODERN PERSPECTIVE

Age and Sex

The condition can appear in individuals as young as 30–35 years, but in the early stages it can often be asymptomatic and so those affected at such ages will be unaware that they have it (Milgram, 1990:869–870).

In North America, PDB is diagnosed during autopsy in approximately 3% of individuals over the age of 40 years (Milgram, 1990). The prevalence of PDB is strongly age-related and after the age of 50 the prevalence doubles every decade (Rousière et al., 2003). Men are affected twice as often as women (Greenfield, 1990:115) and a recent review of individuals in the UK suggests that PDB could affect 2.5% of men and 1.6% of women (Daroszewska and Ralston, 2005). The mortality rates of individuals classified as dying from Paget's disease in Britain have declined in successive cohorts of individuals who have been studied (Barker and Gardner, 1974). Many people actually die from associated tumours, but prior to the availability of recent medical treatments levels of such deaths would have been higher.

Bakwin and Eiger first described a rare condition referred to as Juvenile Paget's Disease (JPD) in 1956. It is characterised by rapidly remodelling woven bone, osteopenia, fractures and progressive skeletal deformity (Whyte et al., 2002) and is linked to a number of inherited genetic mutations (Janssens et al., 2005). This condition has also been referred to as idiopathic hyperphosphatasia (Daroszewska and Ralston, 2005), and familial hyperphosphatasia

(Taneja et al., 1990). There is considerable uncertainty regarding the nature and causes of this condition, as the range of names utilised reflects. It is not certain that JPD is a juvenile version of the adult condition of PDB. Whilst manifestations of this condition appear similar to PDB, several cases have involved the skeletal changes of rickets and Taneja et al. (1990) suggest this may be a rare variant of resistant rickets caused by abnormal serum alkaline levels. Further research is required to better understand the nature of this condition, and caution is necessary before diagnosing potential cases of what appear to be Paget's disease in juvenile skeletal remains.

Geographic Variation

Considerable geographical variations have been reported in the prevalence of PDB (Milgram, 1990:869; Cooper et al., 1999). The condition is rare in Scandinavia, Japan (Cooper et al., 1999) as well as Africa and Asia (Fumio et al., 2004). However, Dahniya (1987) reported that there may be geographical variation in the prevalence of the condition between different African communities, and it may not be rare in some areas. The UK, particularly the northwest of England, has some of the highest prevalence rates in the world (Doyle et al., 2002). The prevalence has been particularly high in some northern British towns, but recent research into the prevalence of the condition indicates that the overall prevalence has declined (Cooper et al., 1999; Doyle et al., 2002).

ANTHROPOLOGICAL PERSPECTIVES

As the causes of PDB are still unclear, interpretation of the condition using a biocultural approach is not possible at present. However, there is a lot of information available on the effects of Paget's disease on individuals. Through a thorough examination of the type and extent of lesions present in the skeleton of affected individuals from archaeological sites, it is possible to gain some idea of the likely impact of the condition on the individual. However, care should be taken with such interpretations as, as stated by Milgram (1990) drawing on experience within clinical practice, the impact and possible consequences of PDB need to be considered on an individual basis.

PALEOPATHOLOGICAL CASES OF PAGET'S DISEASE

The first case of PDB in archaeological bone was reported by Pales (1929), and consisted of a disarticulated right femur of Neolithic date from Lozère, France. Cases currently documented in the published literature are summarised in Table 8.1. The available evidence demonstrates that PDB appears to have considerable antiquity.

TABLE 8.1 Reported Cases of PDB in Paleopathology

Location	Date	Total burials	Age/sex, bones affected	CS or PS	Methods	Sources
England	AD 1700–1900 (PM)	148	1 OAM. Skull, femur	CS	Mac.	Brickley et al. (1999)
England	AD 1700–1900 (PM)	968	9 M, 8 F, all OA where age known Various	CS	Mac. Rad.	Molleson & Cox (1993)
England	AD 900–850 (M)	611	11 M, 3 F, 4 unsexed Various	15 CS, 3 PS	Mac. Rad.	Rogers et al. (2002), Waldron (2004)
England	AD 750–1550 (M)	476	1 OAF. Long	CS	Mac. Rad.	Lilley et al. (1994)
England	AD 750–1550 (M)	436 (MNI)	2 M-OAM. Radius, scapula, humerus, fibula, hand and foot bones	1 CS, 1 PS	Mac. Rad. Mic.	Bell & Jones (1991), Stirland (1991)
England	AD 750–1550 (M)	NS	2 M-OAM. Sacrum, femur, calcaneus	1 CS, 1 PS	Mac. Rad. Mic.	Aaron et al. (1992)
England	AD 1200–1400 (M)	130	6 OAM. Various	CS	Mac. Rad.	Boylston & Ogden (2005)
England	AD 410–1550 (A-S and M)	NS	8 long and 2 skulls	PS	Mac. Rad.	Price (1975)
England	AD 410–1050 (A-S)	NS	1 OAM. Skull, spine, long	CS	Mac. Rad.	Wells & Woodhouse (1975), Anderson et al. (2006)
France	AD 1000 (M)	20	1 OAM. Skull, femur	CS	Mac. Rad. Mic. CT	Roches et al. (2002)

Continued

TABLE 8.1 (*Continued*)

Location	Date	Total burials	Age/sex, bones affected	CS or PS	Methods	Sources
France	AD 750–1550 (M)	NS	NS. Long	CS	Mac.	Heuertz (1957)
France	AD 750–1550 (M)	NS	1 OAM. Tibiae	CS	Mac. Rad.	Morel & Demetz (1961)
Denmark	AD 750–1550 (M)	NS	Femora	CS?	Mac. Rad.	Møller-Christensen (1958)*
Russia	AD 750–1550 (M)	NS	NS	CS	Mac.	Rokhlin (1965)*
Canada	AD 800–1200 (M)	6 (MNI)	NS. 1 Skull	CS	Mac. Rad. Mic.	Gardner et al. (2005)
US	Pre AD 1300 (PE)	NS	NS. Tibiae	2 PS, 3 CS	Mac.	Fisher (1935)
US	Pre AD 1300 (PE)	5 from 3 sites	Where specified OAM. Long	1 PS, 4 CS	Mac. Rad.	Denninger (1933)
France	AD 330–340 (R)	1100	1 OAF. Skull	CS	Mac. Rad. Mic. CT	Roches et al. (2002)
Belgium	650 BC to AD 43 (IA)	NS	Skull, pelvis	PS	Mac.	Jannsens (1963, 1970)*
France	3500–2000 BC (N)	NS	NS. Femora	PS	Mac. Rad.	Pales (1929)
France	3500–2000 BC (N)	NS	NS. Long, innominates	PS	Mac.	Baudouin (1914)*
Italy	3500–2000 BC (N)	NS	NS. Long	PS	Mac. Rad.	Milanesi (1962)
Egypt	NS	NS	NS. Skull (only bone collected)	PS	Mac.	Hutchinson (1889)

Note: Dates given are only approximate to provide a time line. Time periods given by original authors are indicated in brackets: N, Neolithic; R, Roman; A-S, Anglo Saxon; PE, Pre-European; PM, Post Medieval; M, Medieval; MNI, minimum number of individuals; J, juveniles; A, adults; M, male; F, female; MAM, Middle adult male; OA, old adult; OAM, old adult male; OAF, old adult female; NS, not specified; Mac., macroscopic examination; Mic., histological examination; SEM, surface SEM; Rad., radiological examination; BiT, Biochemical test; CS, complete skeleton; PS, partial skeleton or disarticulated bone. Many cases where age and sex were not specified involved isolated bones. Unpublished cases have been omitted

*Indicates cases for which the information reported is derived from Cook (1979).

A substantial number of the possible cases of PDB have come from isolated bones rather than complete articulated skeletons. Human skeletal remains are frequently disarticulated due to a range of cultural and taphonomic factors as discussed in Chapter 2. In the two cases reported by Aaron et al. (1992) many of the bones in the articulated individual were completely normal in appearance, and only certain bones showed radiological evidence of a pathological condition being present. The appearance of individual bones affected by this condition can be very distinctive, and certainly at a histological level once affected bones reach the mixed phase, diagnosis is possible from a small piece of a single bone. Following the careful review of cases undertaken by Cook (1980), in which a number of the early diagnoses of PDB were rejected, many of the more recent reports of PDB have used histological analysis to help confirm a diagnosis. Research currently being undertaken on the WORD project by the Centre for Human Bioarchaeology, Museum of London has identified a number of additional cases of PDB. This work is ongoing, but currently six possible cases have been identified from 1367 individuals examined from medieval sites and seven possible cases from 808 individuals examined from post-Medieval sites. Most of these cases were in older adults, with males being more commonly affected than females (WORD).

One of the most recent reports of PDB in archaeological skeletal material is from Norton Priory, England. A diagnosis of Paget's disease was suggested for six old adult males from this site dating from the mid-thirteenth to fifteenth century (Boylston and Ogden, 2005). One of the individuals from Norton Priory was also reported to have an associated osteosarcoma (Boylston and Ogden, 2005). Just as in clinical cases the possibility of the co-existence of other pathological conditions should be considered when examining human remains from archaeological sites.

Pusch and Czarnetzki (2005) have recently reported the results from a study of more than 8500 archaeological skeletons from central Europe. They found a prevalence of just 0.03%, with no cases before AD 1400, and suggest that the low prevalence is due to lower life expectancy in past populations. These suggestions are discussed by Waldron (2005), who advises PDB can be overlooked in archaeological bone unless all bones are radiographed.

Most of the cases reported from archaeological skeletons are in males, although sex of many of the individuals represented by disarticulated bone is not known, so it is impossible to be certain if the sex distribution of cases has changed. One case has been reported in a female buried at the site of Lisieux-Michelet in Northern France (Roches et al., 2002), dated between AD 330 and 340. The patterns observed in archaeological skeletons appear to reflect the general sex pattern observed in the modern population.

Today, as in the past, some of the highest prevalences are reported from England, with high rates from areas of continental Europe (Resnick, 1988). High levels reported today amongst those in North America and Australia are mainly in individuals with European descent. This makes an early reported case from North America of considerable interest. The case from Ontario in Canada, (AD 800–1200) (Gardner et al., 2005), which has been confirmed through histological analysis is probably the first confirmed case of PDB from North America. Although some of the earlier reported cases of PDB from pre-historic skeletal material were probably inaccurate, it is clear that the assertion made by Ross (1982) that PDB was a disease of people of what could be termed 'Anglo-Saxon origin' is not correct, as there are pre-historic cases reported in individuals from the Americas. Relatively few modern cases have been reported from Japan and China (Resnick, 1988), but to date relatively little paleopathological work has been done in these regions. In the future it would be valuable if paleopathological investigations in areas such as China and Japan where such work is currently limited, were expanded allowing fuller comparisons with modern data.

DIAGNOSIS OF PAGET'S DISEASE IN ARCHAEOLOGICAL BONE

The review of cases of PDB in archaeological skeletons by Cook (1980) found that quite a number of cases were probably not PDB, but may have been treponemal infections, and the advice given regarding exercising caution in making a diagnosis should be heeded (Cook, 1980). Evidence for PDB in archaeological skeletal material can be inconclusive without histological evidence for the condition (Aaron et al., 1992). Histology is rarely required to make a diagnosis in clinical cases, where diagnosis is almost always established radiolographically (Milgram, 1990), but this is not the case in archaeological bone. Due to often incomplete preservation illustrated in the cases presented in Table 8.1 and a variety of taphonomic factors operating on bone, histology is often a valuable aid to diagnosis, see Figures 8.5 and 8.6. Criteria that should be considered when suggesting a diagnosis of PDB are listed in Tables 8.2–8.4. These tables set out the key diagnostic criteria at a macroscopic, radiological and histological level in each of the three phases. It should be remembered that individuals could die at any stage of the condition and not all bones may exhibit the same phase of pathological changes. It may be difficult to fully assess the number of bones involved, as the skeletons of individuals with PDB are frequently poorly preserved and incomplete. This has important implications for the type of sampling and recording strategies that are adopted by those involved in archaeological investigations, a factor recently illustrated by Brickley and Buteux (2006:91–94).

Macroscopic Features of Paget's Disease

Key macroscopic features of Paget's disease seen in archaeological human bone are listed in Table 8.2. Although cases from the first phase may be missed

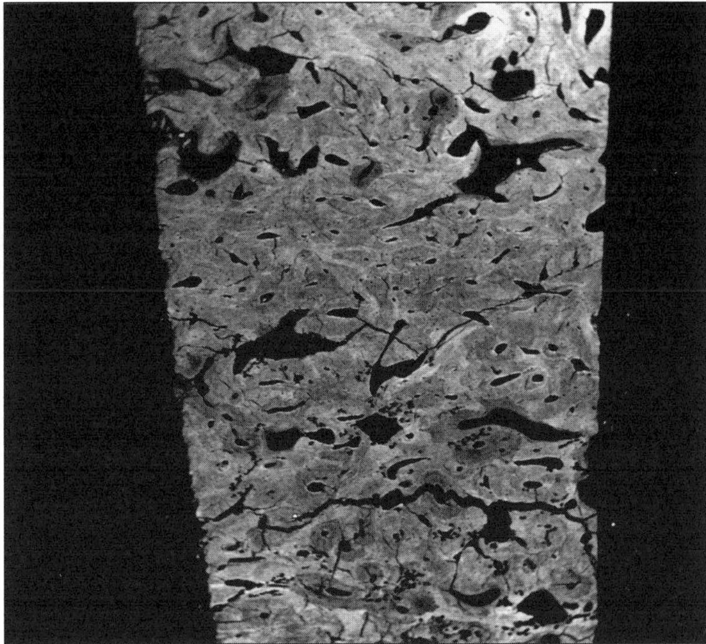

FIGURE 8.5 BSE-SEM image of a sagittal section of occipital bone from an archaeological individual from Sandwell Priory. Increased vascularity with irregularly defined vascular canals evident amongst poorly organised and mineralised bone (less well mineralised bone is darker in tone than relatively more highly mineralised bone) (field width 2625 lμm). Reproduced with permission from Bell and Jones (1991). © John Wiley and Sons Limited.

or not linked to PDB, many of the changes occurring in the second and third phase of the condition are useful for suggesting a diagnosis. Many of the cases listed in Table 8.1 were initially identified by morphological features (although in some cases subsequent radiological analysis revealed additional bones that were affected). For example Stirland (1991) recorded widespread thickening of bones and 'coral like' new bone formation on the external surface of affected bones equating to the type of changes listed in Table 8.2 as being characteristic of the third phase.

Radiological Features of Paget's Disease

The use of radiology in assisting with the diagnosis of PDB has a long history, and was first used in 1896, just one year after the discovery of X-rays (Moore et al., 1994). Key radiological features for the diagnosis of PDB in archaeological human bone are given in Table 8.3. Radiological features of PDB are well known and have been described by a number of authors (e.g. Mirra et al.,

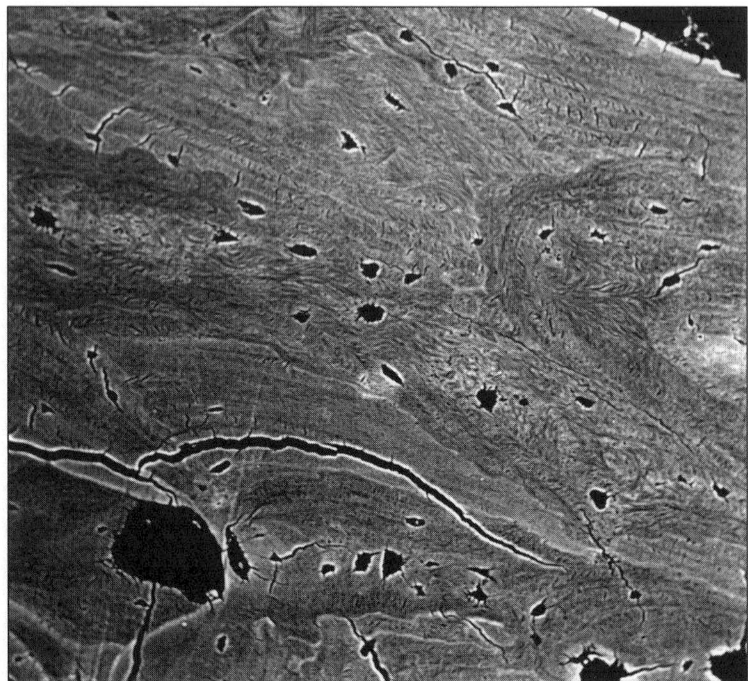

FIGURE 8.6 BSE-SEM image of a specimen from Sandwell Priory. Centre field shows irregular and enlarged osteocyte lacunae with poorly organized collagen (field with 440 μm). Reproduced with permission from Bell and Jones (1991). © John Wiley and Sons Limited.

1995a, 1995b). Some of these features are observable in archaeological bone simply through visual examination. However, features of internal bone structure include: lack of uniformity in areas of increased bone density, cortical thickening – which can encroach on the medullary cavity, very developed pattern of trabecular bone (due to coarsened and accentuated trabecular bone) and a disorganisation of architecture (Doyle et al., 2002). During the first phase when few features are visible macroscopically, radiological analysis will be particularly important for identifying bones affected by PDB.

Although the radiological appearance of PDB generally follows a standard appearance, it is possible in some cases of the condition that the changes to the bones will progress in a different way, and lead to non-standard radiological features. Moore et al. (1994) and Mirra et al. (1995b) provide a review of some less frequent features of the condition. For example, they describe cases where the focus of the change starts in the middle of the shaft of a long bone, rather than at one end. The possibility of variation from the radiological appearance described in Table 8.3 should be considered when obtaining radiographs of archaeological bone from suspected pagetic individuals.

TABLE 8.2 Macroscopic Features of PDB Observable in Archaeological Human Bone

Bones affected	Features	Phase	Code	Differential diagnosis	Sources
Cranium	Platybasia and basilar impression	End 1st	G	Osteomalacia	Milgram (1990)
	Thickened vault – increased dimensions particularly frontal See Figure 8.7	2nd	D		Mirra et al. (1995a, 1995b)
Dentition	Hypercementosis Ankylosis	2nd	G	Age-related change, Trauma, Inflammation	Mirra et al. (1995a)
Vertebrae	Vertebral fractures and compression due to resorption – kyphosis	2nd	G	Osteopenia and Osteoporosis	Mirra et al. (1995a)
	Enlargement and thickening and changes to end plates	3rd	D	Infections	Hadjipavlou et al. (2002)
Ribs	–				
Pelvis	Deformities – e.g. *protrusio acetabuli*	2nd	G	Osteomalacia	Mirra et al. (1995a)
Long-bones	Bowing and widening – particularly lower limbs, e.g. sabre shin	2nd	D	Syphilis Osteomalacia	Mirra et al. (1995a)
	Pathological fractures	2nd	D	Osteomalacia/ Rickets,	
	Coxa vara	2nd	G	Congenital conditions	
	Bones have a pumice-stone-like quality to surface (affects range of skeletal areas)	3rd	D	Osteitis, Osteomyelitis	

Note: Phase, phase of the condition from which features commonly first occur, most such changes will become more marked in later phases. S denotes strongly diagnostic feature, S features are normally required for a diagnosis. D denotes diagnostic feature, the presence of multiple D features are required for a diagnosis. G denotes general changes, can occur in many of the metabolic bone diseases as well as in other conditions. G features can aid a diagnosis with S or D features but cannot be used alone to suggest a diagnosis.

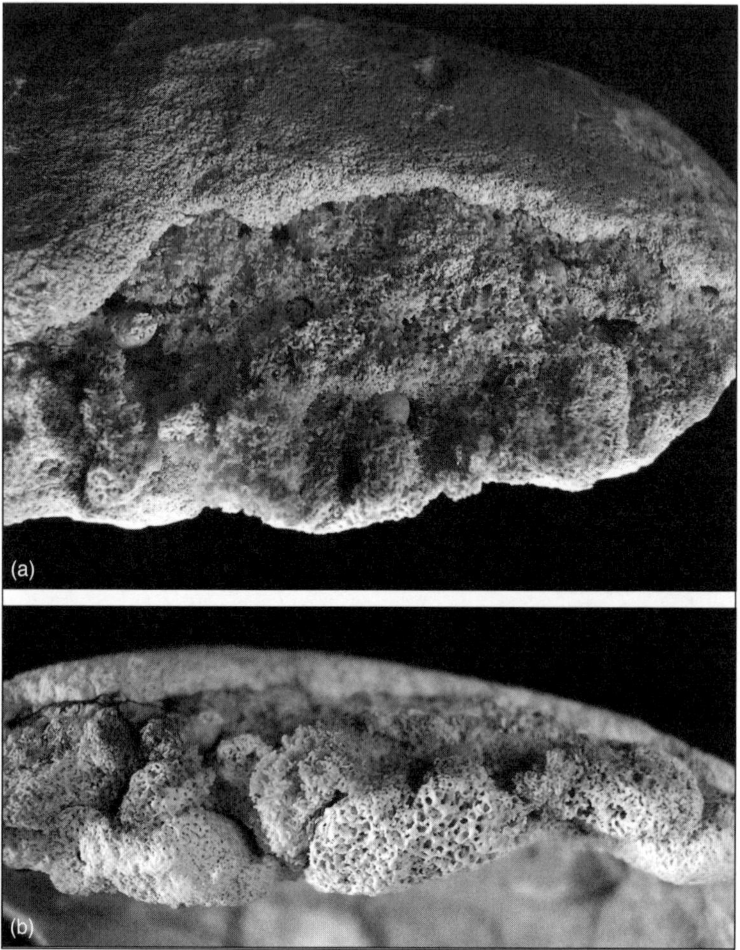

FIGURE 8.7 (a) View of ectocranium with thickened and expanded diploë in an archaeological cranium from Chelsea Old Church (OCU615). (b) Close up of coral-like 'cotton-wool' appearance of the diploë of the cranium. Photographs by Rachel Ives, courtesy of the Museum of London.

Dense and thickened trabecular structures were noted in one of the archaeological calcanei analysed by Aaron et al. (1992), typical of the accentuated trabecular pattern seen in the second phase of PDB. 'V'-shaped areas of radiolucency were noted by Stirland (1991) on radiographs of long bones of the individuals studied. A 'V'-shaped osteolytic focus was also noted on radiographs of one of the femurs analysed by Roches et al. (2002). Another classic radiological feature, 'picture-frame vertebrae' were identified in cases described by Roches et al. (2002) and Boylston and Ogden (2005). Figures 8.8 and 8.9 illustrate some of the radiological features seen in cranial bones of archaeological individuals.

TABLE 8.3 Radiological Features of PDB Observable in Archaeological Human Bone

Bones affected	Features	Stage	Code	Differential diagnosis	Sources
Cranium	Demarcated resorptive changes (affecting outer table)	1st	D		Greenfield (1990), Mirra et al. (1995a, 1995b)
	Cotton wool appearance – very variable density	2nd	D		
	Expansion of skull with coarse sclerotic bone present – particularly inner table	3rd	D		
Dentition	–				
Vertebrae	Widespread resorptive changes – radiolucency or decreased density	1st	G	Other conditions leading to osteopenia see Chapter 7 Osteoporosis	Mirra et al. (1995a, 1995b)
	Increased sclerosis of vertebral margins		D		
Ribs	–				
Pelvis	Thickening of iliopectineal line 'pelvic brim sign'	1st 1st	D		Mirra et al. (1995a, 1995b)
	Bone resorption	2nd	G		
	Marked sclerotic changes and accentuated trabecular pattern. Changes less clearly demarcated than in other skeletal areas.		D		
Long bones	Sharply demarcated resorptive changes V-shaped cutting cone	1st	D	Osteomalacia	Milgram (1990), Mirra et al. (1995a, 1995b), Roodman & Windle (2005)
	Development of coarse or sclerotic bone behind cutting cone	2nd	D		
	Bones wider and thicker than expected	3rd	G		
	Cutting cone no longer visible	3rd			
	Fatigue fractures	3rd	G		

Note: See Table 8.2 for definition of codes used in diagnosis.

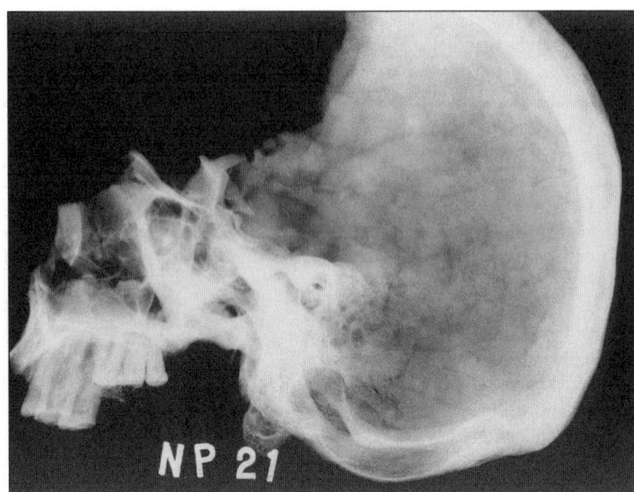

FIGURE 8.8 Skull of the individual from grave 21 at Norton Priory, the skull is dense and sclerotic with some areas of 'cotton-wool' appearance. Courtesy of Anthea Boylston, Alan Ogden and BARC, Archaeological Sciences, University of Bradford.

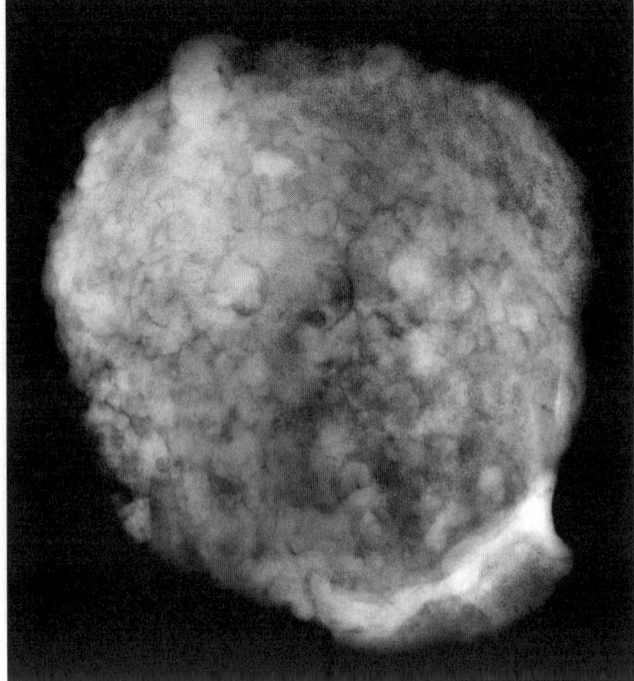

FIGURE 8.9 Radiograph of the skull of an old adult male from Chelsea Old Church (OCU615), London. Sclerotic areas of bone give the characteristic 'cotton-wool' appearance. Radiograph by Rachel Ives, courtesy of the Museum of London.

Histological Features of Paget's Disease

The study by Bell and Jones (1991) was the first to apply histology to the diagnosis of PDB in archaeological bone. Key histological features used in the diagnosis of PDB in archaeological human bone are listed in Table 8.4. Aaron et al. (1992) were subsequently able to identify the presence of 'mosaic' pattern bone (as shown in Figure 8.2) that enabled them to suggest a diagnosis of PDB using plain and polarised light microscopy. The fragility of pagetic bone can make obtaining a sample for histological analysis difficult (Aaron et al., 1992). Although, as reported by Bell and Jones (1991), poor preservation does not always result in poorly preserved microstructure of bones, and they state that this should not deter histological investigation. Due to the nature

TABLE 8.4 Histological Features of PDB Observable in Archaeological Human Bone

Bone affected	Features	Stage	Code	Differential diagnosis	Sources
Cortical	Increased osteoclastic resorption – hypervascularity	1st	G	Hyperparathyroidism, Osteopenia	Gruber et al. (1981), Milgram (1990)) and Mirra et al. (1995a)
	Mosaic pattern, due to scalloping caused by resorption followed by formation	End 1st	S	Metastatic carcinomas	
	Increased amounts lamellar bone	2nd	G		
	Sclerotic bone, with decreased vascularity	3rd	D		
Trabecular	Increased osteoclastic resorption	1st	G	Hyperparathyroidism, Osteopenia	Gruber et al. (1981), Mirra et al. (1995a) and Milgram (1990)
	Focally thinned trabeculae	1st	G		
	Thickened trabeculae at ends of long bones – showing mosaic pattern	2nd	S		

Note: It should be noted that the histological appearance of PDB is very variable, even within a single histological section. See Table 8.2 for definition of codes used in diagnosis.

of PDB only affected bones will show characteristic histological changes, and it may not be permissible to remove samples from the bones with pathological changes. It has been stated that the sensitivity of histology is far greater than that provided by radiology and that changes linked to PDB can be identified in bone in which no radiological features of the condition were visible (Gruber et al., 1981).

DIFFERENTIAL DIAGNOSIS

During the first phase of Paget's disease there are marked similarities with the hypervascularity seen in hyperparathyroidism (Nagant de Deuxchaisnes and Krane, 1964; Mirra et al., 1995a). Confusion can also occur due to the overlap in radiological appearance of primary pagetic lesions and malignant conditions (Hadjipavlou et al., 2002).

There are some rare types of bone metastasis that have been referred to as 'pagetoid', these conditions produce changes in bone that can be confused with PDB and should be considered when making a diagnosis (Roches et al., 2002). It has been suggested that small early lesions of PDB can resemble a tumour (Milgram, 1990:870). In each case the pattern of pathological changes across the skeleton will need to be considered in order to differentiate between the pathological conditions that may have caused the changes seen (see Box Feature 8.2).

Box Feature 8.2. The Contribution of Paleopathology to Modern Medicine

Issues on the current research agenda for those investigating PDB in the modern population include gathering of more data to determine if the prevalence of the condition really is declining. Through greater awareness of skeletal changes with PDB, and gathering of additional data from archaeological skeletal material of different ages, it may be possible for data from paleopathological studies to contribute to this research. Samples of past populations are available from a massive spatial and temporal range, and the data that can be provided by the thorough study of these individuals is enormous.

A recent example of the value of paleopathological investigations is provided in the work of Boylston and Ogden (2005). These researchers were able to state that the recently discovered cases from Norton Priory, England mean that there cannot be a link between arsenic poisoning and PDB, as suggested by Lever (2002). These cases pre-dated the arrival of the cotton industry in the northwest of England.

CONCLUSIONS

There are a number of exiting areas of further research for those involved in the study of PDB to engage with, but wider consideration of all possible types of evidence available including information derived from the study of pathology in animal bone, should assist with the formulation of ideas relating to this fascinating condition. Information derived from paleopathological cases of PDB already provide interesting confirmation of the number of the trends seen in the condition within the modern population. However, as more data are accumulated it will become possible for those engaged in bioarchaeological research to make important contributions to the on-going debate on the aetiology of this condition.

Chapter 9

Miscellaneous Conditions

In the following chapter a number of metabolic bone diseases that have received relatively little attention from bioarchaeologists are discussed. As the information contained within this chapter demonstrates many of these conditions have considerable potential to contribute to a wide range of investigations. Limited work on the conditions to date means that coverage of the conditions here is more restricted and paleopathological cases have not been tabulated as in previous chapters.

FLUOROSIS

Reports of environmental pollution and natural disasters affecting the livelihoods of people around the world are increasingly commonplace. However, such disasters will also have occurred in the past (as discussed in Chapter 4), and investigations of fluorosis in archaeological human bone may assist exploration of such events. The beneficial effects of fluoride have been widely reported in relation to improvements in dental health (Hausen, 2004; Jones et al., 2005), but ingestion or inhalation of high levels of fluoride can have serious health consequences. For example, osteosclerosis can develop (Whyte, 1998). With this condition bones increase in density yet they become prone to fractures due to structural problems. From a review of current literature it is clear that investigations of fluorosis have a potentially important contribution to make to a wide range of bioarchaeological investigations involving environmental and health issues.

Levels of fluoride of <2 ppm can cause tooth enamel to become mottled during tooth formation, levels of 3.15 ppm have been shown to lead to low IQ levels in children, higher levels will be toxic, and at 250–450 ppm can lead to death (Shomar et al., 2004). However, the exact levels at which fluoride becomes toxic vary according to a wide range of factors and there is no global fixed level (Littleton, 1999).

CONSEQUENCES OF FLUOROSIS

Dental Fluorosis

Dental fluorosis will be one of the first indicators of fluorosis. In clinical work one of the most widely used methods of classifying and recording dental fluorosis is the dental fluorosis index (DFI) formulated by Dean (1942). This system allows the amount of enamel alteration on each tooth to be classified. The first evidence of dental fluorosis is a loss of sheen, followed by the development of white patches that later discolour and may be accompanied by pitting (Reddy and Prasad, 2003). The pitting that occurs with fluorosis is a hypoplastic defect (Hillson, 1996:171).

Skeletal Fluorosis

The effects of excessive fluoride on physiology and bone have been fully outlined by Littleton (1999). Toxic levels of fluoride can have an impact on the processes of bone formation and resorption. Resorption can be disrupted either because osteoclasts are affected, or bone mineral is less easily resorbed due to high fluoride concentrations (Greenfield, 1990). Abnormal bone is formed in individuals with fluorosis, which has an increased density (Littleton, 1999). The increase in density tends to be diffuse and could complicate the diagnosis of a number of the other pathological conditions. Recently, there has been particular attention paid to the effects of fluorosis on detecting age-related bone loss in post-menopausal women (Yildiz and Oral, 2003). BMD levels were higher in the women with fluorosis than in the age-matched controls (Yildiz and Oral, 2003). However, bone formed during skeletal fluorosis will not be of such good quality as that formed in healthy conditions, and so individuals with fluorosis may still be at risk of fractures linked to age-related bone loss of the type discussed in Chapter 6.

In cases of skeletal fluorosis there can also be ossification of soft tissues associated with the skeleton. For example, two cases of ossification of the ligamentum flavum, which led to the development of myelopathy, have been reported in recent clinical cases (Muthukumar, 2004). In one individual there were also ossifications of the interosseous membrane. Marked osteophytosis at areas of soft tissue attachment are relatively common features of fluorosis (Greenfield, 1990; see Figure 9.1).

The entire skeleton will be affected by changes linked to the disruption of bone formation and resorption (Littleton, 1999). One important point made in the study by Savas et al. (2001) is that the radiological features of skeletal fluorosis have been found to vary in different parts of the world.

The various skeletal changes attributable to fluorosis often lead to general bone pain, reduction in mobility of joints and stiffness. Figure 9.2 illustrates the new bone formation that can occur around joints of the spine in cases of fluorosis, along with higher levels of bone density. Whyte (1998) provides a fuller list of clinical presentations. It is suggested in the study by Littleton (1999) that skeletal fluorosis may be more marked in individuals undertaking heavy manual labour, and these and other interpretations of the social and economic consequences of the condition are discussed in her paper.

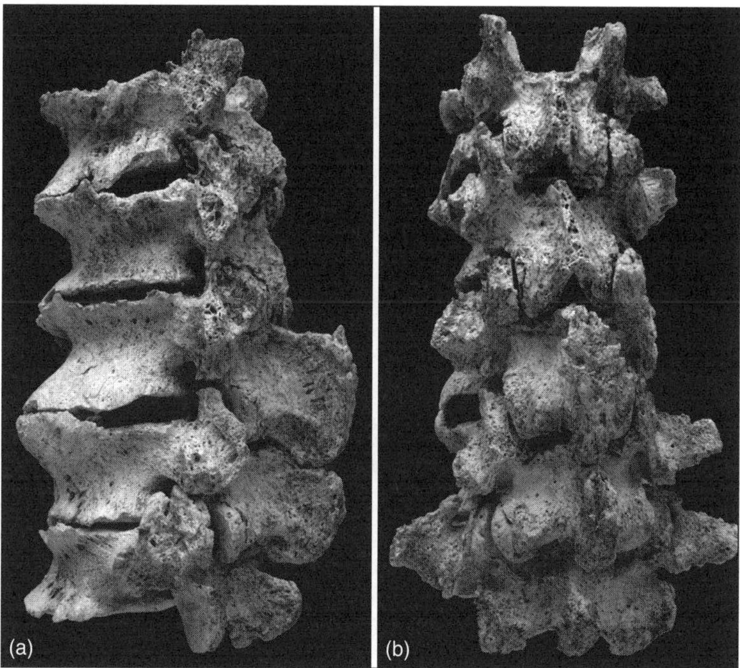

FIGURE 9.1 (a) Severe osteophytosis of lumbar vertebra (L1–L5). Breaks between L1–L2, and L3–L4 post-mortem. Note ankylosis of vertebral arch between L1–L3. Courtesy of Judith Littleton and Wiley-Liss. (b) Posterior view of the same section of spine. Marked new bone formation can be seen at sites of tissue attachment on the spinous processes and transverse processes. Courtesy of Judith Littleton.

CO-MORBIDITIES

A number of conditions can develop alongside fluorosis, particularly where fluorosis has been present since infancy. It is reported by Greenfield (1990:438) that fluorosis affecting children whilst still in the womb can cause features such as bowed legs to develop once they start walking. Co-morbidity with a range of diseases such as osteomalacia and osteoporosis results in a spectrum of bone changes (Greenfield, 1990). It has also been reported that radiological changes can resemble rickets (as discussed in Chapter 5) and hyperparathyroidism, particularly where dietary calcium is inadequate (Littleton, 1999:466).

FLUOROSIS IN THE MODERN PERSPECTIVE

There are a number of regions of the world in which levels of fluorides in water, soil and plants that are consumed will reach levels that can cause fluorosis. Tea in particular has been noted to contain high levels of fluorides

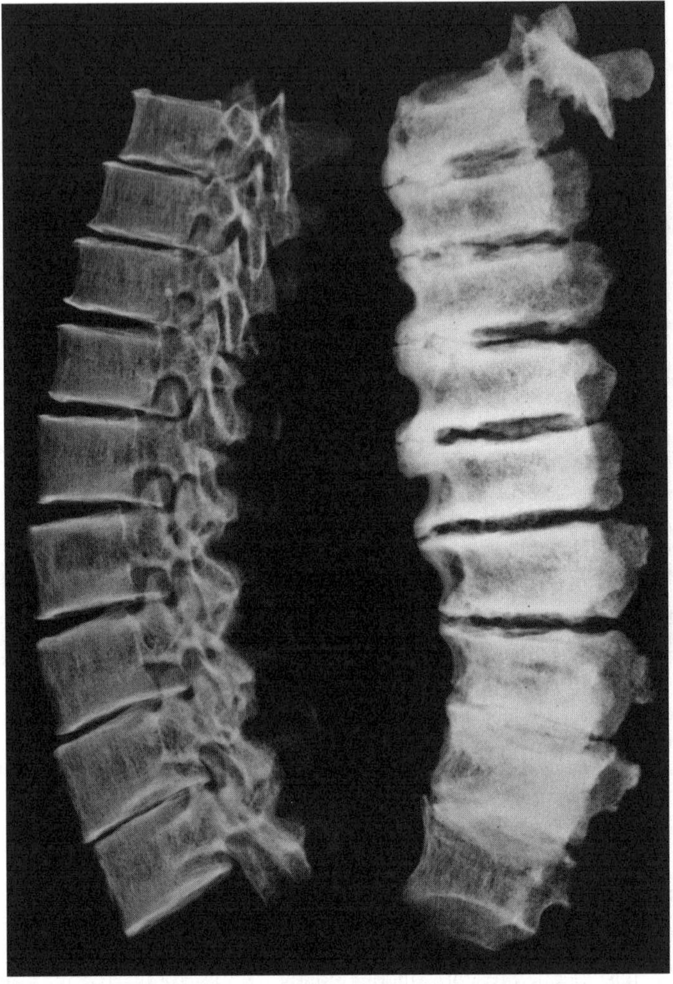

FIGURE 9.2 X-ray of the spine of an individual affected by fluorosis (right) compared to a normal spine (left). Increased density and ossification of sites of soft tissue attachment can be seen on the right. Courtesy of Judith Littleton and Wiley-Liss.

(Yam et al., 1999; Shomar et al., 2004). In affected areas fish can also contain high levels of fluorides (Shomar et al., 2004). The development of fluorosis is a complex area, and a number of factors have been reported to contribute to the development of the condition. For example, altitude may have an effect on the development of fluorosis, with individuals living at higher altitude having a greater prevalence (Martínez-Mier et al., 2004). Although there are many factors that can increase the risk of fluorosis there are also some that can decrease the amount of fluoride absorbed by the body. For example, a study by Khandare et al. (2002) found that ingestion of tamarind (*Tamarindus indicus*) resulted in significantly higher levels of excretion of fluoride. Eating tamarind regularly would probably help delay the effects of fluorosis. It is likely that other foods will also alter the absorption of fluoride.

In areas of Indonesia, the Gaza Strip (Shomar, et al. 2004), parts of India (e.g. Reddy and Prasad, 2003; Bhargavi et al., 2004), many countries in East Africa (Wondwossen et al., 2003), Senegal and other areas of West Africa (Sy et al., 2000), areas of Mexico (Martínez-Mier et al., 2004), southern Turkey (Yildiz and Oral, 2003), Saudi Arabia, some Arab states (Almas et al., 1999) and China have also been reported to have problems with fluorosis. In these regions fluorides probably come from minerals in the bedrock. As a result of levels of fluorides in bedrock certain geographic areas of the world will be more prone to the development of fluorosis within the population. The field of medical geology, in which these types of health issues linked to the environment, is a developing area of research (e.g. Davies et al., 2005a). However, fluorides do not just enter the food chain from the bedrock, and there are reports of fluorosis developing as a result of natural pollution (Löhr et al., 2005). The report by Löhr and co-workers provided a review of the potential effects of volcanic activity on water supply, which include the contamination of water supplies with metals and acidity of water sources. Pollution in East Java caused levels of fluoride in water from wells to exceed recommended levels, and the use of contaminated water for irrigation increased levels of fluoride in crops (Löhr et al., 2005). It is clear that volcanic activity can have very serious consequences for communities in the surrounding region. Such pollution is likely to result in adverse effects on a wide range of plants and animals also living in the affected areas.

Fluorosis can also develop in animals (Patra et al., 2000), leading to a range of pathologies and mortality. It is more likely to occur in domesticated animals, as stored water that has undergone evaporation is liable to contain higher levels of fluoride (Littleton, personal communication). Dependence on stored water sources for drinking and irrigation purposes is a frequent occurrence for individuals in many parts of the world.

ANTHROPOLOGICAL PERSPECTIVES: FLUOROSIS

A number of recent reports have discussed fluorosis in terms of the social and psychological consequences of dental fluorosis (e.g. Wondwossen et al., 2003). It is difficult to evaluate the extent that such changes may have caused similar problems amongst past societies. The fact that individuals in the past would not have had information provided by international media sources would have meant that those in areas with endemic fluorosis might not have been aware to quite the same extent as people today about the lack of normality of their dental defects. Also cosmetic dentistry that would alter the appearance of affected teeth would have been completely unknown. However, it is clear that amongst recent populations in areas affected by fluorosis there is considerable dissatisfaction with the appearance of their dentition among many young individuals, and this has caused significant social problems (Wondwossen et al., 2003). Greater mobility amongst pre-agricultural societies would have meant that problems associated with fluorosis would have been far less common than those seen amongst settled societies.

TABLE 9.1 Macroscopic Features of Fluorosis in Adults

Bones affected	Features	Code	Differential diagnosis	Sources
Cranium	Becomes thick and heavy – gradual obliteration of diploe	D	Paget's	Littleton (1999)
	Irregular margins – foramen magnum	D		
Dentition	Development of white patches that may discolour, possible associated pitting	G	Staining Other hypoplastic defects	Dean (1942), Reddy & Prasad (2003)
Vertebrae	Ossification of ligamentum flavum and other spinal ligaments	D	DISH, AS	Littleton (1999)
	Marked osteophyte formation at areas of soft tissue attachment	G	Paget's	
	Fusion of vertebrae	G		
	Enlarged vertebrae	G		
Ribs	Larger bones than normal	G	DISH	Greenfield (1990), Littleton (1999)
	Rough bone surfaces	G		
	Bony projections into intercostal muscles	D		
	Ankylosis costovertebral and costosternal joints	D		
Pelvis	Ossification of insertions around iliac spine	G		Avioli and Krane (1998)
Long bones	Bones become heavy and irregular – development of enthesophytes at muscle and tendon insertions	D		Littleton (1999)

Note: AS, ankylosing spondylitis; DISH, diffuse idiopathic skeletal hyperostosis. Note all sites of tendon and ligament insertion can be affected and only some areas are dealt with in this table. The features listed demonstrate the skeletal changes useful for pathological identification. Diagnosis can be determined on the strength of each feature: S denotes strongly diagnostic feature, S features are normally required for a diagnosis. D denotes diagnostic feature, the presence of multiple D features are required for a diagnosis. G denotes general changes, can occur in many of the metabolic bone diseases, as well as in other conditions. G features can aid a diagnosis together with S or D features but cannot be used alone to suggest a diagnosis. Commonly occurring conditions that display these pathological features are listed to aid differential diagnosis, although this list is not exhaustive.

PALEOPATHOLOGICAL CASES OF FLUOROSIS

Few studies of fluorosis have been undertaken on archaeological human bone (and to the authors knowledge, none have been undertaken on animal bone). One of the few studies undertaken was that by Judith Littleton (1999) from the island of Bahrain in the Arabian Gulf. As Littleton (1999) points out, in areas with high levels of fluorides, fluorosis needs to be considered in the differential diagnosis of pathological conditions. Due to the interest of Littleton and her co-workers there have been a number of published cases of fluorosis from the Arabian Gulf (Littleton and Frolich, 1989; Frolich et al., 1989; Littleton, 1999). Key features of fluorosis at the macroscopic, radiological and histological level in adult individuals from archaeological sites are presented in Tables 9.1–9.3.

Dental fluorosis will be far more common than skeletal fluorosis, which occurs where levels are <5 ppm (Reddy and Prasad, 2003). The difficulty with archaeological dental material is that staining can occur post-mortem and dental plaque, which researchers would be reluctant to remove from teeth, could obscure markers linked to fluorosis. Hillson (1996) suggests that

TABLE 9.2 Radiological Features of Fluorosis in Adults

Bones affected	Features	Code	Differential diagnosis	Sources
Cranium	Sclerosis of skull base and posterior clinoids	D		Greenfield (1990), Littleton (1999)
Dentition	–			
Vertebrae	Narrowing of intervertebral disk space between fused vertebrae	D	Osteomalacia	Greenfield (1990), Avioli & Krane (1998), Littleton (1999)
	Thickening of trabeculae followed by uniform sclerosis	D		
Ribs	Thickening of trabeculae followed by uniform sclerosis	D		Greenfield (1990), Littleton (1999)
Pelvis	Thickening of trabeculae followed by uniform sclerosis	D		Greenfield (1990), Choubisa et al. (2001)
Long bones	Osteopenia at ends of long bones – early sign particularly in younger individuals	G	See Chapter 7 Scurvy	Greenfield (1990), Littleton (1999)

Note that with fluorosis there is marked radiodensity of many bones. See Table 9.1 for definition of codes used in diagnosis.

TABLE 9.3 Histological Features of Fluorosis in Adults

Tissue type	Features	Code	Differential diagnosis	Source
Cortical bone	Poorly mineralised osteons	G	Osteomalacia	Frolich et al. (1989)
Trabecular bone	Thickening of trabeculae	G		

Note: See Table 9.1 for definition of codes used in diagnosis.

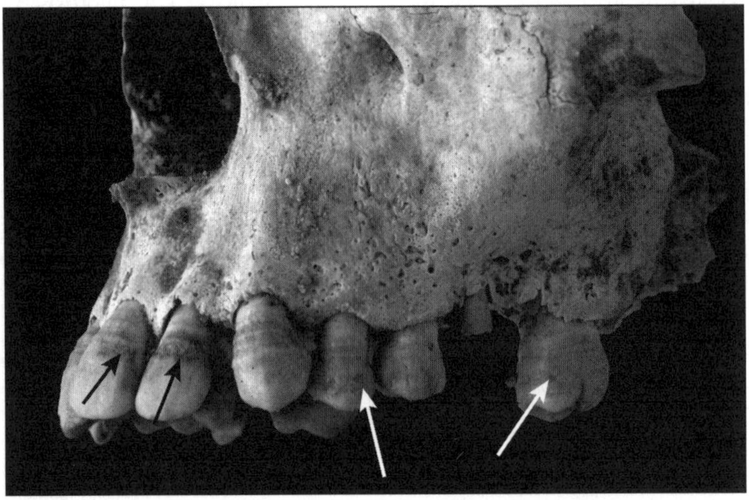

FIGURE 9.3 Ridges of hypoplastic defects in the enamel (indicated with white arrows) of the teeth of an individual from Bahrain (250 BC to AD 250). Staining on the incisors (indicated with black arrows) is probably linked to fluorosis. Courtesy of Judith Littleton.

it would be difficult to diagnose fluorosis based entirely on the presence of enamel defects in archaeological teeth. However, in a number of cases changes consistent with fluorosis have been reported from archaeological sites (Lukacs et al., 1985; Littleton and Frolich, 1989; Littleton, 1999). Littleton (1999) took care to exclude the possibility of post-mortem staining when analysing archaeological material, and provides full information on procedures used to investigate dental fluorosis in archaeological skeletal material (see Figures 9.3 and 9.4). If skeletal fluorosis can be identified in archaeological human bone then it is almost certain that dental fluorosis will also be present.

One reason why there have been so few reports of cases of fluorosis in past populations is that to date relatively little paleopathological work has been undertaken in areas of the world where high levels of fluoride are a problem.

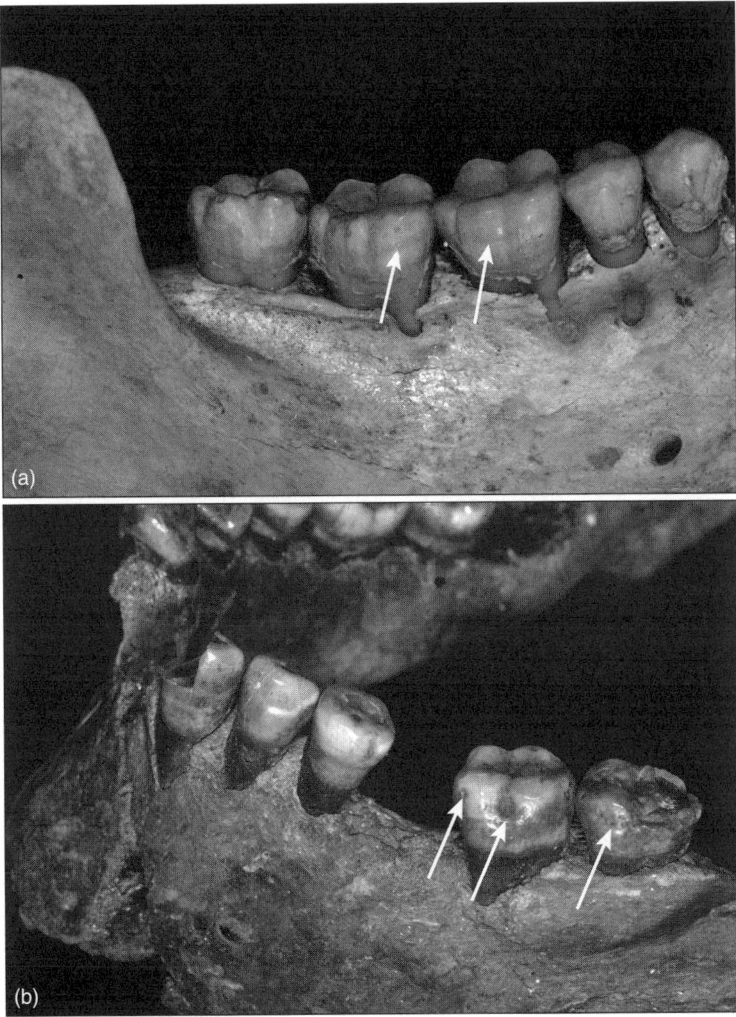

FIGURE 9.4 (a) Bands of discolouration on the molars of an individual from an archaeological site in Bahrain (250 BC to 250 AD). (b) Hypoplastic defects in the form of pitting on molars of an individual from the site in Bahrain. Courtesy of Judith Littleton.

However, as the brief review presented here demonstrates diagnosis of this condition has the potential to contribute considerable amounts of information on a number of important aspects of past societies.

Other Conditions Linked to Intoxication

There are quite a wide range of substances that, like fluoride, can cause intoxication, and coverage of a range of conditions including lead, mercury, arsenic poisoning and mycotoxicosis are provided by Aufderheide and Rodríguez-Martín (1998).

HYPERPARATHYROIDISM

Hyperparathyroidism can occur when the amount of PTH in the body is no longer adequately regulated by the thyroid glands. As discussed in Chapter 3, PTH plays an important role in the regulation of calcium, so a disruption in these processes will have important consequences for the skeleton. In hyperparathyroidism intestinal absorption of calcium is reduced and there are elevated PTH levels circulating in the body (see further, Civitelli et al., 1998).

CAUSES OF HYPERPARATHYROIDISM

Various factors can cause hyperparathyroidism, and the condition is usually described as being primary, secondary or tertiary depending upon the causative factors. However, some forms of hyperparathyroidism are only relevant to studies involving contemporary groups, as they are a side affect of modern medical treatments (e.g. taking thyroid hormone supplements). A full discussion of the range of such conditions is provided by Ross (1998).

Primary Hyperparathyroidism

Primary hyperparathyroidism is caused by excess production of PTH, by one or more of the parathyroid glands, but most cases are caused by a single adenoma (Milgram, 1990; Bilezikian, 1999). Although the condition is usually asymptomatic in its early stages, with time a number of symptoms may develop. In advanced cases of primary hyperparathyroidism the skeleton will be affected, as excess levels of PTH will lead to an increase in osteoclastic activity and bone resorption (Milgram, 1990).

Secondary Hyperparathyroidism

There are a number of possible causes of secondary hyperparathyroidism including renal problems. In the modern population renal insufficiency is the most common cause of secondary hyperparathyroidism (Kahn, 2005), but without medical treatments it is not clear how many of these individuals would have lived long enough to develop the condition in past communities. Secondary hyperparathyroidism can also occur with vitamin D deficiency (Kahn, 2005) as is illustrated in a case reported by Mays et al. (2007). As discussed in Chapter 5, there are a number of past communities in which vitamin D deficiency could have been a problem and secondary hyperparathyroidism should certainly be considered when examining evidence of vitamin D deficiency in archaeological skeletal material. Tertiary hyperparathyroidism occurs in cases of secondary hyperparathyroidism where PTH secretion continues even after the underlying cause has been eliminated (Klein, 2006).

CONSEQUENCES OF HYPERPARATHYROIDISM

As discussed by Mays et al. (2001) the effects of hyperparathyroidism are quite variable. Some individuals will experience limited symptoms, but others will suffer from symptomatic bone disease, and may also develop related problems, such as kidney damage (Milgram, 1990). A wide variety of neuromuscular and neuropsychiatric symptoms have been linked to primary hyperparathyroidism (Freitas, 2002). Bone pain that can progress to a point where the patient is bedridden has been reported in secondary hyperparathyroidism (Khan, 2005).

ANTHROPOLOGICAL PERSPECTIVES: HYPERPARATHYROIDISM

The potential contribution of diet to causing secondary hyperparathyroidism is complex and incompletely understood, and requires further investigation in the future (Gannage-Yared et al., 2005; Mehrotra et al., 2006). A recent clinical study has linked obesity and vitamin D deficiency in black individuals to an associated increased risk of developing secondary hyperparathyroidism (Yanoff et al., 2006).

Following a number of recent studies, the characteristic features of hyperparathyroidism (both primary and secondary) in archaeological human bone are now far better understood. This will undoubtedly make identification of cases in the future easier. What bioarchaeologists now need to do is give further consideration to possible interpretations of finding cases of hyperparathyroidism from different social and cultural contexts. To date there has been little interpretive work, but as more cases are reported it should become easier to expand on this area of research. There are a number of biological, social and cultural factors that will impact on the expression of these conditions, all of which could be of interest to those in a number of fields of anthropology. The study by Gannage-Yared et al. (2005) found a link between urban living, gender, diet and risk of hyperparathyroidism. In particular the link between secondary hyperparathyroidism and vitamin D deficiency has a number of implications for interpretations of social and cultural aspects of past societies, and there are important issues relating to co-morbidities suffered by those who have primary hyperparathyroidism. However, non-human animals can also be affected by hyperparathyroidism (Feldman et al., 2005; Mehrotra et al., 2006) and to date little work has been undertaken considering this condition in studies of animal bone.

PALEOPATHOLOGICAL CASES OF HYPERPARATHYROIDISM

There has been a growing awareness amongst researchers of the skeletal changes that could result from the various types of hyperparathyroidism and a number of cases have now been reported in the literature (e.g. Mays et al., 2001; Mays et al., 2007). The small numbers of reported cases are likely to be linked to the subtle nature of many of the skeletal changes associated with hyperparathyroidism, particularly during the earlier stages of the condition.

A case of possible primary hyperparathyroidism was reported in an adult individual from an Egyptian site in the Dakhleh Oasis by Cook et al. (1988), and Zink et al. (2005) have now reported on a Neolithic case from Germany. The first possible case of secondary hyperparathyroidism was described in a spontaneously mummified body from South America (Blackman et al., 1991). More recently Mays et al. (2007) have reported on a possible case of secondary hyperparathyroidism linked to vitamin D deficiency in a juvenile individual from a site in England.

Hyperparathyroidism was considered as part of the differential diagnosis in the case of a woman from a sixth century burial context in the Negev Desert (Foldes et al., 1995). It is probable that the skeletal changes represent hyperparathyroidism secondary to vitamin D deficiency, potentially caused by a number of social and cultural practices (Foldes et al., 1995). Foods with high phytate content, combined with types of clothing worn and activities of women outside the home could have contributed to vitamin D deficiency (see Chapter 5 for details).

DIAGNOSIS OF HYPERPARATHYROIDISM IN ARCHAEOLOGICAL BONE

Key features of hyperparathyroidism at a macroscopic, radiological and histological level are provided in Tables 9.4–9.6. Hyperparathyroidism is likely to result in bone loss and an overall deterioration in the quality of bone present (Ross, 1998). As discussed previously in relation to osteomalacia, this could contribute to poor preservation of archaeological skeletons of individuals who suffered from these conditions. In the skeleton with possible hyperparathyroidism reported on by Mays et al. (2001), the bones present were friable and poorly preserved.

The subtle nature of hyperparathyroidism means that radiological investigation and histological examination are likely to be required in order to suggest a diagnosis (Mays et al., 2001). Zink et al. (2005) and Mays et al. (2007) confirmed that histology was a useful means of identifying the patterns of osteoclastic resorption diagnostic of hyperparathyroidism.

PELLAGRA

Pellagra is a nutritional deficiency disease caused by a niacin deficiency and has recently received renewed attention, in part due to its appearance in refugees and

TABLE 9.4 Macroscopic Features of Hyperparathyroidism

Bones affected	Features	Code	Differential diagnosis	Sources
Cranium	–			
Dentition	–			
Vertebrae	Wedging and biconcavity	G	Osteoporosis, osteomalacia	Jaffe (1972), Milgram (1990)
Ribs	Sub-periosteal bone resorption – ribs (upper border)	G	Osteomalacia	Greenfield (1990)
Pelvis	Sub-periosteal bone resorption – ischial tuberosity, pubic symphysis Sclerosis of subchondral bone, erosive changes sacroiliac joint Triradiate deformity	G	Osteomalacia	Greenfield (1990), Mays et al. (2001)
Long bones	Sub-periosteal bone resorption – femur, tibia and humerus (proximal medial aspects) Sclerosis of subchondral bone	G G	Osteomalacia	Greenfield (1990), Mays et al. (2001)
Metacarpals, phalages and clavicles	Sub-periosteal bone resorption – middle phalanges of 2nd and 3rd finger particularly affected	D		Greenfield (1990), Potts (1998)

Note: See Table 9.1 for definition of codes used in diagnosis.

displaced peoples (Mason, 2002 see discussion in Box Feature 9.1). However, there are a number of reports of pellagra from historical contexts. Sub-clinical pellagra has been suggested to have been widespread in Ireland following the use of Indian meal (a type of cornmeal) distributed as relief rations during the famine of 1845–1848 (Crawford, 1988). As was discussed in Chapter 4, on scurvy, micronutrient deficiency diseases are relatively common in certain situations. Pellagra and scurvy are suggested to have co-occurred in African-Americans in the Southwest of America in the twentieth century (Davidson et al., 2002). These conditions

TABLE 9.5 Radiological Features of Hyperparathyroidism

Bones affected	Features	Code	Differential diagnosis	Source
Cranium	'Salt and pepper' appearance of skull	D		Greenfield (1990), Potts (1998), Bilezikian (1999)
	Both tables loose sharp outline	G	Osteoporosis	
Vertebrae	Generalised osteopenia with relatively dense appearance of endplates	G	Chapters 6 & 7	Greenfield (1990) Aufderheide & Rodríguez-Martín (1998)
	Increased volume of trabecular bone	G		
Ribs	–			
Pelvis	Increased volume of trabecular bone	G	Osteomalacia	Potts (1998)
Long bones	Tunnelling resorption in cortical bone	D	Osteomalacia	Jaffe (1972), Resnick & Niwayama (1988), Greenfield (1990), Mays et al. (2001)
	Osteopenia	G		
	Reduced cortical thickness – increased volume trabecular bone	G		
	Loss of cortical definition	G		
	Fractures	G		
	Brown tumours may be visible (well-defined radiolucencies)	D		
Metacarpals and phalanges	Subperiosteal bone resorption	D		Greenfield (1990)

Note: Brown tumours may be present at a variety of locations. See Table 9.1 for definition of codes used in diagnosis. Specific radiological feaures are rarely noted clinically, certainly in the early stages.

frequently lead to high mortality rates and nutritional deficiency diseases often go unrecognised even today. It is stated by Mason (2002) that the quantity of micronutrients in diets provided to refugees and displaced people is still not considered when thinking about provision of food for refugees. This is despite the now substantial information on the importance of this aspect of diet, and conditions that result from a deficiency of such nutrients. In the past such deficiency diseases would almost certainly have been present when populations suffered adverse conditions or were displaced.

In a brief history of pellagra by Rajakumar (2003), it is reported that the Spanish physician Don Gasper Casal, first described the condition in 1735 in

TABLE 9.6 Histological Features of Adult Hyperparathyroidism

Bones affected	Features	Code	Differential diagnosis	Sources
Cortical bone	Resorption of Haversian canals	D	Acromegaly, Hyperthyroidism	Bogumill & Schwamm (1984), Resnick & Niwayama (1995), Mays et al. (2007)
	Numerous reversal lines	G		
	Eroded perimeters, bite-like areas of resorption – 'cookie-bite' defects	D		
Trabecular bone	Removal of central portions of trabecular elements	D		Bogumill & Schwamm (1984), Resnick & Niwayama (1995), Mays et al. (2007)
	Numerous reversal lines	G		

Note: See Table 9.1 for definition of codes used in diagnosis.

poor peasants who consumed significant amounts of maize. Pellagra was first recognised in the U.S. in 1902, where it was seen amongst poorer sections of the community who would have eaten maize as a staple part of their diet (Rajakumar, 2003). To date very little attention has been paid to pellagra in the paleopathological literature, but the available evidence, particularly relating to the transition to maize-based diets, has been reviewed by Paine and Brenton (2006a).

Paine and Brenton (2006b) noted that pellagra has an important impact on bone remodelling. They observed very few fragmentary or secondary osteons in older individuals with these conditions (Figure 9.5). As will be apparent from Chapter 3, the pattern seen in individuals with pellagra is very unusual. Paine and Brenton (2006b) demonstrated that histological aging techniques underestimated the age at death of individuals with pellagra, and to a lesser extent those with general malnutrition.

This work raises a number of important points for those engaged in the study of paleopathology and past human health. It is clear that a number of the metabolic bone diseases and dietary deficiencies can cause a variation in bone turnover rates (producing either higher or lower turnover than would otherwise be expected). This has important implications for the application of histological age determination techniques to archaeological human remains, or those from forensic contexts (e.g. Maat and Bond, 2007). Such remains have a high chance of including individuals who may have had a deficient diet.

Pellagra is not a condition that is covered by any of the standard texts on paleopathology, probably because the macroscopic skeletal changes produced by this condition are not specific, and suggesting a diagnosis in human remains

Box Feature 9.1. Anthropological Investigations of Displaced Peoples

There are a number of metabolic bone diseases that are frequently observed in displaced peoples. Pellagra and starvation covered in this chapter have been reported in various recent studies, but conditions such as scurvy covered in Chapter 4 are also a problem in such groups (e.g. WHO, 1999b).

Reports of displaced peoples are now quite common in the media, both those that are displaced within a country and those who are forced by a range of circumstances to leave their home countries. Many of the issues involved with such communities have been investigated by social, cultural and medical anthropologists (e.g. Johnston et al., 1985; Cameron, 1991). However, it is likely that natural disasters and conflicts in past communities would also have led to situations in which there were significant numbers of displaced peoples. Such individuals often have to live in poor quality housing and have poor access to adequate nutrition. In addition to these health issues, where wars and conflicts were the reason for their displacement people may also face psychosocial problems.

From looking at a recent reports of displaced peoples (e.g. Charnley, 1997; Kett, 2005) it can be seen that such individuals face a complex set of problems that will have a significant impact on their long-term health and well being. Occurrence of high levels of osteopenia and co-occurrence of conditions such as scurvy in archaeological collections could be an indication that the individuals involved may have experienced some form of socio-cultural disruption. There have been a number of reports that have used paleopathological evidence to suggest the possible occurrence of some kind of natural disaster or event. However, to date such reports have not fully integrated the range of possible information available on modern displaced peoples, or considered the wider socio-cultural factors that large numbers of displaced people bring. This is a potentially valuable area of research, which deserves further consideration in the future.

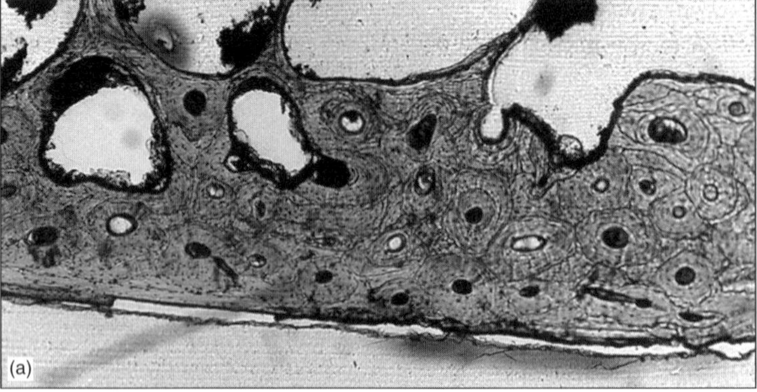

FIGURE 9.5 (a) Rib microanatomy, intact secondary osteons seen in a 70-year-old female who died from pellagra. Note that there are very few fragmentary osteons seen in this photo. A grey filter was used to take this photo (Paine and Brenton, 2006a: Figure 1).

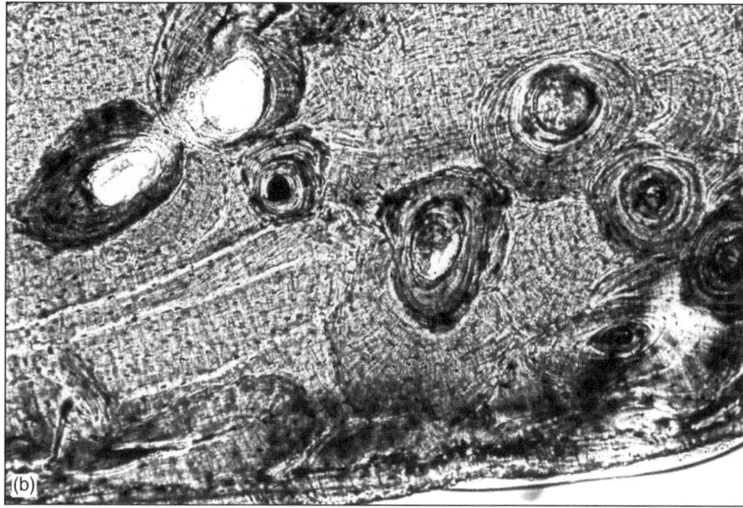

FIGURE 9.5 (*Continued*) (b) Rib microanatomy, a low density of secondary osteons seen in a 45-year-old female who died from pellagra. Blue and grey filters were used to create this photo (Paine and Brenton, 2006a: Figure 2). Courtesy of Robert Paine and Barrett Brenton.

would be difficult, even with histological analysis. However, the condition has important consequences for human health and it should be considered in investigations of the health of people from regions where it is common. There have been numerous past communities that will have been affected by poverty and poor nutrition; factors such as alcoholism, malabsorptive states and food fads can also have an impact on the appearance of pellagra (Rajakumar, 2003). There are few reports of pellagra from areas of Central American and Mexico despite a heavy reliance on maize in the diets of individuals from these areas and poverty within these regions. It is suggested (Rajakumar, 2003) that this might be related to a tradition of pre-soaking maize in limejuice, which makes more niacin available. As in any study involving archaeological human remains, it is clear that it is also important to consider a wide range of social and cultural factors that may impact on health. For example agricultural practices, technologies available for food preparation and traditions pertaining to food combinations or avoidance, and other cultural and social factors relating to the activities of individuals within a society and the risks and activities that they would be exposed to.

STARVATION

Quite a lot of space is devoted to the effects of starvation upon bone by Ortner (2003), and it is likely that such dietary factors were important in the health of past populations. Deficiencies are unlikely to occur singularly and people deficient in one aspect of their diet are likely to also be deficient in other areas.

Box Feature 9.2. Malnutrition, Starvation and Osteoporosis

Malnutrition, disordered eating and/or starvation can cause osteopenia. Starvation can result in amenorrhoea, or premature cessation of menstruation removing the inhibition of bone resorption (Riggs et al., 1998; Pacifici, 2001; see Chapter 6). Under-nutrition also triggers a decline in osteoblast function and therefore bone formation (see Shires et al., 1980; Grinspoon et al., 1995; Orden et al., 2002).

Extremes of food availability are well documented in recent times. One out of every five people in the developing world is currently chronically undernourished, with an estimated 777 million individuals affected (FAO, 2004). The risk of amenorrhoea and osteopenia may be exacerbated by lack of energy and decreased physical activity. Many eating disorders have also been linked with amenorrhoea, for example anorexia nervosa. Where such conditions have a peak occurrence during puberty, there may be detrimental impacts on the attainment of peak bone mass (LaBan et al., 1995; Rome and Ammerman, 2003:422; Stone et al., 2006:835).

Skeletal evidence of osteoporosis has been causatively linked with dietary deficiencies or malnutrition in past populations (see Martin et al., 1985; Ortner et al., 2006:100, 2007:192). Periods of malnutrition may be cyclical, reflecting the seasonal influences of climatic or environmental changes on subsistence practices (Cohen, 1989:97; deMenocal, 2001; Ubelaker and Newson, 2002:351; Fagan, 2005; Allen, 2006). Severe environmental damage (e.g. natural disasters) may have long-lasting disruptions to food supplies (see also Chapter 4). For example, extensive sea-water flooding by the Asian tsunami in 2004 destroyed over 61,000 ha of agricultural land in Indonesia, affecting 92,000 farms and small enterprises (FAO, 2005:2). In the Maldives, over 73% of agricultural land became unusable owing to salted mud, and nearly 12,000 people were displaced to other islands (FAO, 2005:2; see further Box Feature 9.1). Severe effects on health, including osteopenia, may result in individuals affected by such disastrous events. Evidence for natural disasters in affecting past populations does exist (e.g. Ubelaker and Newson, 2002:349; Fagan, 2005) and without the modern aid intervention, the consequences would have been very severe for many past communities.

The ranges of possible effects of starvation are discussed in more detail in Box Feature 9.2.

RARE METABOLIC BONE DISEASES

The following conditions are very rare and are unlikely to be reported very frequently in anthropological investigations of either past or present communities. However, a range of such conditions will be briefly listed here, as they may occasionally be encountered.

Hyperostosis

The term 'hyperostosis' is used to describe conditions in which there is enlargement of the outer portion of bones. However, a range of conditions in which there is thickening of the endosteal surface can also occur. Endosteal hyperostosis has various causes and these are reviewed by Whyte (1998), but these are rare conditions. Various types of localised hyperostosis have been recognised including internal frontal hyperostosis, leontiasis ossea and infantile cortical hyperostosis (Caffey's disease). Information on all these conditions is provided by Ortner (2003), but apart from internal frontal hyperostosis and Caffey's disease (e.g. Roberts and Cox, 2003:127; Rogers and Waldron, 1988) none of these conditions have been identified in archaeological human bone.

In a review of the literature undertaken by Aufderheide and Rodríguez-Martín (1998) only one case of generalised hyperostosis with pachydermia was identified, that was reported by Allison et al. (1976) from a Peruvian skeleton dating to approximately AD 1000. No additional cases of hyperostosis were identified when the authors undertook a review of recently published literature.

Hypophosphatasia

Hypophosphatasia is a rare, genetically determined metabolic bone disease. In this condition the mineralisation of newly formed bone is abnormal, and generally deficient. The condition can become apparent during childhood, although occasionally may not be discernable until later in life (Grech et al., 1985). The clinical presentation of hypophosphatasia varies widely, but is linked to the age of onset (Anadiotis, 2003). All patients with the condition are affected by premature loss of the dentition, and also frequently suffer from bone pain associated with stress fractures. Such complications often lead to gait difficulties for affected individuals (Anadiotis, 2003).

Due to failure of osteoid to be properly mineralised, some of the skeletal changes associated with the condition can resemble those produced in rickets. Less severe forms of the condition may be mistaken for rickets, but characteristic 'punched-out' defects in the metaphyses should be detectable radiographically in hypophosphatasia (Grech et al., 1985). Aufderheide and Rodríguez-Martín (1998:64) provide additional information on characteristics of the condition. To date the authors have only found one published report in which hypophosphatasia was considered in the differential diagnosis for an archaeological skeleton with pathological changes. The case reported was that of a middle-aged adult male from an Italian site dating to the twelfth millennium BC (Formicola, 1995). Hypophosphatasia is rare and so few cases would be expected from past communities, but consideration should be given to the possibility of misdiagnosis of the condition as rickets or osteomalacia when studying archaeological skeletal material. In individuals who have the condition it is likely that there will be a range of other pathological changes present in the skeleton, that are uncharacteristic of changes caused by vitamin D deficiency (covered in Chapter 5). These additional pathological changes should assist in making a differential diagnosis.

Osteogenesis Imperfecta

Osteogenesis imperfecta is a rare heritable disease of the connective tissues and has been referred to as 'brittle bone disease' in the past (Whyte, 1999). Whyte provides a clear review of classification of the severity of osteogenesis imperfecta and the clinical presentations of the condition (Whyte, 1999). There are in fact a variety of conditions that are grouped under this heading and a clear description of each of these conditions is provided by Ortner (2003). There are a couple of examples of the condition that have been identified in archaeological human bone (e.g. Wells, 1965; Gray, 1970), and Ortner (2003) provides information on further possible archaeological examples of osteogenesis imperfecta. The levels of morbidity and mortality associated with this group of conditions depends on the exact form of the condition and ranges from forms that may result in still birth, to those in which the condition is almost asymptomatic (Ramachandran, 2006).

Osteopetrosis

Osteopetrosis is a genetic condition in which osteoclasts do not resorb bone normally, and as a result bone formation and remodelling are impaired (Bharagava and Blank, 2006). The abnormal structure of bone that develops in this condition result in the occurrence of pathological fractures (Ortner, 2003). There are at least four types of osteopetrosis and these have been described by Ortner (2003). The exact frequency of these conditions in the modern population is unknown, but they are not common. Individuals who have infantile osteopetrosis rarely survive to adulthood, but individuals who develop the condition later in life generally have a good life expectancy (Bharagava and Blank, 2006). To date few cases have been identified in archaeological human remains, but Aufderheide and Rodríguez-Martín (1998:363) discuss the couple of cases that have been reported.

CONCLUSIONS

The various chapters within this book clearly illustrate that far more research has been undertaken on the bioarchaeology of some of the metabolic bone diseases than others. However, in the future as more diagnostic case reports and clearly undertaken studies are carried out, it is likely that interpretive studies will be carried out on a wider range of metabolic bone diseases. Some of those discussed within this chapter may well receive additional interest and study, although the coverage is not exhaustive and some conditions not covered may also receive increased attention. Some of the rarer conditions will only provide limited information that helps the understanding of wider anthropological issues. However, identification of cases of these conditions will assist in piecing together the history and development of a number of pathological conditions in humans.

Chapter 10

Overview and Directions for Future Research

The research presented throughout this book has illustrated how the investigation of the metabolic bone diseases can contribute to an enormous range of fundamental and challenging debates within bioarchaeology. Analysis of these conditions has expanded considerably over the last ten years, most notably within paleopathology. Although significant advances have been made within this field, it is evident that future research can achieve much more in developing meaningful interpretations in the study of health and disease across past populations. The current research particularly demonstrates how an integrated understanding of many of these conditions within modern contexts can broaden the interpretative frameworks that are considered within bioarchaeology.

The metabolic bone diseases comprise conditions that are affected by complex interactions between many causative variables, including the diet, environment, climate, living conditions, status, cultural practices and heritability. Manifestations of these conditions can be further modified depending on an individual's age, sex and ancestry, or through co-existence with many diverse pathological conditions. Whilst complex, analyses of the metabolic bone diseases can provide an intriguing and significant insight into health and the quality of life across many past and present contexts. This chapter reviews the key developments highlighted from the various metabolic conditions investigated, and discusses directions for future research into the diseases.

BONE BIOLOGY

Metabolic bone diseases are characterised by defects in the processes of bone resorption, formation and mineralisation. As such, accurate interpretations of skeletal pathological changes are enhanced by detailed realisation of the processes evident at the underlying bone cell level. Chapter 3 presented a brief review of the processes inherent in bone growth (modelling) and maintenance (remodelling), as well as the functions of individual cells and the systemic and local factors that can influence cell behaviour. The introduction to mineral metabolism highlighted the complex inter-relationship of calcium metabolism not only with other mineral elements in the skeleton, but also with various organs and hormonal factors.

This chapter aimed to establish a useful resource of recent literature that can substantially enhance the understanding of bone biology for consideration within many fields of bioarchaeology. Interactions between the many factors that can act on bone are complex and, despite substantial advances in recent knowledge, are still incompletely understood. Future research in bone biology will probably better determine how bone cells respond to mechanical loading for example, and will continue to document the extensive influence that hormones, such as oestrogen and vitamin D, have in bone cell regulation.

VITAMIN C DEFICIENCY, SCURVY

Only ten years ago the study of scurvy in both past and present populations was limited. It was widely, but probably mistakenly believed not to be present in modern populations, and diagnostic criteria for archaeological bone were poorly developed. However, as discussed in Chapter 4, there has been a huge development in the diagnosis of scurvy in archaeological human remains. Information on scurvy amongst archaeological juveniles is starting to reach a point where prevalence rates can provide valuable evidence regarding a range of aspects of past societies. Research into the development of diagnostic criteria for scurvy in adults is being undertaken by a number of researchers and this along with greater awareness of the condition by paleopathologists should improve its recognition in past communities.

Alongside the advances in the paleopathology of scurvy there have also been significant developments in appreciating the extent to which scurvy is affecting modern populations from around the world. Recent reports by organisations such as the World Health Organization demonstrate that scurvy occurs in a broad range of situations, but is particularly common amongst refugees and displaced people (WHO, 1999b). Scurvy can be a substantial factor underlying many other conditions, and future investigations need to consider the potential for such co-morbidities to exist. These disease interactions will provide an important means in enabling the social and cultural causes and aspects of this disease in both modern and past contexts to be better understood. It is increasingly apparent that scurvy should be recognised as a prominent health factor in a broad spectrum of social and cultural situations.

VITAMIN D DEFICIENCY, RICKETS AND OSTEOMALACIA

The manifestations of vitamin D deficiency have recently received increasing attention and as a result this condition is becoming more widely recognised from a range of contexts in both present and past communities. Bioarchaeological investigations have demonstrated that it is inherently inter-linked to the development of industrialisation, with causative variables including air pollution, architecture associated with urban centres, as well as increasing levels of food adulteration. Vitamin D deficiency can stem from illnesses that limit mobility and therefore sunlight exposure, as well as underlying episodes of malnutrition or calcium deficiency. These latter causes indicate that this disease may be evident in situations where food supplies are limited, including after natural disasters or bouts of warfare, or where food is deliberately withheld at an individual level of abuse (as debated within Box Features 5.1 and 5.2).

The impact of the geographical habitat and the ancestry of groups affected by vitamin D deficiency are significant considerations, which together with analysis of social conditions may enable a greater interpretation of health in both past and present communities. For example, dark skin pigmentation in individuals residing at high latitudes and experiencing conditions of limited food stuffs, are factors which likely explain the prevalence of vitamin D deficiency in certain demographic groups in the modern world. These factors are also likely to have been underlying variables that impacted on the health and lives of those affected by slavery in the relatively recent past.

Investigation into the presence of vitamin D deficiency can offer a particularly valuable insight into the manner in which cultural practices may influence health. Such factors are evident in the occurrence of vitamin D deficiency in infants living in countries with virtually year-round sunlight exposure. Better understanding of cultural precedents which manifest in clothing styles or nutritional quality are an important mechanism through which more meaningful interpretations of disease presence can be made. The occurrence of this disease in elderly females may exacerbate bone fatigue and failure occurring in osteoporosis. Determining the extent to which this disease may co-exist with other conditions represents an important future challenge for research into vitamin D deficiency.

AGE-RELATED OSTEOPOROSIS

Aging has a significant impact on skeletal tissue. Whilst the outcome of degenerative changes can be manifest in osteoporosis-related fracture, the mechanisms underlying bone failure still require further investigation. Paleopathological study of these changes can contribute substantially to knowledge in this field. However, improvements in the reporting of evidence are essential as are large-scale analyses drawing on considerations of sex- and ancestry-specific bone variation in response to age. The potentially protective effects that mechanical loading through heightened physical activity may have in the accumulation of bone needs further investigation. Intra-skeletal variability of bone has been investigated recently, and needs to be a prominent consideration in future investigations. Continued analysis of variation evident in bone remodelling at the micro-structural level, possibly combined with the determination of alterations in long bone morphology and strength are also likely to play a significant role in developing understanding of bone response to age.

Age-related imbalances in hormonal levels, particularly oestrogen, are a vital factor in the pathogenesis of osteoporosis and related fractures. Broader examination of bioarchaeological evidence may demonstrate that this condition is not simplistically related only to the domain of the post-menopausal female. While prevalence rates of males affected by this disease are likely to remain less than those of females, the range of potential factors that could impact on both sexes particularly in past communities should not be overlooked. In particular, factors that may influence the development of peak bone mass prior to the onset of age-related bone loss may be significant in highlighting population-level osteoporosis-risk and deserve careful consideration in assessments of this disease. Past perspectives on this condition may illustrate a different experience of this condition than is evident in modern communities.

The inter-relationship between the diverse factors that may mediate behaviour, cultural practices, reproduction, dietary quality and health at various stages across the life course is of particular concern when investigating osteoporosis. Examination of such interacting variables may be more suited to multi-faceted investigations involving macroscopic and microscopic evaluation of bone structure together with stable isotope analysis, as has been demonstrated in research on past populations from Nubia (such as White and Armelagos, 1997).

Bioarchaeological evidence exists for this condition, but estimating its occurrence in terms of fracture prevalence is difficult to ascertain from the literature at present. Age- and sex-specific data, which would facilitate interpretations, is also not always presented. Significantly, analyses of trauma in past populations less frequently consider vertebral fracture in relation to cranial and long bone trauma. Consistent and accurate reporting of locations affected and bone element prevalence are required for meaningful future analyses. Future data could significantly enhance our current knowledge of this disease in the past and will contribute to our understanding of the condition in the modern population.

SECONDARY OSTEOPENIA AND OSTEOPOROSIS

Significant advances in the interpretation of disease impact on life in both present and past communities could be made through consideration of secondary pathologies and adaptations to illness. Chapter 7 demonstrated that the investigation of secondary osteopenia could play an important role in furthering such understanding. Osteopenia can develop as a secondary feature to a diverse range of conditions, including immobility due to injury as well as disease and trauma. The consequences of injury, trauma and pathology on subsequent health are often overlooked in relation to the diagnosis and quantification of the primary insult.

A well-balanced diet is important in preventing osteopenia, but unfortunately there are many circumstances in which this balance cannot be maintained. The range of nutritional factors in modern and past diets that can lead to dietary-related osteopenia are documented in Chapter 7, although the complex interaction between age, sex, ancestry and lifestyle, including physical activity habits, can make interpretations difficult. Determining the composition of past diets is becoming increasingly viable through utilising a wide array of techniques and approaches and the association between diet and health remains a fundamental aspect of bioarchaeological and anthropological investigation.

PAGET'S DISEASE OF BONE

Discussion within Chapter 8 illustrates the many aspects of PDB that currently remain poorly understood; in particular, there is presently insufficient knowledge regarding the specific details of the disease process itself. Continued documentation of cases from archaeological contexts would substantially enhance the understanding of the variety of processes and manifestations inherent within this enigmatic condition. As with osteoporosis, determination of PDB can be synergistically enhanced by the continued improvements in accuracy of aging techniques within physical anthropology. Radiological and histological evaluation (where permissible) in future studies will significantly improve the diagnosis of this condition.

Anthropological investigations presently raise more questions concerning the origins and causes of this disease than current clinical and archaeological evidence can answer. As discussed in Box Feature 8.2, this is one condition where paleopathological investigations may provide important information on the causative factors of this disease. For example, modern trends of the geographical occurrence of the disease can be compared with bioarchaeological evidence, potentially illuminating evolutionary trends of this disease.

MISCELLANEOUS METABOLIC BONE DISEASES

The less frequently investigated conditions covered towards the end of this book, illustrate the broad series of conditions that come under the heading of metabolic bone diseases. At present many of the diseases covered in this chapter are not well known. However, over the next ten years some of these conditions are likely to receive more investigation. As illustrated by the Box Feature 9.1, on displaced peoples, several of these disorders have the potential to develop our knowledge of how both past and present populations cope with environmental disruption.

CONCLUSIONS

This book aimed to develop an enriched understanding of past human health through an explicit examination of the factors that can mediate the expression of the metabolic bone diseases. The variability in causes and consequences of this disease group are significant elements that can enable greater insights into health across many communities. Future investigations will yield fundamental opportunities to better determine the interaction between genetic factoring and environmental influence on disease etiologies. Moreover, variation in human health at different stages of the life course can be augmented in both past and present contexts through the medium of this disease group. The many issues and questions addressed within the Box Features throughout this book highlight the diverse range of directions that exist for future research. Reviews of the archaeological evidence for many of these conditions provide a salient perspective on the past experience of these diseases, although collating this data remains an on-going challenge. We hope that the study of the metabolic bone diseases over the next ten years will establish many exciting new avenues of enhanced understanding within the continued development of bioarchaeology.

Bibliography

Aaron JE, Gallagher JC, Anderson J, Stasiak L, Longton EB, Nordin BEC, and Nicholson M. 1974. Frequency of osteomalacia and osteoporosis in fractures of the proximal femur. Lancet 1:229–233.

Aaron JE, Rogers J, and Kanis JA. 1992. Paleohistology of Paget's disease in two Medieval skeletons. American Journal of Physical Anthropology 89:325–331.

Adams J. 1997. Radiology of rickets and osteomalacia. In: Feldman D, Glorieux F, Pike J, editors. Vitamin D. San Diego: Academic Press. pp. 619–641.

Adams P, and Berridge FR. 1969. Effects of kwashiorkor on cortical and trabecular bone. Archives of Disease in Childhood 44:705–709.

Agarwal SC. 2001. The effects of pregnancy and lactation on the maternal skeleton: A historical perspective. American Journal of Physical Anthropology 114:30.

Agarwal SC, and Grynpas MD. 1996. Bone quantity and quality in past populations. Anatomical Record 264:423–432.

Agarwal SC, and Stout SD, editors. 2003. Bone loss and osteoporosis. An anthropological perspective. New York: Kluwer Academic/Plenum Publishers.

Agarwal SC, and Stuart-Macadam P. 2003. An evolutionary and biocultural approach to understanding the effects of reproductive factors on the female skeleton. In: Agarwal SC, Stout SD, editors. Bone loss and osteoporosis. An anthropological perspective. New York: Kluwer Academic/Plenum Publishers. pp. 105–119.

Agarwal SC, Dumitriv M, Tomlinson GA, and Grynpas MD. 2004. Medieval trabecular bone architecture: The influence of age, sex and lifestyle. American Journal of Physical Anthropology 124:33–44.

Aiello L, and Wheeler P. 1995. The expensive tissue hypothesis. Current Anthropology 36:199–222.

Akikusa JD, Garrick D, and Nash MC. 2003. Scurvy: Forgotten but not gone. Journal of Paediatric Child Health 39:75–77.

Albright F, and Reifenstein EC. 1948. Parathyroid glands and metabolic bone disease. Baltimore: Williams & Wilkins.

Albright F, Burnett C, Parson W, Reifenstein EC, and Roos A. 1946. Osteomalacia and late rickets. Medicine 2:399–479.

Allen M. 2006. New ideas about late Holocene climate variability in the Central Pacific. Current Anthropology 47:521–535.

Allen MR, and Burr DB. 2005. Human femoral neck has less cellular periosteum, and more mineralized periosteum than femoral diaphyseal bone. Bone 36:311–316.

Allen MR, Hock JM, and Burr DB. 2004. Periosteum: Biology, regulation, and response to osteoporosis therapies. Bone 35:1003–1012.

Allison MJ, Gerszten E, Sontil R, and Pezzia A. 1976. Primary generalized hyperostosis in ancient Peru. Medical College of Virginia Quarterly 12:49–51.

Almas K, Shakir ZF, and Afzal M. 1999. Prevalence and severity of dental fluorosis in Al-Qaseem province – Kingdom of Saudia Arabia. Tropical Dental Journal 22:44–47.

Alvarez H. 2000. Grandmother hypothesis and primate life histories. American Journal of Physical Anthropology 113:435–450.

Ammann P, Bourrin S, Bonjour JP, Meyer JM, and Rizzoli R. 2000. Protein undernutrition-induced bone loss is associated with decreased IGF-1 levels and estrogen deficiency. Journal of Bone and Mineral Research 15:683–690.

Anadiotis GA. 2003. Hypophosphatasia. eMedicine. Accessed on 20th June 2006: www.emedicine.com/ped/topic1126.htm.

Anderson S. 1993. The human skeletal remains from Caister-on-Sea. East Anglian Archaeology 60:261–268.

Anderson JJB. 1999. Nutritional mechanisms of age-related bone loss. In: Rosen C, Glowacki J, Bilezikian JP, editors. The aging skeleton. San Diego: Academic Press. pp. 229–234.

Anderson S, Wells C, and Birkett D. 2006. The human skeletal remains. In: Cramp R, editor. Wearmouth and Jarrow monastic sites. Vol. 2. London: English Heritage. pp. 481–624.

Angel JL, Kelley JO, Parrington M, and Pinter S. 1987. Life stresses of the Free Black community as represented by the First African Baptist Church, Philadelphia, 1823–1841. American Journal of Physical Anthropology 74:213–229.

Anon. 1890. Cruel neglect of children. Lancet 136:460–461.

Anon. 1906. A sad case of neglect of children. Lancet 168:1088.

Apley AG, and Solomon L. 1994. Concise system of orthopaedics and fractures. Second edition. Oxford: Butterworth-Heinermann.

Arden N, Nevitt M, Lane N, Gore R, Hochberg M, Scott J, Pressman A, and Cummings S. 1999. Osteoarthritis and risk of falls, rates of bone loss and osteoporotic fractures. Arthritis and Rheumatism 42:1378–1385.

Armelagos GJ, Leatherman T, Ryan M, and Sibley L. 1992. Biocultural synthesis in medical anthropology. Medical Anthropology 14:35–52.

Arnett T. 2003. Regulation of bone cell function by acid-base balance. Proceedings of the Nutrition Society 62:511–520.

Arnstein AR, Frame B, and Frost HM. 1967. Recent progress in osteomalacia and rickets. Annals of Internal Medicine 67:1296–1330.

Aspin RK. 1993. Illustrations from the Wellcome Institute Library: The papers of Sir Thomas Barlow (1845–1945). Medical History 37:333–340.

Atoyebi W, Brown M, Wass J, Littlewood T, and Hatton C. 2002. Lymphoplasmacytoid lymphoma presenting as severe osteoporosis. American Journal of Hematology 70:77–80.

Aufderheide AC, and Rodríguez-Martín C. 1998. The Cambridge encyclopedia of human paleopathology. Cambridge: Cambridge University Press.

August M, and Kaban LB. 1999. The aging maxillofacial skeleton. In: Rosen C, Glowacki J, Bilezikian JP, editors. The aging skeleton. San Diego: Academic Press. pp. 359–371.

Avioli LV, and Krane SM, editors. 1998. Metabolic bone disease and clinically related disorders. Third edition. San Diego: Academic Press.

Baker J, and Brothwell D. 1980. Animal diseases in archaeology. London: Academic Press.

Banse X, Devogelaer JP, Delloye C, Lafosse A, Holmyard D, and Grynpas MD. 2003. Irreversible perforations in vertebral trabeculae? Journal of Bone and Mineral Research 18:1247–1253.

Baraz D. 2003. Medieval cruelty. Changing perceptions, late antiquity to the early modern period. New York: Cornell University Press.

Barker DJP, and Gardner MJ. 1974. Distribution of Paget's disease in England, Wales and Scotland and a possible relationship with vitamin D deficiency in childhood. British Journal of Preventative and Social Medicine 28:226–232.

Barlow T. 1883. On cases described as 'acute rickets' which are probably a combination of scurvy and rickets, the scurvy being an essential, and the rickets a variable element. Medico-Chirurgical Transactions 66:159–219.

Barlow T. 1894. Infantile scurvy and its relation to rickets. British Medical Journal 2:1029–1034.

Barnes E. 1994. Developmental defects of the axial skeleton in palaeopathology. Boulder: University Press of Colorado.

Barnes E. 2005. Disease and human evolution. Albuquerque: University of Mexico Press.

Barnes E, and Ortner D. 1997. Multifocal eosinophilic granuloma with a possible trepanation in a fourteenth century Greek young skeleton. International Journal of Osteoarchaeology 7:542–547.

Barnes C, Mercer G, and Shakespeare T. 1999. Exploring disability. A sociological introduction. Cambridge: Polity Press.

Baron R. 1999. Anatomy and ultrastructure of bone. In: Favus M, editor. Primer on the metabolic bone diseases and disorders of mineral metabolism. Fourth edition. Philadelphia: Lippincott William & Wilkins. pp. 3–10.

Bartley W, Krebs HA, and O'Brien JRP. 1953. Vitamin C requirements of human adults. A report by the vitamin C subcommittee of the accessory food factors committee and Barnes AE, Bartley W, Frankau IM, Higgins GA, Pemberton J, Roberts GL, and Vickers HR. London: Her Majesty's Stationary Office.

Baudouin M. 1914. L'ostéite déformante chronique dans l'ossuaire de Bazoges-en-Pareds. Bulletin de la Société Française d'Histoire de la Médecine 13:96–102.

Bell LS. 1990. Palaeopathology and diagenesis: An SEM evaluation of structural changes using backscattered electron imaging. Journal of Archaeological Science 17:85–102.

Bell N. 1998. Sarcoidosis and related disorders. In: Avioli LV, Krane SM, editors. Metabolic bone disease and clinically related disorders. Third edition. San Diego: Academic Press. pp. 607–619.

Bell LS, and Jones SJ. 1991. Macroscopic and microscopic evaluation of archaeological pathological bone: Backscattered electron imaging of putative pagetic bone. International Journal of Osteoarchaeology 1:179–184.

Bell JC, and Palmer SR. 1983. Control of zoonoses in Britian: Past, present and future. British Medical Journal 287:591–593.

Bell LS, and Piper K. 2000. An introduction to palaeohistopathology. In: Cox M, Mays S, editors. Human osteology in archaeology and forensic science. London: Greenwich Medical Media. pp. 255–274.

Bell KL, Loveridge N, Power J, Garrahan N, Stanton M, Lunt M, Meggitt BF, and Reeve J. 1999a. Structure of the femoral neck in hip fracture: Cortical bone loss in the inferoanterior to superoposterior axis. Journal of Bone and Mineral Research 14:111–119.

Bell KL, Loveridge N, Power J, Garralan N, Megitt BF, and Reeve J. 1999b. Regional differences in cortical porosity in the fractured femoral neck. Bone 24:57–64.

Benyshek DC, and Watson JT. 2006. Exploring the thrifty genotype's food-shortage assumptions: A cross cultural comparison of ethnographic accounts of food security among foraging and agricultural societies. American Journal of Physical Anthropology 131:120–126.

Berlyne GM, Ari JB, Nord E, and Shainkin R. 1973. Bedouin osteomalacia due to calcium deprivation caused by high phytic acid content of unleavened bread. American Journal of Clinical Nutrition 26:910–911.

Berner YN, Stern F, Polyak Z, and Dror Y. 2002. Dietary intake analysis in institutionalized elderly: A focus on nutrient density. Journal of Nutrition and Health in Aging 6:237–242.

Berry J, Davies M, and Mee A. 2002. Vitamin D metabolism, rickets and osteomalacia. Seminars in Musculoskeletal Radiology 6:173–181.

Bertoli S, Battezzati A, Merati G, Margonato V, Maggioni M, Testolin G, and Veicsteinas A. 2006. Nutritional status and dietary patterns in disabled people. Nutrition, Metabolism and Cardiovascular Diseases 16:100–112.

Bharagava A, and Blank R. 2006. Osteopetrosis. eMedicine. Accessed on 20th June 2006: www.emedicine.com/med/topic1692.htm.

Bhargavi V, Khandare AL, Venkaiah K, and Sarojini G. 2004. Mineral content of water and food in fluorotic villages and prevalence of dental fluorosis. Biological Trace Element Research 100:195–203.

Bhat BV, and Srinivasan S. 1989. Neonatal scurvy. Indian Pediatrics 23:1258–1260.

Bickle DD, Halloran BP, and Morey-Holton E. 1997. Spaceflight and the skeleton: Lessons for the earthbound. Endocrinologist 7:10–22.

Bilezikian JP. 1999. Primary hyperparathyroidism. In: Favus M. editor. Primer on the metabolic bone diseases and disorders of mineral metabolism. Fourth edition. Philadelphia: Lippincott William & Wilkins. pp. 187–192.

Bilezikian JP. 2006. Editorial: What's good for the goose's skeleton is good for the gander's. Journal of Clinical Endocrinology and Metabolism 91:1223–1225.

Bilezikian JP, and Silverberg SJ. 2001. The role of parathyroid hormone and vitamin D in the pathogenesis of osteoporosis. In: Marcus R, Feldman D, Kelsey J, editors. Osteoporosis. Volume 2. Second edition. San Diego: Academic Press. p 71–84.

Bilezikian JP, Raisz LG, and Rodan GA, editors. 2002. Principles of bone biology. Second edition. San Diego: Academic Press.

Bisel C. 1991. The human skeletons from Herculaneum. International Journal of Anthropology 6:1–20.

Bizzari G. 2004. The right to food. Food and Agriculture Organization Fact Sheet FAO 18679. Accessed on 18th January 2007:http://www.fao.org/worldfoodsummit/english/fsheets/food.pdf.

Black A, Tilmont E, Handy A, Scott W, Shapses S, Ingram D, Roth G, and Lane M. 2001. A nonhuman primate model of age-related bone loss: A longitudinal study in male and premenopausal female rhesus monkeys. Bone 28:295–302.

Blackman J, Allison MJ, Aufderheide AC, Oldroyd N, and Steinbock RT. 1991. Secondary hyperparathyroidism in an Andean mummy. In: Ortner DJ, Aufderheide AC, editors. Palaeopathology: Current syntheses and future options. Washington: Smithsonian Institution Press. pp. 291–296.

Blau S. 2001. Limited yet informative: Pathological alterations observed on human skeletal remains from Third and Second Millennium BC collective burials in the United Arab Emirates. International Journal of Osteoarchaeology 11:173–205.

Blom DE, Buikstra JE, Keng L, Tomczak PD, Shoreman E, and Stevens-Tittle D. 2005. Anemia and childhood mortality: Latitudinal patterning along the coast of pre-Columbian Peru. American Journal of Physical Anthropology 127:152–169.

Blondiaux J, Cotten A, Fontaine C, Hänni C, Bera A, and Flipo RM. 1997. Two Roman and Medieval cases of symetrical erosive polyarthropathy from Normandy: Anatomico-pathological and radiological evidence for rheumatoid arthritis. International Journal of Osteoarchaeology 7:451–466.

Blondiaux G, Blondiaux J, Secousse F, Cotten A, Danze P, and Flipo R. 2002. Rickets and child abuse: The case of a two year old girl from the 4th century in Lisieux (Normandy). International Journal of Osteoarchaeology 12:209–215.

Blurton Jones NG, Hawkes K, and O'Connell J. 2002. Antiquity of postreproductive life: Are there modern impacts on hunter-gatherer life spans? American Journal of Human Biology 14:184–205.

Boddington A, Garland AN, and Janaway RC, editors. 1987. Death decay and reconstruction. Manchester: Manchester University Press.

Bogumill GP, and Schwamm HA. 1984. Orthopaedic pathology: A synopsis with clinical and radiographic correlation. Philadelphia: WB Saunders.

Bollet AJ. 1992. Scurvy and chronic diarrhea in Civil War troops: Were they both nutritional deficiency syndromes? Journal of the History of Medicine and Allied Sciences 47:49–67.

Bonjour JP, Chevalley T, Ferrari S, and Rizzoli R. 2003. Peak bone mass and its regulation. In: Glorieux FH, Pettifor JM, Jüppner H, editors. Pediatric bone. Biology and Diseases. San Diego: Academic Press. pp. 235–248.

Boonen S, Vanderschueren D, Guang Cheng X, Verbeke G, Dequeker J, Geusens P, Broos P, and Bouillon R. 1997. Age-related (Type II) femoral neck osteoporosis in men: Biochemical evidence for both hypovitaminosis D- and androgen deficiency-induced bone resorption. Journal of Bone and Mineral Research 12:2119–2126.

Borden IV S. 1975. Roentgen recognition of acute plastic bowing of the forearm in children. American Journal of Roentgenology 125:524–530.

Bourbou C. 2003a. Health patterns of proto-byzantine populations (6th–7th centuries AD) in South Greece: The cases of Eleutherna (Crete) and Messene (Peloponnesse). International Journal of Osteoarchaeology 13:303–313.

Bourbou C. 2003b. The interaction between a population and its environment: Probable case of subadult scurvy from proto-byzantine Greece. Eres Arquelogía/ Bioantroplogía 11:105–114.

Bourne GH. 1942a. Vitamin C and repair of injured tissues. Lancet 240:661–664.

Bourne GH. 1942b. The effect of graded doses of vitamin C upon the regeneration of bone in guinea-pigs on a scorbutic diet. Journal of Physiology 101:327–336.

Boyde A, Maconnachie E, Reid S, Delling G, and Mundy G. 1986. Scanning electron microscopy in bone pathology: Review of methods, potential and applications. Scanning Electron Microscopy 6:1537–1554.

Boyle IT. 1991. Bones for the future. Acta Pediatrica Scandinavia 373 (Supplement):58–65.

Boylston A, and Ogden A. 2005. A study of Paget's disease at Norton Priory, Cheshire. A Medieval religious house. In: Zakrzewski SR, Clegg M, editors. Proceedings of the fifth annual conference of the British Association for Biological Anthropology and Osteoarchaeology. British Archaeological Reports International Series 1383. Oxford: Archeopress. pp. 69–76.

Boylston A, and Roberts CA. 2004. The Roman inhumations. In: Dawson M, editor. Archaeology in the Bedford region. Bedfordshire Archaeology Monograph Series 4. British Archaeological Reports British Series 373. Oxford: Archeopress. pp. 322–370.

Boylston A, Knüsel CJ, Roberts CA, and Dawson M. 2000. Investigation of a Romano-British rural ritual in Bedford, England. Journal of Archaeological Science 27:241–254.

Brailsford JF. 1929. Deformities of the lumbo-sacral region of the spine. British Journal of Surgery 16:562–627.

Braude-Heller A, Rotbalsam I, and Elbinger R. 1979. Clinical aspects of hunger disease in children. Current Concepts in Nutrition 7:45–68.

Brennan RJ, and Waldman RJ. 2006. The South Asian Earthquake six months later – an ongoing crisis. New England Journal of Medicine 354:1769–1771.

Briançon D, de Gaudemar JB, and Forestier R. 2004. Management of osteoporosis in women with peripheral osteoporotic fractures after 50 years of age: A study of practices. Joint Bone Spine 71:128–130.

Brickley M. 2002. An investigation of historical and archaeological evidence for age-related bone loss and osteoporosis. International Journal of Osteoarchaeology 12:364–371.

Brickley M. 2006. Rib fractures in the archaeological record: A useful source of sociocultural information? International Journal of Osteoarchaeology 16:61–75.

Brickley M, and Agarwal SC. 2003. Techniques for the investigation of age-related bone loss and osteoporosis in archaeological bone. In: Agarwal S, Stout SD, editors. Bone loss and osteoporosis. An anthropological perspective. New York: Kluwer Academic/Plenum Publishers. pp. 157–172.

Brickley M, and Ferllini R, editors. 2007. Forensic anthropology: Case studies from Europe. Springfield: Charles C Thomas.

Brickley M, and Howell PGT. 1999. Measurement of changes in trabecular bone structure with age in an archaeological population. Journal of Archaeological Science 26:151–157.

Brickley M, and Ives R. 2006. Skeletal manifestations of infantile scurvy. American Journal of Physical Anthropology 129:163–172.

Brickley M, and McKinley J, editors. 2004. Guidance to standards for recording human skeletal remains. University of Reading: Institute of Field Archaeologists/British Association of Biological Anthropology and Osteoarchaeology.

Brickley M, and Thomas R. 2004. The young woman and her baby or the juvenile and their dog: Reinterpreting osteological material from a Neolithic long barrow. Archaeological Journal 161:1–10.

Brickley M, Miles A, and Stainer H. 1999. The Cross Bones burial ground, Redcross Way, Southwark, London. Archaeological excavations (1991–1998) for the London Underground Limited Jubilee line extension project. London: Museum of London Archaeology Service Monograph 3.

Brickley M, Mays S, and Ives R. 2005. Skeletal manifestations of vitamin D deficiency osteomalacia in documented historical collections. International Journal of Osteoarchaeology 15:389–403.

Brickley M, and Buteux S, Adams J, & Cherrington R. 2006. St. Martin's uncovered. Investigations in the churchyard of St. Martin's-in-the-Bull Ring, Birmingham, 2001. Oxford: Oxbow Books.

Brickley M, Mays S, and Ives R. 2007. An investigation of skeletal indicators of vitamin D deficiency in adults: Effective markers for interpreting past living conditions and pollution levels in eighteenth and nineteenth century Birmingham, England. American Journal of Physical Anthropology 132:67–79.

Bridges PS. 1989. Bone cortical area in the evaluation of nutrition and activity levels. American Journal of Human Biology 1:785–792.

Brimblecombe P. 1987. The big smoke: A history of air pollution in London since Medieval times. London: Meuthen.

Broadus AE. 1999. Mineral balance and homeostasis. In: Favus M, editor. Primer on the metabolic bone diseases and disorders of mineral metabolism. Fourth edition. Philadelphia: Lippincott William & Wilkins. pp. 74–80.

Brothwell DR, editor. 1968. The skeletal biology of earlier human populations. Symposia of the society for the study of human biology VIII. Oxford: Pergamon.

Brothwell DR, and Browne S. 1994. Pathology. In: Lilley J, Stroud G, Brothwell DR, Williamson M, editors. The Jewish Burial Ground at Jewbury. The Archaeology of York Volume 12/3. The Medieval Cemeteries. York Archaeological Trust for Excavation and Research. York: Council for British Archaeology. pp. 457–494.

Brothwell DR, and Møller-Christensen V. 1963a. A possible case of amputation dated to c. 2000 BC. Man 63:192–194.

Brothwell DR, and Møller-Christensen V. 1963b. Medico-historical aspects of a very early case of mutilation. Danish Medical Bulletin 10:21–25.

Brothwell DR, and Sandison A, editors. 1967. Diseases in antiquity: A survey of the diseases, injuries and surgery of early populations. Springfield: Charles C Thomas.

Brousseau TJ, Kissoon N, and McIntosh B. 2005. Vitamin K deficiency mimicking child abuse. Journal of Emergency Medicine 29:283–288.

Brown L, Streeten E, Shuldiner A, Almasy L, Peyser P, and Mitchell B. 2004. Assessment of sex-specific genetic and environmental effects on bone mineral density. Genetic Epidemiology 27:153–161.

Buckberry J, and Chamberlain A. 2002. Age estimation from the auricular surface of the ilium: A revised method. American Journal of Physical Anthropology 119:231–239.

Buckley HR. 2000. Subadult health and disease in prehistoric Tonga, Polynesia. American Journal of Physical Anthropology 113:481–505.

Buckley HR, and Tayles NG. 2003. The functional cost of tertiary yaws (*Treponema pertenue*) in a prehistoric Pacific Island skeletal sample. Journal of Archaeological Science 30:1301–1314.

Buikstra JE, and Cook D. 1980. Palaeopathology: An American account. Annual Review of Anthropology 9:433–470.

Buikstra JE, and Ubelaker D, editors. 1994. Standards for data collection from human skeletal remains. Fayetteville: Archaeological Survey Research Seminar Series 44.

Buikstra JE, and Beck LA, editors. 2006. Bioarchaeology. The contextual analysis of human remains. San Diego: Elsevier.

Bunn HT. 2007. Meat made us human. In: Ungar PS, editor. Evolution of the human diet. Oxford: Oxford University Press. pp. 191–211.

Burland C. 1918. An historical case of rickets. The Practitioner 100:391–395.

Burr DB, Forwood MR, Fyhrie DP, Martin RB, Schaffler MB, and Turner CH. 1997. Bone microdamage and skeletal fragility in osteoporotic and stress fractures. Journal of Bone and Mineral Research 12:6–15.

Buzhilova AP. 2005. The environment and health condition of the Upper Palaeolithic Sunghir people of Russia. Journal of Physiological Anthropology and Applied Human Science 24:413–418.

Byrne CD, and Wild SH. 2006. The metabolic syndrome. Chichester: Wiley.

Cachel S. 1997. Dietary shifts and the European Upper Palaeolithic transition. Current Anthropology 38:579–603.

Cadranel JL, Garabedian M, Milleron B, Guilozzo R, Valeyre D, Paillard F, Akoun G, and Hance AJ. 1994. Vitamin D metabolism by alveolar immune cells in tuberculosis: Correlation with calcium metabolism and clinical manifestations. European Respiratory Journal 7:1103–1110.

Cail WS, Keats TS, and Sussman MD. 1978. Plastic bowing fracture of the femur in a child. American Journal of Roentgenology 130:780–782.

Cameron N. 1991. Human growth, nutrition, and health status in sub-Saharan Africa. American Journal of Physical Anthropology 34:211–250.

Cameron N. 1996. Antenatal and birth factors and their relationship to child growth. In: Henry CJ, Ulijaszek SJ, editors. Long term consequences of early environment. Growth, development and the lifespan. A developmental perspective. Cambridge: Cambridge University Press. pp. 69–90.

Cameron N, and Demerath EW. 2002. Critical periods in human growth and their relationship to diseases of aging. Yearbook of Physical Anthropology 45:159–184.

Capecchi V, and Rabino Massa E, editors. 1984. Proceedings of the fifth European meeting of the Paleopathology Association. Sienna: Tipografia Senese.

Cardwell P. 1995. Excavation of the hospital of St. Giles by Brompton Bridge, North Yorkshire. Archaeological Journal 152:109–245.

Care A. 1997. Vitamin D in pregnancy, the fetoplacental unit and lactation. In: Feldman D, Glorieux F, Pike J, editors. Vitamin D. San Diego: Academic Press. pp. 437–446.

Carli-Thiele P. 1995. Scurvy: Investigations on the human skeleton using macroscopic, radiological and microscopic methods. Journal of Paleopathology 7:88.

Carli-Thiele P. 1996. Spuren von mangelerkrankungen an steinzeitlichen kinderskelete. Göttingen: Verlag Erich Goltze.

Carlson DS, Armelagos GJ, and Van Gerven D. 1976. Patterns of age-related cortical bone loss (osteoporosis) within the femoral diaphysis. Human Biology 48:295–314.

Carpenter K. 1986. The history of scurvy and vitamin C. Cambridge: Cambridge University Press.

Carpenter K. 1987. The problem of land scurvy, 1820–1910. Society for the Social History of Medicine 40:20–22.

Castelo-Branco C, Reina F, Montivero AD, Colodron M, and Vannell JA. 2006. Influence of high intensity training and of dietic and anthropometric factors on menstrual cycle disorders in ballet dancers. Gynecology and Endocrinology 22:31–35.

Center JR, Nguyen TV, Schneider D, Sambrook PN, and Eisman J. 1999. Mortality after all major types of osteoporotic fracture in men and women: An observational study. Lancet 353:878–882.

Centre for Human Bioarchaeology. Museum of London. Launched May 2007 http://www.museumoflondon.org.uk/English/Collections/OnlineResources/CHB/Home.htm.

Cerroni A, Tomlinson G, Turnquist J, and Grynpas MD. 2000. Bone mineral density, osteopenia and osteoporosis in the rhesus macaques of Cayo Santiago. American Journal of Physical Anthropology 113:389–410.

Chalmers J. 1970. Subtrochanteric fractures in osteomalacia. Journal of Bone and Joint Surgery 52B:509.

Chapuy MC, and Meunier PJ. 1997. Vitamin D insufficiency in adults and the elderly. In: Feldman D, Glorieux F, Pike J, editors. Vitamin D. San Diego: Academic Press. pp. 679–693.

Charlton W, Kharazzi D, Alpert S, Gloussman R, and Chandler R. 2003. Unstable non-union of the scapula: A case report. Journal of Shoulder and Elbow Surgery 12:517–519.

Charnley S. 1997. Environmentally-displaced peoples and the cascade effect: Lessons from Tanzania. Human Ecology 25:593–618.

Chatproedprai S, and Wananukul S. 2001. Scurvy: A case report. Journal of the Medical Association of Thailand 84:S106–S110.

Cheadle W. 1878. Three cases of scurvy supervening on rickets in young children. Lancet 2:685–687.

Cheng YL, Wright JM, Walstad WR, and Finn MD. 2002. Osteosarcoma arising in Paget's disease of the mandible. Oral Oncology 38:785–792.

Cheung E, Mutahar R, Assefa F, Ververs M, Nasiri SM, Borrel A, and Salama P. 2003. An epidemic of scurvy in Afghanistan: Assessment and response. Food and Nutrition Bulletin 24:247–255.

Cho H, and Stout SD. 2003. Bone remodelling and age-associated bone loss in the past: A histomorphometric analysis of the Imperial Roman skeletal population of Isola Sacra. In: Agarwal SC, Stout SD, editors. Bone loss and osteoporosis. An anthropological perspective. New York: Kluwer Academic/Plenum Publishers. pp. 207–228.

Cho H, Stout SD, and Bishop TA. 2006. Cortical bone remodelling rates in a sample of African American and European American descent groups from the American Midwest: Comparisons of age and sex in ribs. American Journal of Physical Anthropology 130:214–226.

Choubisa SL, Choubisa L, and Choubisa DK. 2001. Endemic fluorosis in Rajasthan. Indian Journal of Environmental Health 43:177–189.

Chundun Z, and Roberts CA. 1995. Human skeletal remains. In: Cardwell P, editor. Excavation of the hospital of St. Giles by Brompton Bridge, North Yorkshire. Archaeological Journal 152:109–245.

Churchill SE, and Formicola V. 1997. A case of marked bilateral asymmetry in the upper limbs of an Upper Palaeolithic male from Barma Grande (Liguria), Italy. International Journal of Osteoarchaeology 7:18–38.

Civitelli R, Ziambaras K, and Leelawattana R. 1998. Pathophysiology of calcium, phosphate and magnesium absorption. In: Avioli LV, Krane SM, editors. Metabolic bone disease and clinically related disorders. Third edition. San Diego: Academic Press. pp. 165–206.

Clark NG, Sheard NF, and Kelleher JF. 1992. Treatment of iron-deficiency anemia complicated by scurvy and folic acid deficiency. Nutrition Review 50:134–137.

Clemens T, Henderson S, Adams J, and Holick M. 1982. Increased skin pigment reduces the capacity of skin to synthesise vitamin D_3. Lancet 1:74–76.

Clowes JA, Riggs III BL, and Khosla S. 2005. The role of the immune system in the pathophysiology of osteoporosis. Immunological Reviews 208:207–227.

Cobb K, Bachrach L, Greendale G, Marcus R, Neer R, Nieves J, Sowers F, Brown M, Byron W, Gopalakrishman G, Luetters C, Tanner H, Ward B, and Kelsey J. 2003. Disordered eating, menstrual irregularity and bone mineral density in female runners. Medicine and Science in Sports and Exercise 35:711–719.

Cockburn E, editor. 1984. Papers on paleopathology. Sienna: Fifth European Members Meeting.

Cohen MN. 1989. Health and the rise of civilisation. New Haven: Yale University Press.

Cohen MN, and Armelagos GJ, editors. 1984. Paleopathology at the origins of agriculture. Orlando: Academic Press.

Coleman RE. 2001. Metastatic bone disease: Clinical features, pathophysiology and treatment strategies. Cancer Treatment Reviews 27:165–176.

Colman R, Kemnitz J, Lane M, Abbtt D, and Binkley N. 1999a. Skeletal effects of aging and menopausal status in female rhesus macaques. Journal of Clinical Endocrinology and Metabolism 84:4144–4148.

Colman R, Lane M, Binkley N, Wegner F, and Kemnitz J. 1999b. Skeletal effects of aging in male rhesus monkeys. Bone 24:17–23.

Compston J. 1999. Histomorphometric manifestations of age-related bone loss. In: Rosen C, Glowacki J, Bilezikian JP, editors. The aging skeleton. San Diego: Academic Press. pp. 251–261.

Cook D. 1979. Subsistence base and health in prehistoric Illinois Valley: Evidence from the human skeleton. Medical Anthropology 3:109–124.

Cook D. 1980. Paget's disease and treponematosis in prehistoric Midwestern Indians: The case for misdiagnosis. OSSA 7:41–63.

Cook D. 1984. Subsistence and health in the lower Illinois Valley: Osteological evidence. In: Cohen MN, Armelagos GJ, editors. Paleopathology at the origins of agriculture. New York: Academic Press. pp. 235–269.

Cook GC, and Bjelland JC. 1979. Acute bowing fracture of the fibula in an adult. Radiology 131:637–638.

Cook GC. 2005. Scurvy in the British mercantile marine in the 19th century and the contribution of the Seamen's hospital society. Postgraduate Medical Journal 80:224–229.

Cook M, Molto E, and Anderson C. 1988. Possible case of hyperparathyroidism in a Roman period skeleton from the Dakhleh Oasis, Egypt, diagnosed using bone histomorphometry. American Journal of Physical Anthropology 75:23–30.

Cooper L. 1996. A Roman cemetery in Newarke Street, Leicester. Leicestershire Archaeology 70:33–49.

Cooper C. 2005. Epidemiology of osteoporotic fracture: Looking to the future. Rheumatology 44 (Supplement 4):iv 36–iv 40.

Cooper C, Schafheutle K, Dennison E, Kellingray S, Guyer P, and Barker D. 1999. The epidemiology of Paget's disease in Britain: Is the prevalence decreasing? Journal of Bone and Mineral Research 14:192–197.

Cordain L. 1999. Cereal grains: Humanity's double-edged sword. World Review of Nutrition and Dietetics 84:19–73.

Cordain L. 2007. Implications of Plio-Pleistocene hominin diets for modern humans. In: Ungar PS, editor. Evolution of the human diet. Oxford: Oxford University Press. pp. 363–383.

Cordain L, Brand Miller J, Eaton SB, Mann N, Holt S, and Speth J. 2000. Plant-animal subsistence ratios and macronutrient energy estimations in worldwide hunter-gatherer diets. American Journal of Clinical Nutrition 71:682–692.

Corruccini RS, Jacobi KP, Handler JS, and Aufderheide AC. 1987. Implications of tooth root hypercementosis in a Barbados slave skeletal collection. American Journal of Physical Anthropology 74:179–184.

Corsi A, Collins MT, Riminucci M, Howell PG, Boyde A, Robney PG, and Bianco P. 2003. Osteomalacic and hyperparathyroid changes in fibrous dysplasia of bone: Core biopsy and clinical correlations. Journal of Bone and Mineral Research 18:1235–1246.

Corwin R. 2003. Effects of dietary fats on bone health in advanced age. Prostaglandins, Leukotrienes and Essential Fatty Acids 68:379–386.

Corwin R, Hartman T, Maczuga S, and Graubard B. 2006. Dietary saturated fat intake is inversely associated with bone density in humans: Analysis of NHANES III. Journal of Nutrition 136:159–165.

Costeff H, and Breslaw Z. 1962. Rickets in Southern Israel. Tropical Pediatrics 61:919–924.

Cox M, editor. 1998. Grave concerns. Death and burial in England 1700–1850. York: Council of British Archaeology.

Cox M, and Mays S, editors. 2000. Human osteology in archaeology and forensic science. London: Greenwich Medical Media Ltd.

Crabtree N, Loveridge N, Parker M, Rushton N, Power J, Bell KL, Beck TJ, and Reeve J. 2001. Intracapsular hip fracture and the region-specific loss of cortical bone: Analysis by peripheral quantitative computed tomography. Journal of Bone and Mineral Research 16:1318–1328.

Cramp R, editor. 2006. Wearmouth and Jarrow monastic sites. Vol. 2. London: English Heritage.

Crawford M. 1988. Scurvy in Ireland during the great famine. Society for the Social History of Medicine 1:281–300.

Crellin JK. 2000. Early settlements in Newfoundland and the scourge of scurvy. Canadian Bulletin of Medical History 17:127–136.

Crowe JE, and Swischuck LE. 1977. Acute bowing fractures of the forearm in children: A frequently missed injury. American Journal of Roentgenology 128:981–984.

Cummings SR, and Melton LJ III. 2002. Epidemiology and outcomes of osteoporotic fractures. Lancet 359:1761–1767.

Cundiff DJ, and Harris W. 2006. Case report of 5 siblings. Nutrition Journal 5:1–8.

Currey J. 2002. Bones: Structure and mechanics. Princeton: Princeton University Press.

Curtis T, Ashrafi S, and Weber D. 1985. Canalicular communication in the cortices of human long bones. Anatomical Record 212:336–344.

Cutforth RH. 1958. Adult scurvy. Lancet 271:454–456.

Czerny V. 1873. Eine locale Malacie des unterschenkels. Wiener Medizinische Wochenschrift 23:895.

Dahniya MH. 1987. Paget's disease of bone in Africans. British Journal of Radiology 60:113–116.

Dai LY, and Jiang LS. 2004. Loss of bone mass after Colles' fractures: A follow-up study. Chinese Medical Journal 117:327–330.

Dalinka MK, Aronchick JM, and Haddard JG. 1983. Paget's disease. Orthopedic Clinics of North America 14:3–19.

Damany DS, and Parker MJ. 2005. Varus impacted hip fracture. Injury. International Journal of Care of Injuries 36:627–629.

Danzeiser Wols HD, and Baker JE. 2004. Dental health of elderly confederate veterans: Evidence from the Texas State Cemetery. American Journal of Physical Anthropology 124:59–72.

Daroszewska A, and Ralston SH. 2005. Genetics of Paget's disease of bone. Clinical Science 109:257–263.

Davidson JM, Rose JC, Gutmann MP, Haines MR, Condon K, and Condon C. 2002. The quality of African-American life in the old Southwest near the turn of the twentieth century. In: Steckel RH, Rose JC, editors. The backbone of history. Health and nutrition in the Western Hemisphere. New York: Cambridge University Press. pp. 226–277.

Davies BE. 1994. Trace elements in the human environment: Problems and risks. Environmental Geochemistry and Health 16:97–106.

Davies BE, Bowman C, Davies TC, and Selinus O. 2005a. Medical geology: Perspectives and prospects. In: Selinus O, Alloway B, Centeno JA, Finkelman RB, Fuge R, Lindh U, Smedley P, editors. Essentials of medical geology. Impacts of the natural environment on public health. San Diego: Elsevier. pp. 1–14.

Davies J, Fabiš M, Mainland I, Richards M, and Thomas R. 2005b. Diet and health in past animal populations: Current research and future directions. Oxford: Oxbow Books.

Dawes JD. 1980. The human bones. In: Dawes JD, Magilton JR, editors. The cemetery of St. Helen-on-the-walls, Aldwark. The archaeology of York 12/1. York: York Archaeological Trust. pp. 19–120.

Dawes JD, and Magilton JR, editors. 1980. The cemetery of St. Helen-on-the-walls, Aldwark. The archaeology of York 12/1. York: York Archaeological Trust.

Dawson M, editor. 2004. Archaeology in the Bedford region. Bedfordshire Archaeology Monograph Series 4. British Archaeological Reports British Series 373. Oxford: Archeopress.

Dean HT. 1942. The investigation of physiological effects by the epidemiological method. In: Moulton FR, editor. Fluoride and dental health. Washington: American Association for Advancement of Science. pp. 23–31.

De Bruin E, Frey-Rindova P, Herzog RE, Dietz V, Dambacher MA, and Stüssi E. 1999. Changes of tibia bone properties after spinal cord injury: Effects of early intervention. Archives of Physiotherapy and Medical Rehabilitation 80:214–220.

DeLuca HF. 1976. Vitamin D endocrinology. Annals of Internal Medicine 85:367–377.

DeMenocal P. 2001. Cultural responses to climate change during the late Holocene. Science 292:667–673.

Demissie D, Dupuis J, Cupples LA, Beck TJ, Kiel DP, and Karasik D. 2007. Proximal hip geometry is linked to several chromosomal regions: Genome-wide linkage results from the Framingham osteoporosis study. Bone 40:743–750.

Denninger HS. 1933. Paleopathological evidence of Paget's disease. Annals of Medical History 5:73–81.

Dent CE, and Hodson CJ. 1954. Radiological changes associated with certain metabolic bone diseases. British Journal of Radiology 27:605–618.

Dequeker J, Ortner DJ, Stix AI, Cheng XG, Brys P, and Boonen S. 1997. Hip fracture and osteoporosis from Lisht. Upper Egypt. Journal of Bone and Mineral Research 16:881–888.

De Torrenté de la Jara G, Pécoud A, and Favrat B. 2006. Female asylum seekers with musculoskeletal pain: The importance of diagnosis and treatment of hypovitaminosis D. BMC Family Practice 7. Accessed online on 21st March 2007: http://www.biomedcentral.com/1471-2296/7/4www.biomedcentral.com/1471-2296/7/4.

Dettwyler KA. 2005. Can paleopathology provide evidence for 'compassion'? American Journal of Physical Anthropology 84:375–384.

Dewey K. 2005. Guiding principles for feeding non-breastfed children 6–24 months of age. Geneva: World Health Organization.

Diab T, Condon KW, Burr DB, and Vashishth D. 2006. Age-related change in the damage morphology of human cortical bone and its role in bone fragility. Bone 38:427–431.

Dickel D, and Doran G. 1989. Severe neural tube defect syndrome from the early Archaic of Florida. American Journal of Physical Anthropology 80:325–334.

Dickinson CJ. 2000. The possible role of osteoclastogenic oral bacterial products in etiology of Paget's disease. Bone 26:101–102.

Domett KM, and Tayles N. 2006. Adult fracture patterns in prehistoric Thailand: A biocultural interpretation. International Journal of Osteoarchaeology 16:185–199.

Donahue S, McGee M, Harvey K, Vaughan M, and Robbins C. 1999. Hibernating bears as a model for preventing disuse osteoporosis. Journal of Biomechanics 39:1480–1488.

Donahue S, Vaughan M, Demers L, and Donahue H. 2003. Bone formation is not impaired by hibernation (disuse) in black bears. Journal of Experimental Biology 206:4233–4239.

Downey P, and Siegel M. 2006. Bone biology and the clinical implications for osteoporosis. Physical Therapy 86:77–91.

Doyle T, Gunn J, Anderson G, Gill M, and Cundy T. 2002. Paget's disease in New Zealand: Evidence for declining prevalence. Bone 31:616–619.

Draper J, editor. 1993. Excavations at Poundbury 1966–1980. Dorset Natural History and Archaeology Society Monograph Series 11. Dorchester: Dorset County Museum.

Draper H, editor. 1994. Advances in nutritional research. Vol. 9. New York: Plenum Press.

Drusini AG, Bredariol S, Carrara N, and Rippa Bonati M. 2000. Cortical bone dynamics and age-related osteopenia in a Longobard archaeological sample from three graveyards of the Veneto Region (Northeast Italy). International Journal of Osteoarchaeology 10:268–279.

Duan Y, Turner CH, Kim BT, and Seeman E. 2001a. Sexual dimorphism in vertebral fragility is more the result of gender differences in age-related bone gain than bone loss. Journal of Bone and Mineral Research 16:2267–2275.

Duan Y, Seeman E, and Turner CH. 2001b. The biomechanical basis of vertebral body fragility in men and women. Journal of Bone and Mineral Research 16:2276–2283.

Dufour DL, Staten LK, Reina JC, and Spurr GB. 1998. Living on the edge: Dietary strategies of economically impoverished women in Cali, Columbia. American Journal of Physical Anthropology 102:5–15.

Duggeli O, and Trendezenberg F. 1961. Spinal tuberculosis. Acta Rheumatological 11:9–84.

Duintjer Tebbens RJ, Pallansch MA, Kew OM, Caceres VM, Sutter RW, and Thompson KM. 2005. A dynamic model of poliomyelitis outbreaks: Learning from the past to help inform the present. American Journal of Epidemiology 162:358–372.

Dunnigan MG, and Henderson JB. 1997. An epidemiological model of privational rickets and osteomalacia. Proceedings of the Nutrition Society 56:939–956.

Eastell R, and Riggs BL. 1997. Vitamin D and osteoporosis. In: Feldman D, Glorieux F, Pike J, editors. Vitamin D. San Diego: Academic Press. pp. 695–711.

Eaton SB. 2007. Pre-agricultural diets and evolutionary health promotion. In: Ungar PS, editor. Evolution of the human diet. Oxford: Oxford University Press. pp. 384–394.

Eaton SB, and Nelson DA. 1991. Calcium in the evolutionary perspective. American Journal of Clinical Nutrition 54:281S–287S.

Eaton SB, Eaton SB III, and Konner MJ. 1997. Paleolithic nutrition revisited: A twelve-year retrospective on its nature and implications. European Journal of Clinical Nutrition 51:207–216.

Einhorn TA. 1992. Bone strength: The bottom line. Calcified Tissue International 51:333–339.

Eisman JA. 1998. Relevance of pregnancy and lactation to osteoporosis? Lancet 352:504–505.

Eisman JA. 1999. Genetics of osteoporosis. Endocrine Reviews 20:788–804.

Eisman JA, Sambrook PN, Kelly PJ, and Pocock NA. 1991. Exercise and its interaction with genetic influences in the determination of bone mineral density. American Journal of Medicine 91 (Supplement):5B–5S.

Ekenman S, Eriksson AV, and Lindgren JU. 1995. Bone density in Medieval skeletons. Calcified Tissue International 56:355–358.

El Maghraoui A. 2004. Osteoporosis and ankylosing spondylitis. Joint Bone Spine 71:291–295.

El-Najjar MY. 1979. Human treponematosis and tuberculosis: Evidence from the New World. American Journal of Physical Anthropology 51:599–618.

Epstein S, Inzerillo A, Caminis J, and Zaidi M. 2003. Disorders associated with acute rapid and severe bone loss. Journal of Bone and Mineral Research 18:2083–2094.

Erdal YS. 2006. A pre-Columbian case of congenital syphilis from Anatolia (Nicaea, 13th century AD). International Journal of Osteoarchaeology 16:16–33.

Eren E, and Yilmaz N. 2005. Biochemical markers of bone turnover and bone mineral density in patients with beta-thalassaemia major. International Journal of Clinical Practice 59:46–51.

Ericksen MF. 1976. Cortical bone loss with age in three Native American populations. American Journal of Physical Anthropology 45:443–452.

Ericksen MF. 1978a. Aging in the lumbar spine II L1 and L2. American Journal of Physical Anthropology 48:241–246.

Ericksen MF. 1978b. Aging in the lumbar spine III L5. American Journal of Physical Anthropology 48:247–250.

Ericksen MF. 1982. Aging changes in thickness of the proximal femoral cortex. American Journal of Physical Anthropology 59:121–130.

Eshed V, Gopher A, Gage TB, and Hershkovitz I. 2004. Has the tranisition to agriculture reshaped the demographic structure of prehistoric populations? New evidence from the Levant. American Journal of Physical Anthropology 124: 315–329.

Estes JW. 1997. Naval surgeon: Life and death at sea in the age of sail. Canton: Science History Publications.

Everts V, Delaissé JM, Korper J, Jansen DC, Tigchelaar-Gutter W, Saftig P, and Beertsen W. 2002. The bone lining cell: It's role in cleaning Howship's lacunae and initiating bone formation. Journal of Bone and Mineral Research 17:77–90.

Faerman M, Nebel A, and Filon D. 2000. From a dry bone to a genetic portrait: A case study of sickle cell anaemia. American Journal of Physical Anthropology 111:153–163.

Faerman M, Horwitz LK, Kahana T, and Zilberman U, editors. 2007. Faces from the past. Papers in honor of Patricia Smith. British Archaeological Reports International Series 1603. Oxford: Archeopress.

Fagan B. 2005. The long summer. How climate changed civilization. London: Granta Books.

Fain O. 2005. Musculoskeletal manifestations of scurvy. Joint Bone Spine 72:124–128.

Fairgrave SI, and Molto JE. 2000. Cribra orbitalia in two temporally distinct population samples from the Dakhleh Oasis, Egypt. American Journal of Physical Anthropology 111:319–331.

Falys CG, Schutkowski H, and Western DA. 2006. Auricular surface aging: Worse than expected? A test of the revised method on a documented historic skeletal assemblage. American Journal of Physical Anthropology 130:508–513.

Farquharson MJ, and Brickley M. 1997. Determination of mineral makeup in archaeological bone using energy dispersive low angle X-ray scattering. International Journal of Osteoarchaeology 7:95–99.

Farwell DE, and Molleson TI. 1993. Poundbury volume 2: The cemeteries. In: Draper J, editor. Excavations at Poundbury 1966–1980. Dorset Natural History and Archaeology Society Monograph Series 11. Dorchester: Dorset County Museum.

Favus MJ, editor. 1996. Primer on the metabolic bone diseases and disorders of mineral metabolism. Third edition. Philadelphia: Lippincott-Raven Press.

Favus M, editor. 1999. Primer on the metabolic bone diseases and disorders of mineral metabolism. Fourth edition. Philadelphia: Lippincott William & Wilkins.

Favus MJ. 1999. Vitamin D. In: Rosen C, Glowacki J, Bilezikian JP, editors. The aging skeleton. San Diego: Academic Press. pp. 613–622.

Feeney RE. 1992. Food technology and polar exploration: Problems faced by polar explorers in the 19th and 20th centuries stimulated improvements in subsequent food supplies. Arctic Medical Research 51:35–46.

Fehrsen GS. 1974. Communicating the concept of nutritional disease in a rural area. South African Medical Journal 48:2521–2522.

Feldman D, Glorieux F, and Pike J, editors. 1997. Vitamin D. San Diego: Academic Press.

Feldman EC, Hoar B, Pollard R, and Nelson RW. 2005. Pretreatment and laboratory findings in dogs with primary hyperparathyroidism: 210 cases (1987–2004). Journal of American Veterinary Medical Association 227:756–761.

Fermesrdorf-Koln F. 1927. Gräber der einheimischen Bevölkerung römischer Zeit in Köln. Preehistorische Zeitschrift 18:255–293.

Ferreira MT. 2002. A scurvy case in an infant from Monte da Cegonha (Vidigueria – Portugal). Antropologia Portuguesa 19:57–63.

Feskanich D, Willett WC, and Colditz GA. 2003. Calcium, vitamin D, milk consumption, and hip fractures: A prospective study among postmenopausal women. American Journal of Clinical Nutrition 77:504–511.

Fildes VA. 1986. Breasts, bottles and babies. A history of infant feeding. Edinburgh: Edinburgh University Press.

Fiorato V, Boylston A, and Knüsel C. 2000. Blood red roses: The archaeology of a mass grave from the battle of Towton AD 1461. Oxford: Oxbow Books.

Fisher AK. 1935. Additional paleopathological evidence of Paget's disease. Annals of Medical History 7:197–198.

FitzGerald C, Saunders S, Bondioli L, and Macchiarelli R. 2006. Health of infants in an Imperial Roman skeletal sample: Perspective from dental microstructure. American Journal of Physical Anthropology 130:179–189.

Foldes AJ, Moscovici A, Popovtzer MM, Mogle P, Urman D, and Zias J. 1995. Extreme osteoporosis in a sixth century skeleton from the Negev Desert. International Journal of Osteoarchaeology 5:157–162.

Follis BH, Park EA, and Jackson D. 1950. The prevalence of scurvy at autopsy during the first two years of age. Bulletin of the John Hopkins Hospital 87:569–591.

Food and Agriculture Organization of the United Nations. 2004. The spectrum of malnutrition. Rome: Food and Agriculture Organization.

Food and Agriculture Organization of the United Nations. 2005. Report of the regional workshop on salt-affected soils from sea-water intrusion: Strategies for rehabilitation and management. Thailand: Regional Office for Asia and Pacific, Food and Agricultural Organization.

Formicola V. 1995. X-linked hypophosphatemic rickets: A probable upper paleolithic case. American Journal of Physical Anthropology 98:403–409.

Formicola V, and Buzhilova AP. 2004. Double child burial from Sunghir (Russia): Pathology and inferences for Upper Paleolithic funerary practices. American Journal of Physical Anthropology 124:189–198.

Formicola V, Pontrandolfi A, and Svoboda J. 2001. The Upper Paleolithic triple burial of Dolní Věstonice: Pathology and funerary behaviour. American Journal of Physical Anthropology 115:372–379.

Forsius H. 1993. Medical problems connected with wintering in the arctic during A.E. Nordenskiöld's expeditions in 1872–73 and 1878–79. Arctic Medical Research 52:131–136.

Francis R, and Selby P. 1997. Osteomalacia. In: Reid I, editor. Metabolic bone disease in clinical endocrinology and metabolism. London: W.B. Saunders. pp. 145–163.

Frayer D, and Martin D. 1997. Troubled times: violence and welfare in the past war and society research series. Volume 3. Amsterdam: Gordon & Breach Publishers.

Freeman S. 1989. Mutton and oysters: Food, cooking and eating in Victorian times. London: Gollancz.

Freitas B. 2002. Hyperparathyroidism, primary. eMedicine. Accessed on 18th May 2006: www.emedicine.com/radio/topic355.htm.

Friedlander AL, Genant HK, Sadowksy S, Byle NN, and Glüer CC. 1995. A 2 year program of aerobics and weight-training enhances bone mineral density of young women. Journal of Bone and Mineral Research 10:574–585.

Frigo P, and Lang C. 1995. Osteoporosis in a woman of the early Bronze Age. New England Journal of Medicine 333:1468.

Frohlich B, Ortner D, and Al Khalifa H. 1989. Human disease in the ancient Middle East. Dilmun 14:61–73.

Frost HM. 1963. Bone remodelling dynamics. Springfield: Charles C Thomas.

Frost HM. 2001. From Wolff's Law to the Utah Paradigm: Insights about bone physiology and its clinical applications. Anatomical Record 262:398–419.

Frost HM. 2003. On changing views about age-related bone loss. In: Agarwal S, Stout SD, editors. Bone loss and osteoporosis. An anthropological perspective. New York: Kluwer Academic/Plenum Publishers. pp. 19–31.

Frost ML, Fogelman I, Blake GM, Marsden PK, and Cook GJR. 2004. Dissociation between global markers of bone formation and direct measurement of spinal bone formation in osteoporosis. Journal of Bone and Mineral Research 19:1797–1804.

Fumio O, Kenichi I, Jiro S, Toshio O, and Hirofumi M. 2004. Skull Paget's disease developing into chiari malformation. Endocrine Journal 51:391–392.

Fürst CM. 1920. När de döda vittna. Stockholm: Svenska Teknolog Förlag.

Gage T. 2005. Are modern environments really bad for us?: Revisiting the demographic and epidemiologic transitions. Yearbook of Physical Anthropology 128:96–117.

Gandevia B, and Cobley J. 1974. Mortality at Sydney Cove, 1788–1792. Australian and New Zealand Journal of Medicine 4:111–125.

Gannage-Yared MH, Chemali R, Sfeir C, Maalouf G, and Halaby G. 2005. Dietary calcium and vitamin D intake in an adult Middle Eastern population: Food sources and relation to lifestyle and PTH. International Journal for Vitamin and Nutrition Research 75:281–289.

Gardner J, Beauchesne P, and Spence M. 2005. The identification of Paget's disease in a prehistoric specimen from Ontario, Canada. 32nd annual meeting of the Paleopathology Association, Milwaukee, Wisconsin. pp. 12.

Garn SM, Guzman MA, and Wagner B. 1969. Subperiosteal gain and endosteal loss in protein-calorie malnutrition. American Journal of Physical Anthropology 30:153–156.

Garn SM, Sullivan TV, Decker SA, Larkin FA, and Hawthorne VM. 1992. Continuing bone expansion and increasing bone loss over a two-decade period in men and women from a total community sample. American Journal of Human Biology 4:57–67.

Garner A, and Ball J. 1966. Quantitative observations on mineralised and unmineralised bone in chronic renal azotaemia and intestinal malabsorption syndrome. Journal of Pathology and Bacteriology 91:545–561.

Gartland JJ, and Werley CW. 1951. Evaluation of healed Colles' fractures. Journal of Bone and Joint Surgery 33A:895–907.

Gejvall NG. 1960. Westerhus. Medieval population and church in light of skeletal remains. Lund: Hakan Ohlssons Boktryckeri.

Genant HK. 1988. Quantitative bone mineral analysis. In: Resnick D, Niwayama G, editors. Diagnosis of bone and joint disorders. Volume 8. Second edition. Philadelphia: W.B. Saunders. pp. 1999–2019.

Genant HK, Volger JB, and Block JE. 1988. Radiology of osteoporosis. In: Riggs BL, Melton LJ III, editors. Osteoporosis. Etiology, diagnosis and management. New York: Raven Press. pp. 181–220.

Genant HK, Wu CY, Van Kuijk C, and Nevitt MC. 1993. Vertebral fracture assessment using a semi-quantitative technique. Journal of Bone and Mineral Research 8:1137–1148.

Gerstenfeld L, and Einhorn T. 2003. Developmental aspects of fracture healing and the use of pharmacological agents to alter healing. Journal of Musculoskeletal Neuronal Interactions 3:297–303.

Giangregorio L, and Blimkie CJ. 2002. Skeletal adaptations to alterations in weight-bearing activity: A comparison of models of disuse osteoporosis. Sports Medicine 32:459–476.

Gilbert RI, and Mielke JH, editors. 1985. The analysis of prehistoric diets. San Diego: Academic Press.

Gilbert MT, Bandelt HJ, Hofreiter M, and Barnes I. 2005. Assessing ancient DNA studies. Trends in Ecology and Evolution 20:541–544.

Glencross B, and Stuart-Macadam P. 2000. Childhood trauma in the archaeological record. International Journal of Osteoarchaeology 10:198–209.

Glisson F. 1650. Rachitide sive morbo puerili, qui vulgo. The Rickets dicitur. London.

Glorieux FH. 1999. Hypophosphatemic vitamin D-resistant rickets. In: Favus M, editor. Primer on the metabolic bone diseases and disorders of mineral metabolism. Fourth edition. Philadelphia: Lippincott William & Wilkins. pp. 328–331.

Glorieux FH, Pettifor JM, and Jüppner H, editors. 2003. Pediatric bone. Biology and diseases. San Diego: Academic Press.

Goldbloom EB, Cummings EA, and Yhap M. 2005. Osteoporosis at presentation of childhood ALL: Management with pamidronate. Pediatric Hematology and Oncology 22:543–550.

Goldring SR. 2001. Osteoporosis associated with rheumatologic disorders. In: Marcus R, Feldman D, Kelsey J, editors. Osteoporosis. Volume 2. Second edition. San Diego: Academic Press. pp. 351–362.

Goodman AH. 1993. On the interpretation of health from skeletal remains. Current Anthropology 34:281–288.

Goodman LR, and Warren MP. 2005. The female athlete and menstrual function. Current Opinion Obstetrics and Gynecology 17:466–470.

Goodman AH, Thomas RB, Swedlund AC, and Armelagos GJ. 1988. Biocultural perspectives on stress in prehistoric, historical, and contemporary population research. Yearbook of Physical Anthropology 31:169–202.

Grauer AL, and Roberts CA. 1996. Paleoepidemiology, healing, and possible treatment of trauma in the Medieval cemetery population of St. Helen-on-the-Walls, York, England. American Journal of Physical Anthropology 100:531–544.

Grauer AL, editor. 1995. Bodies of evidence. Chichester: Wiley.

Gray P. 1970. Osteogenesis imperfecta associated with dentinogenesis imperfecta dating from antiquity. Clinical Radiology 21:106–108.

Grech P, Martin TJ, Barrington NA, and Ell PJ. 1985. Diagnosis of metabolic bone disease. London: Chapman & Hall Medical.

Greenfield GB. 1990. Radiology of bone diseases. Fifth edition. Philadelphia: JB Lippincott Company.

Grewar D. 1965. Infantile scurvy. Clinical Pediatrics 4:82–89.

Grinspoon SK, Baum BA, Kim V, Coggins C, and Klibanski A. 1995. Decreased bone formations and increased mineral dissolutions during acute fasting in young women. Journal of Clinical Endocrinology and Metabolism 80:3628–3633.

Groen JJ, Balogh M, Levy M, Yaron E, Zemach R, and Benaderet S. 1964. Nutrition of the Bedouins in the Negev Desert. American Journal of Clinical Nutrition 14:37–46.

Groen JJ, Eshchar J, Ben-Ishay D, Alkan WJ, and Assa BI. 1965. Osteomalacia among the Bedouin of the Negev Desert. Archives of Internal Medicine 116:195–204.

Gruber HE, Stauffer ME, Thompson ER, and Baylink DJ. 1981. Diagnosis of bone disease by core biopsies. Seminars in Hematology 18:258–278.

Grynpas MD. 2003. The role of bone quality on bone loss and fragility. In: Agarwal S, Stout SD, editors. Bone loss and osteoporosis. An anthropological perspective. New York: Kluwer Academic/Plenum Publishers. pp. 33–44.

Grynpas MD, Chachra D, and Lundon K. 2000. Bone quality in animal models of osteoporosis. Drug Development Research 49:146–158.

Güllü S, Erdoğan MF, Uysal AR, Başkal N, Kamel AN, and Erdoşan G. 1998. A potential risk for osteomalacia due to socio-cultural lifestyle in Turkish women. Endocrine Journal 45:675–678.

Hacking P, Allen T, and Rogers J. 1994. Rheumatoid arthritis in a Medieval skeleton. International Journal of Osteoarchaeology 4:251–255.

Hadjipavlou AG, Gaitanis IN, and Kontakis GM. 2002. Paget's disease of the bone and its management. Journal of Bone and Joint Surgery 84-B:160–169.

Haffter C. 1968. The changeling: History and psychodynamics of attitudes to handicapped children in European folklore. Journal of the History of Behavioural Sciences 4:55–61.

Haibach H, Farrell C, and Dittric FJ. 1985. Neoplasms arising in Paget's disease of bone: A study of 82 cases. American Journal of Clinical Pathology 83:594–600.

Hallel T, Malkin C, and Garti R. 1980. Epiphyseometaphyseal cupping of the distal femoral epiphysis following scurvy in infancy. Clinical Orthopaedics and Related Research 153:166–168.

Halloran BP, and Portale AA. 1997. Vitamin D metabolism: The effects of aging. In: Feldman D, Glorieux F, Pike J, editors. Vitamin D. San Diego: Academic Press. pp. 541–554.

Hampl JS, Johnston CS, and Mills RA. 2001. Scourge of the black-leg (scurvy) on the Mormon trail. Nutrition 17:416–418.

Haneveld GT, and Perizonius WRK. editors. 1982. Proceedings of the fourth European members meeting of the Paleopathology Association, Middelburg-Antwerpen, 16th–19th September 1982. Utrecht: Eliinkwijk.

Hannan M, Tucker K, Dawson-Hughes B, Cupples A, Felson D, and Kiel D. 2000. Effect of dietary protein on bone loss in elderly men and women: The Framingham osteoporosis study. Journal of Bone and Mineral Research 15:2504–2512.

Hardy A. 1993. The epidemic streets. Infectious disease and the rise of preventive medicine 1856–1900. Oxford: Clarendon Press.

Harkess JW, Ramsey WC, Ahmadi B. 1984. In: Rockwood CA, Green DP, editors. Fractures in adults. Volume 1. Second edition. Philadelphia: Lippincott Williams & Wilkins. pp. 1–36.

Harlow M, and Laurence R. 2002. Growing up and growing old in ancient Rome. A life course approach. London: Routledge.

Harman M, Molleson TI, and Prince JL. 1981. Burials, bodies and beheadings in Romano-British and Anglo-Saxon cemeteries. Bulletin of the British Museum of Natural History (Geology) 35:145–188.

Harper KD, and Weber TJ. 1998. Secondary osteoporosis. Endocrinology and Metabolism Clinicals of North America 27:325–348.

Hauge EM, Qvesel D, Eriksen EF, Mosekilde L, and Melsen F. 2001. Cancellous bone remodelling occurs in specialised compartments lined by cells expressing osteoblastic markers. Journal of Bone and Mineral Research 16:1575–1582.

Hausen H. 2004. Benefits of topical fluorides firmly established. Evidence Based Dentistry 5:36–37.

Havill L, Mahaney M, Czerwinski S, Carey K, Rice K, and Rogers J. 2003. Bone mineral density reference standards in adult baboons (*Papio hamadryas*) by sex and age. Bone 33:877–888.

Hawkey D. 1998. Disability, compassion and the skeletal record. Using musculoskeletal stress markers (MSM) to construct an osteobiography from early New Mexico. International Journal of Osteoarchaeology 8:326–340.

Hayes W, and Ruff C. 1986. Biomechanical compensatory mechanisms for age-related changes in cortical bone. In: Uhthoff HK, editor. Current concepts in bone fragility. Berlin: Springer-Verlag. pp. 371–377.

Heaney RP. 1997a. Vitamin D: Role in the calcium economy. In: Feldman D, Glorieux F, Pike J, editors. Vitamin D. San Diego: Academic Press. pp. 485–497.

Heaney RP. 1997b. The roles of calcium and vitamin D in skeletal health: An evolutionary perspective. Food Nutrition and Agriculture 20. Food and Agricultural Organization Corporate Document Depository. Accessed on 15th February 2007: http://www.fao.org/docrep/W7336T/w7336t03.htm.

Heaney RP. 1999. Aging and calcium balance. In: Rosen C, Glowacki J, Bilezikian JP, editors. The aging skeleton. San Diego: Academic Press. pp. 19–26.

Heaney RP. 2002. Calcium. In: Bilezikian JP, Raisz LG, Rodan GA, editors. Principles of bone biology. Second edition. San Diego: Academic Press. pp. 1007–1018.

Henry A, and Bowler L. 2003. Fracture of the neck of the femur and osteomalacia in pregnancy. British Journal of Obstetrics and Gynaecology 110:329–330.

Henry CJ, and Ulijaszek SJ, editors. 1996. Long term consequences of early environment. Growth, development and the lifespan. A developmental perspective. Cambridge: Cambridge University Press.

Herm F, Killguss H, and Stewart A. 2005. Osteomalacia in Hazara District, Pakistan. Tropical Doctor 35:8–10.

Hernandez J, and Baca D. 1998. Effect in tuberculosis of milk production in dairy cows. Journal of the American Veterinary Medical Association 213:851–854.

Hershkovitz I, Rothschild BM, Latimer B, Dutour O, Leonetti G, Greenwald CM, Rothschild C, and Jellema LM. 1997. Recognition of sickle cell anemia in skeletal remains of children. American Journal of Physical Anthropology 104:213–226.

Hess AF. 1930. Rickets including osteomalacia and tetany. London: Henry Kimpton.

Heuertz M. 1957. Etude des squelettes du cimetière franc d'Ennery (Moselle). Bulletin Mémoires Societie Anthropologie 10e Séries 8:81–141.

Higgins RL, Haines MR, Walsh L, and Sirianni JE. 2002. The poor in the mid-nineteenth century Northeastern United States. Evidence from the Monroe County Almshouse, Rochester, New York. In: Steckel RH, Rose JC, editors. The backbone of history. Health and nutrition in the Western Hemisphere. New York: Cambridge University Press. pp. 162–184.

Hillson S. 1996. Dental anthropology. Cambridge: Cambridge University Press.

Hirsch M, Mogle P, and Barkli Y. 1976. Neonatal scurvy report of a case. Pediatric Radiology 4:251–253.

Hochberg Z. 2003. Rickets – past and present. In: Hochberg Z, editor. Vitamin D and rickets. Basel: Karger. pp. 1–13.

Hochberg Z, editor. 2003. Vitamin D and rickets. Basel: Karger.

Hock JM, Fitzpatrick LA, and Bilezikian JP. 2002. Actions of parathyroid hormone. In: Bilezikian JP, Raisz LG, Rodan GA, editors. Principles of bone biology. Second edition. San Diego: Academic Press. pp. 463–481.

Hoenderop JGJ, Willems PGM, and Bindels RJM. 2000. Toward a comprehensive molecular model of active calcium reabsorption. American Journal of Physiology and Renal Physiology 278:F352–F360.

Holck P. 1984. Scurvy – a paleopathological problem. In: Cockburn E, editor. Papers on paleopathology. Sienna: Fifth European Members Meeting. pp. 163–171.

Holick MF. 1994. McCollum award lecture, 1994: Vitamin D new horizons for the 21st century. American Journal of Clinical Nutrition 60:619–630.

Holick MF. 2002a. Sunlight and vitamin D. Journal of General Internal Medicine 17:733–735.

Holick MF. 2002b. Too little vitamin D in pre-menopausal women: Why should we care?. American Journal of Clinical Nutrition 76:3–4.

Holick MF. 2003. Vitamin D: A millennium perspective. Journal of Cellular Biochemistry 88:296–307.

Holick MF. 2004. Vitamin D: Importance in the prevention of cancers, type 1 diabetes, heart disease and osteoporosis. American Journal of Clinical Nutrition 79:362–371.

Holick MF. 2005. The vitamin D epidemic and its health consequences. Journal of Nutrition 135:2739S–2748S.

Holick MF. 2006. Resurrection of vitamin D deficiency and rickets. Journal of Clinical Investigation 116:2062–2072.

Holck P. 2006. Bone mineral densities in the prehistoric, viking-age and medieval populations of Norway. International Journal of Osteoarchaeology 17:199–206.

Holick MF, and Adams J. 1998. Vitamin D metabolism and biological function. In: Avioli L, Krane S, editors. Metabolic bone disease and clinically related disorders. Third edition. San Diego: Academic Press. pp. 123–164.

Holt B. 2003. Mobility in upper paleolithic and mesolithic Europe: Evidence from the lower limb. American Journal of Physical Anthropology 122:200–215.

Hoppa RD, and Vaupel JW, editors. 2002. Paleodemography: Age distributions from skeletal samples. Cambridge: Cambridge University Press.

Houillier P, Nicolet-Barouse L, Maruani G, and Paillard M. 2003. What keeps serum calcium levels stable? Joint Bone Spine 70:407–413.

Houston CS. 1990. Scurvy and the Canadian exploration. Canadian Bulletin of Medical History 7:161–167.

Howe G. 1997. People, environment, disease and death. A medical geography throughout the ages. Cardiff: University of Wales Press.

Huang QY, and Wai Chee Kung A. 2006. Genetics of osteoporosis. Molecular Genetics and Metabolism 88:295–306.

Hubert J, editor. 2000. Madness, disability and social exclusion: The archaeology and anthropology of difference. One World Archaeology Series Volume 40. London: Routledge.

Hughes RE. 1990. The rise and fall of the 'antiscorbutics': Some notes on the traditional cures for 'land scurvy'. Medical History 34:52–64.

Hutchinson J. 1889. 'Osteitis deformans'. Illustrated Medical News 2:177.

Hutchinson J. 1987. The age-sex structure of the slave population in Harris County, Texas: 1850 and 1860. American Journal of Physical Anthropology 74:231–238.

Huuskonen J, Vaisanen SB, Kroger H, Jurvelin JS, Alhava E, and Rauramaa R. 2001. Regular physical exercise and bone mineral density. A four year controlled randomized trail in middle aged men. The DNASCO study. Osteoporosis International 12:349–355.

Illarvamendi X, Nery JAC, Vieira LMM, and Sarno EN. 2002. Acral bone resorption in multibacillary patients. A retrospective clinical study. In: Roberts CA, Lewis ME, Manchester K, editors. The past and present of leprosy. Archaeological, historical, paleopathological and clinical approaches. Proceedings of the international congress on the evolution and paleoepidemiology of the infectious diseases 3 (ICEPID). University of Bradford 26–31st July 1999. British Archaeological Reports International Series 1054. Oxford: Archeopress. pp. 43–50.

Illés T, Halmai V, de Jonge T, and Dubousset J. 2001. Decreased bone mineral density in neurofibromatosis 1 patients with spinal deformities. Osteoporosis International 12:823–827.

Inoue K, Hukuda S, Nakai M, Katayama K, and Huang J. 1999. Erosive peripheral polyarthritis in ancient Japanese skeletons: A possible case of rheumatoid arthritis. International Journal of Osteoarchaeology 9:1–7.

İsçan MY, and Kennedy KAR, editors. 1989. Reconstruction of life from the skeleton. New York: Wiley-Liss.

Ivanhoe F. 1994. Osteometric scoring of adult residual rickets skeletal plasticity in two archaeological populations from Southeastern England: Relationship to sunshine and calcium deficits and demographic stress. International Journal of Osteoarchaeology 4:97–120.

Ivanhoe F, and Chu PW. 1996. Cranioskeletal size variation in San Francisco Bay prehistory: Relation to calcium deficit in the reconstructed high-seafoods diet and demographic stress. International Journal of Osteoarchaeology 6:346–381.

Ivanhoe F. 1995. Secular decline in cranioskeletal size over two millennia of interior central California prehistory: Relation to calcium deficit in the reconstructed diet and demographic stress. International Journal of Osteoarchaeology 5:213–253.

Ives R. 2005. Vitamin D deficiency osteomalacia in a historic urban collection. An investigation of age, sex and lifestyle-related variables. Paleopathology Association Newsletter 130:6–15.

Ives R. in preparation. Investigating the prevalence of adult vitamin D deficiency osteomalacia in five post-Medieval urban sites from England.

Ives R, and Brickley M. 2004. A procedural guide to metacarpal radiogrammetry in archaeology. International Journal of Osteoarchaeology 14:7–17.

Ives R, and Brickley M. 2005. Metacarpal radiogrammetry: A useful indicator of bone loss throughout the skeleton? Journal of Archaeological Science 32:1552–1559.

Iwaniec UT. 1997. Effects of dietary acidity on cortical bone remodeling: A histomorphometric assessment. Unpublished doctoral dissertation, University of Wisconsin-Madison, US.

Jablonski NG, and Chaplin G. 2000. The evolution of human skin colouration. Journal of Human Evolution 39:57–106.

Jackson D, and Park EA. 1935. Congenital scurvy a case report. Journal of Pediatrics 7:741–753.

Jacob RA, and Sotoudeh G. 2002. Vitamin C function and status in chronic disease. Nutrition in Clinical Care 5:66–74.

Jacobsen DM, Itani K, Digre KB, Ossoinig KC, and Varner MW. 1988. Maternal orbital hematoma associated with labor. American Journal of Ophthalmology 105:547–553.

Jaffe HL. 1972. Metabolic degenerative and inflammatory diseases of bones and joints. Philidelphia: Lea & Febiger.

Jannsens PA. 1963. De crematieresten uit het Urnenveld te Grote-Brogel. In: Roosens H, Beex G, Bonenfant P, editors. Een urnenveld Grote-Brogel. Limberg 42. Archaeologia Belgica 67. pp. 261–300.

Jannsens PA. 1970. Paleopathology: Diseases and injuries of prehistoric man. London: John Baker.

Janssens K, de Vernejoul MC, de Freitas F, Vanhoenacker F, and Van Hul W. 2005. An intermediate form of juvenile Paget's disease caused by truncating *TNFRSF11B* mutation. Bone 36:542–548.

Järup L. 2003. Hazards of heavy metal contamination. British Medical Bulletin 68:167–182.

Järup L, and Alfvén T. 2004. Low level cadmium exposure, renal and bone effects – the OSCAR study. BioMetals 17:505–509.

Jee W, and Ma Y. 1999. Animal models of immobilization osteopenia. Morphologie 83:25–34.

Jeffcoat MK, Reddy MS, and DeCarlo AA. 2001. Oral bone loss and systemic osteopenia. In: Marcus R, Feldman D, Kelsey J, editors. Osteoporosis. Volume 2. Second edition. San Diego: Academic Press. pp. 363–381.

Jelliffe DB. 1955. Infant feeding in the subtropics and tropics. Geneva: World Health Organization.

Jergas MD, and Genant HK. 1999. Radiology of osteoporosis. In: Favus M, editor. Primer on the metabolic bone diseases and disorders of mineral metabolism. Fourth edition. Philadelphia: Lippincott William & Wilkins. pp. 160–169.

Jergas MD, and Genant HK. 2001. Imaging of osteoporosis. In: Marcus R, Feldman D, Kelsey J, editors. Osteoporosis. Volume 2. Second edition. San Diego: Academic Press. pp. 411–431.

Jin T, Nordberg G, Ye T, Bo M, Wang H, Zhu G, Kong Q, and Bernard A. 2004. Osteoporosis and renal dysfunction in a general population exposed to cadmium in China. Environmental Research 96:353–359.

Jobling MA, Hiures ME, and Tyler-Smith C. 2004. Human evolutionary genetics: Origins, peoples and disease. New York: Garland Science.

Joffe N. 1961. Some radiological aspects of scurvy in the adult. British Journal of Radiology 34:429–437.

Johansson SR, and Owsley D. 2002. Welfare history on the Great Plains. Mortality and skeletal health 1650–1900. In: Steckel RH, Rose JC, editors. The backbone of history. Health and nutrition in the Western Hemisphere. New York: Cambridge University Press. pp. 524–560.

Johnson YJ, Kaneene JB, Gardiner JC, Lloyd JW, Sprecher DJ, and Coe PH. 2001. The effect of subclinical mycobacterium paratuberculosis infection on milk production in Michigan dairy cows. Journal of Dairy Science 84:2188–2194.

Johnston FE, Low SM, de Baessa Y, and MacVean RB. 1985. Growth status of disadvantaged urban guatemalan children of a resettled community. American Journal of Physical Anthropology 68:215–224.

Jones A, editor. 2003. Settlement, burial and industry in Roman Godmanchester. Birmingham University Field Archaeology Unit Monograph Series 6. British Archaeological Reports British Series 346. Oxford: Archeopress.

Jones SD. 2004. Mapping a zoonotic disease: Anglo-American efforts to control bovine TB before WWI. Osiris 19:133–148.

Jones SJ, and Boyde A. 1999. Development and structure of teeth and periodontal tissues. In: Favus M, editor. Primer on the metabolic bone diseases and disorders of mineral metabolism. Fourth edition. Philadelphia: Lippincott William & Wilkins. pp. 455–458.

Jones S, Burt BA, Petersen PE, and Lennon MA. 2005. The effective use of fluorides in public health. Bulletin of the World Health Organization 83:670–677.

Jónsdóttir B, Ortner DJ, and Frohlich B. 2003. Probable destructive meningioma in an archaeological adult male skull from Alaska. American Journal of Physical Anthropology 122:232–239.

Jowsey J, Kelly PJ, Riggs BL, Bianco AJ, Scholz DA, and Gershon-Cohen J. 1965. Quantitative microradiographic studies of normal and osteoporotic bone. Journal of Bone and Joint Surgery 47A:785–806.

Joyce JM, and Keats TE. 1986. Disuse osteoporosis: Mimic of neoplastic disease. Skeletal Radiology 15:129–132.

Jüppner H, Brown EM, and Kronenberg HM. 1999. Parathyroid hormone. In: Favus M, editor. Primer on the metabolic bone diseases and disorders of mineral metabolism. . Fourth edition. Philadelphia: Lippincott William & Wilkins. pp. 80–87.

Khan AN. 2005. Secondary Hyperparathyroidism. eMedicine. Accessed May 18th 2006. www.emedicine.com/radio/topic356.htm.

Kanis J. 1994. Osteoporosis. Oxford: Blackwell Science.

Kanis J. 2002. Diagnosis of osteoporosis and assessment of fracture risk. Lancet 359:1929–1936.

Kaplan JM. 1986. Pseudoabuse – the misdiagnosis of child abuse. Journal of Forensic Sciences 31:1420–1428.

Karlsson M, Ahlborg H, and Karlsson C. 2005. Maternity and bone mineral density. Acta Orthopaedica 76:2–13.

Katzenberg MA, and Lovell NC. 1999. Stable isotope variation in pathological bone. International Journal of Osteoarchaeology 9:316–324.

Katzenberg MA, and Saunders S, editors. 2000. Biological anthropology of the human skeleton. New York: Wiley-Liss.

Keesing RM, and Strathern AJ. 1998. Cultural anthropology. A contemporary perspective. Third edition. Fort Worth: Harcourt Brace College Publishers.

Kelley JO, and Angel JL. 1987. Life stresses of slavery. American Journal of Physical Anthropology 74:199–211.

Kelley MA, and El-Najjar MY. 1980. Natural variation and differential diagnosis of skeletal changes in tuberculosis. American Journal of Physical Anthropology 52:153–167.

Kellogg N, and Lukefahr J. 2005. Criminally prosecuted cases of child starvation. Pediatrics 116:1309–1316.

Kennedy K. 1984. Growth, nutrition, and pathology in changing paleodemographic settings in South Asia. In: Cohen MN, Armelagos G, editors. Paleopathology at the origins of agriculture. New York: Academic Press. pp. 169–192.

Kerley ER. 1978. The identification of battered infant skeletons. Journal of Forensic Sciences 23:163–168.

Kerstetter JE, O'Brien KO, and Insogna KL. 2003a. Dietary protein, calcium metabolism and skeletal homeostasis revisited. American Journal of Clinical Nutrition 78 (Supplement):584S–592S.

Kerstetter JE, O'Brien KO, and Insogna KL. 2003b. Low protein intake: Impact on calcium and bone homeostasis in humans. American Society for Nutritional Sciences 133:855s–861s.

Kerstetter JE, Gaffney ED, O'Brien KO, Caseria DM, and Insogna KL. 2007. Dietary protein increases intestinal calcium absorption and improves bone balance: A hypothesis. International Congress Series 1297:204–216.

Kertzer DI, and Laslett P. 1995. Aging in the past: Demography, society and old age. Berkeley: University of California Press.

Kett ME. 2005. Internally displaced peoples in Bosnia-Herzegovina: Impacts of long-term displacement on health and well-being. Medicine, Conflict and Survival 21:199–215.

Khandare AL, Rao GS, and Lakshmaiah N. 2002. Effect of tamarind ingestion on fluoride excretion in humans. European Journal of Clinical Nutrition 51:82–85.

Kilgore L, Jurmain R, and Van Gerven D. 1997. Paleoepidemiological patterns of trauma in a Medieval Nubian skeletal population. International Journal of Osteoarchaeology 7:103–114.

Kiple KF, editor. 1993. The Cambridge world history of human disease. Cambridge: Cambridge University Press.

Kiratli BJ. 2001. Immobilization osteopenia. In: Marcus R, Feldman D, Kelsey J, editors. Osteoporosis. Volume 2. Second edition. San Diego: Academic Press. pp. 207–227.

Klein GL. 2006. Hyperparathyroidism. eMedicine. Accessed on 14th June 2007: www.emedicine.com/ped/topic1086.htm.

Kneissel M, Roschger P, Steiner W, Schamall D, Kalchhauser G, Boyde A, and Teschler-Nicola M. 1997. Cancellous bone structure in the growing and aging lumbar spine in a historic Nubian population. Calcified Tissue International 61:95–100.

Knick SG. 1981. Linear enamel hypoplasia and tuberculosis in the pre-Columbian North America. OSSA 8:131–138.

Knüsel CJ, and Bowman JE. 1996. A possible case of neurofibromatosis in an archaeological skeleton. International Journal of Osteoarchaeology 6:202–210.

Komara JS, Kottamasu L, and Kottamasu S. 1986. Acute plastic bowing fractures in children. Annals of Emergency Medicine 15:585–588.

Kosaryan M, Zadeh M, and Shahi V. 2004. The bone density of thalassemic patients of Boo Ali Sina Hospital, Sari, Iran, in 2002, does hydroxyurea help?. Pediatric Endocrinology Review 2 (Supplement 2):303–306.

Krane SM, and Simon LS. 1987. Metabolic consequences of bone turnover in Paget's disease of bone. Clinical Orthopedics 217:26–36.

Krishnamachari KAVR, and Iyengar L. 1975. Effect of maternal malnutrition on bone density of the neonates. American Journal of Clinical Nutrition 28:482–486.

Krohel GB, and Wright JE. 1979. Orbital hemorrhage. American Journal of Ophthalmology 88:254–258.

Kuh D, and Hardy R, editors. 2002. A life course approach to women's health. Oxford: Oxford University Press.

Kuhns J, and Wilson PD. 1928. Major amputations. Analysis and study of end-results in four hundred and twenty cases. Archives of Surgery (New York) 16:887–921.

Kuorilehto T, Pöyhönen M, Bloigu R, Heikkinen J, Väänänen K, and Peltonen J. 2005. Decreased bone mineral density and content in neurofibromatosis type 1: Lowest local values located in the load-carrying parts of the body. Osteoporosis International 16:928–936.

LaBan M, Wilkins J, Sackeyfio A, and Taylor R. 1995. Osteoporotic stress fractures in anorexia nervosa: Etiology, diagnosis and review of four cases. Archives of Physical Medicine and Rehabilitation 76:884–887.

Ladegaard Jakobsen AL. 1978. A cripple from the late middle ages. OSSA 5:17–24.

Låg J. 1987. Soil properties of special interest in connection with health problems. Cellular and Molecular Life Sciences 43:63–67.

Lagia A, Eliopoulos C, and Manolis S. 2007. Thalassemia: Macroscopic and radiological study of a case. International Journal of Osteoarchaeology 17:269–285.

Lahariya C, and Pradhan SK. 2007. Prospects of eradicating poliomyelitis by 2007: Compulsory vaccinaion may be a strategy. Indian Journal of Pediatrics 74:61–63.

Lamberg-Allardt CJE, Outila TA, Karkkainen MUM, Rita HJ, and Valsta LM. 2001. Vitamin D deficiency and bone health in healthy adults in Finland: Could this be a concern in other parts of Europe? Journal of Bone and Mineral Research 16:2066–2073.

Lambert P, editor. 2000. Bioarchaeological studies of life in the age of agriculture. A view from the Southeast. Tuscaloosa: University of Alabama Press.

Lambert P. 2002. Rib lesions in a prehistoric Puebloan sample from Southwestern Colorado. American Journal of Physical Anthropology 117:281–292.

Lambert PM. 2006. Infectious disease among enslaved African Americans at Eaton's Estate, Warren County, North Carolina, ca. 1830–1850. Memorias do Instituto Oswaldo Cruz 101 (Supplement II):107–117.

Lambert JE. 2007. Seasonality, fallback strategies, and natural selection: A chimpanzee and Cercopithecoid model for interpreting the evolution of hominin diet. In: Ungar PS, editor. Evolution of the human diet. Oxford: Oxford University Press. pp. 324–343.

Lanyon L, and Skerry T. 2001. Postmenopausal osteoporosis as a failure of bone's adaptation to functional loading: A hypothesis. Journal of Bone and Mineral Research 16:1937–1947.

Larsen CS. 1997. Bioarchaeology. Interpreting behaviour from the human skeleton. Cambridge: Cambridge University Press.

Larsen CS. 2003. Animal source foods and human health during evolution. Journal of Nutrition 133:3893S–3897S.

Larsen CS, Craig J, Sering LE, Schoeninger MJ, Russell KF, Hutchinson DL, and Williamson MA. 1995. Cross homestead: Life and death on the Midwestern frontier. In: Grauer AL, editor. Bodies of evidence. Chichester: Wiley. pp. 139–159.

Larsen CS, Crosby AW, Griffin MC, Hutchinson DL, Ruff CB, Russell KF, Schoeninger MJ, Sering LE, Simpson SW, Takács JL, and Teaford MF. 2002. A biohistory of health and behaviour in the Georgia Bight. In: Steckel RH, Rose JC, editors. The backbone of history. Health and nutrition in the Western Hemisphere. New York: Cambridge University Press. pp. 406–439.

Laskey MA, Prentice A, Hanratty LA, Jarjou LMA, Dibba B, Beavan SR, and Cole TJ. 1998. Bone changes after 3 mo of lactation: Influence of calcium intake, breast-milk output, and vitamin D-receptor genotype. American Journal of Clinical Nutrition 67:685–692.

Lazenby RA. 1990. Continuing periosteal apposition I: Documentation, hypotheses, and interpretation. American Journal of Physical Anthropology 82:451–472.

Lazenby RA. 1997. Bone loss, traditional diet and cold adaptation in artic populations. American Journal of Human Biology 9:329–341.

Lazenby RA, and Pfeiffer SK. 1993. Effects of a nineteenth century below the knee amputation and prosthesis on femoral morphology. International Journal of Osteoarchaeology 3:19–28.

Lazo MG, Shirazi P, Sam M, Giobbie-Hurder A, Blacconiere MJ, and Muppidi M. 2001. Osteoporosis and risk of fracture in men with spinal cord injury. Spinal Cord 39:208–214.

Le May M, and Blunt JW. 1949. A factor determining the location of pseudofractures in osteomalacia. Journal of Clinical Investigation 28:521–525.

Lee T. 1940. Historical notes on some vitamin deficiency diseases in China. Chinese Medical Journal 58:314–323.

Lee R, and DeVore I, editors. 1968. Man the hunter. Chicago: Aldine.

Lees B, Molleson T, Arnett TR, and Stevenson JC. 1993. Differences in proximal femur bone density over two centuries. Lancet 341:673–675.

Lee-Thorp J, and Sponheimer M. 1993. Contributions of biogeochemistry to understanding hominin dietary ecology. Yearbook of Physical Anthropology 49:131–148.

Lehmann-Nitsche R. 1903. La arthritis deformans de los antiguos Patagones. Revta Museum La Plata 11:199.

Lemann J Jr., and Favus MJ. 1999. The intestinal absorption of calcium, magnesium and phosphate. In: Favus M, editor. Primer on the metabolic bone diseases and disorders of mineral metabolism. Fourth edition. Philadelphia: Lippincott William & Wilkins. pp. 63–67.

Lever JH. 2002. Paget's disease of bone in Lancashire and arsenic pesticide in cotton mill wastewater: A speculative hypothesis. Bone 31:434–436.

Levy H. 1945. Scurvy in London. Lancet 246:92.

Lewis M. 2002. Urbanisation and child health in Medieval and post-Medieval England. An assessment of the morbidity and mortality of non-adult skeletons from the cemeteries of two urban and two rural sites in England (AD 850–1859). British Archaeological Reports British Series 339. Oxford: Archaeopress.

Lewis S, Baker I, and Davey-Smith G. 2005. Meta-analysis of vitamin D receptor polymorphisms and pulmonary tuberculosis risk. International Journal of Tuberculous Lung Disease 9:1174–1177.

Lewis M. 2007. The bioarchaeology of children: Perspectives from biological and forensic anthropology. Cambridge: Cambridge University Press.

Li G, White G, Connolly C, and Marsh D. 2002. Cell proliferation and apoptosis during fracture healing. Journal of Bone and Mineral Research 17:791–799.

Lian J, Stein G, Canalis E, Gehron Robey P, and Boskey A. 1999. Bone formation: Osteoblast lineage cells, growth factors, matrix proteins and the mineralisation process. In: Favus M, editor. Primer on the metabolic bone diseases and disorders of mineral metabolism. Fourth edition. Philadelphia: Lippincott William & Wilkins. pp. 14–29.

Liang HW, Chen SY, Hsu JH, and Chang CW. 2004. Work-related upper limb amputations in Taiwan 1999–2001. American Journal of Industrial Medicine 46:649–655.

Liberman UA, and Marx SJ. 1999. Vitamin D-dependent rickets. In: Favus M, editor. Primer on the metabolic bone diseases and disorders of mineral metabolism. Fourth edition. Philadelphia: Lippincott William & Wilkins. pp. 323–328.

Lilley J, Stroud G, Brothwell DR, and Williamson M, editors. 1994. The Jewish burial ground at Jewbury. The archaeology of York volume 12/3. The Medieval cemeteries. York archaeological trust for excavation and research. York: Council for British Archaeology.

Lindsay R, and Cosman F. 1999. Estrogen. In: Rosen C, Glowacki J, Bilezikian JP, editors. The aging skeleton. San Diego: Academic Press. pp. 495–505.

Linkhart TA, Mohan S, and Baylink DJ. 1996. Growth factors for bone growth and repair: IGF, TGF beta and BMP. Bone 19 (Supplement 1):1S–12S.

Lips P. 2001. Vitamin D deficiency and secondary hyperparathyroidism in the elderly: Consequences for bone loss and fractures and therapeutic implications. Endocrine Reviews 22:477–501.

Littleton J. 1998. A middle eastern paradox: Rickets in skeletons from Bahrain. Journal of Paleopathology 10:13–30.

Littleton J. 1999. Paleopathology of skeletal fluorosis. American Journal of Physical Anthropology 109:465–483.

Littleton J, and Frolich B. 1989. An analysis of dental pathology from historic Bahrain. Paléorient 15:59–75.

Livshits G, Karasik D, Otremski I, and Kobyliansky E. 1998. Genes play an important role in bone aging. American Journal of Human Biology 10:421–438.

Lloyd T, and Cardamone Cusatis D. 1999. Nutritional determinants of peak bone mass. In: Rosen C, Glowacki J, Bilezikian J, editors. The aging skeleton. San Diego: Academic Press. pp. 95–103.

Löhr AJ, Bogaard TA, Heikens A, Hendriks MR, Sumarti S, Van Bergen MJ, Van Gestel CAM, Van Straalen NM, Vroon PZ, and Widianarko B. 2005. Natural pollution caused by the extremely acidic crater lake Kawah Ijen, East Java, Indonesia. Environmental Science and Pollution Research 12:85–95.

Lomax E. 1986. Difficulties in diagnosing infantile scurvy before 1878. Medical History 30:70–80.

Looser E. 1920. Über pathologische formen von infraktionen und callusbildungen bei rachitis und osteomalakie und anderen knochenerkrankungen. Zentralblatt für Chirurgie 47:1470.

Lovejoy CO, and Heiple KG. 1981. The analysis of fractures in skeletal populations with an example from the libben site, Ottawa County, Ohio. American Journal of Physical Anthropology 55:529–541.

Lovell NC. 1997a. Trauma analysis in paleopathology. Yearbook of Physical Anthropology 40:139–170.

Lovell NC. 1997b. Anaemia in the ancient Indus Valley. International Journal of Osteoarchaeology 7:115–123.

Loveridge N, Power J, Reeve J, and Boyde A. 2004. Bone mineralization density and femoral neck fracture. Bone 35:929–941.

Luckin B. 2003. 'The heart and home of horror': The great London fogs of the late nineteenth century. Social History 28:31–48.

Luk KDK. 1999. Tuberculosis of the spine in the new millennium. European Spine Journal 8:338–345.

Lukacs JR, Retief DH, and Jarrige JF. 1985. Dental disease in prehistoric Baluchistan. National Geographic Research 1:184–197.

Lyman RL. 1994. Vertebrate taphonomy. Cambridge: Cambridge University Press.

Lynnerup N, and Von Wowern N. 1997. Bone mineral content in Medieval Greenland Norse. International Journal of Osteoarchaeology 7:235–240.

Maat GJR. 1982. Scurvy in Dutch whalers buried at Spitsbergen. In: Haneveld GT, Perizonius WRK, editors. Proceedings of the fourth European members meeting of the Paleopathology Association, Middelburg-Antwerpen, 16th–19th September 1982. Utrecht: Eliinkwijk. pp. 82–93.

Maat GJR. 2004. Scurvy in adults and youngsters: The Dutch experience. A review of the history and pathology of a disregarded disease. International Journal of Osteoarchaeology 14:77–81.

Maat GJR, and Bond J. 2007. The repudiated angle: Identification of a male corpse found where it simply couldn't be. In: Brickley M, Ferllini R, editors. Forensic anthropology: Case studies from Europe. Springfield: Charles C Thomas. pp. 58–68.

Maat GJR, and Mastwijk R. 2000. Avulsion injuries of vertebral endplates. International Journal of Osteoarchaeology 10:142–152.

Maat GJR, and Uytterschaut HT. 1984. Microscopic observations on scurvy in Dutch whalers buried at Spitsbergen. In: Capecchi V, Rabino Massa E, editors. Proceedings of the fifth European meeting of the Paleopathology Association. Sienna: Tipografia Senese. pp. 211–218.

MacCurdy GG. 1923. Human skeletal remains from the highlands of Peru. American Journal of Physical Anthropology 6:217–329.

Magnus O. 1555. Historia de Gentibus Septenrionalbus, Rome.

Malabanan A, and Holick M. 2003. Vitamin D and bone health in postmenopausal women. Journal of Women's Health 12:151–156.

Manchester K. 1983. The archaeology of disease. Leeds: Arthur Wrigley & Sons.

Mankin H. 1974. Rickets, osteomalacia and renal osteodystrophy. Part I. Journal of Bone Joint Surgery 56A:101–128.

Mann RW, Roberts CA, Thomas MD, and Davy DT. 1991. Pressure erosion of the femoral trochlea, patella baja, and altered patellar surfaces. American Journal of Physical Anthropology 85:321–327.

Marcsik A, Szentgyörgyi R, Gyetvai A, Finnegan M, and Pálfi G. 1999. Probable Pott's paraplegia from the 7th–8th century AD. In: Pálfi G, Dutour O, Deak J, Hutas I, editors. Tuberculosis past and present. Hungary: Golden Book Publisher Ltd/ Tuberculosis Foundation. pp. 333–336.

Marcus R, and Majumder S. 2001. The nature of osteoporosis. In: Marcus R, Feldman D, Kelsey J, editors. Osteoporosis. Volume 2. Second edition. San Diego: Academic Press. pp. 3–18.

Marcus R, Feldman D, and Kelsey J, editors. 2001. Osteoporosis. Volume 2. Second edition. San Diego: Academic Press.

Marks S, and Odgren P. 2002. Structure and development of the skeleton. In: Bilezikian JP, Raisz LG, Rodan GA, editors. Principles of bone biology. Second edition. San Diego: Academic Press. pp. 3–15.

Martin RB. 2000. Toward a unifying theory of bone remodelling. Bone 26:1–6.

Martin RB. 2003. Functional adaptation and fragility of the skeleton. In: Agarwal S, Stout SD, editors. Bone loss and osteoporosis. An anthropological perspective. New York: Kluwer Academic/Plenum Publishers. pp. 121–138.

Martin DL, and Armelagos GJ. 1979. Morphometrics of compact bone: An example from Sudanese Nubia. American Journal of Physical Anthropology 51:571–578.

Martin W III, and Roddervold HO. 1979. Acute plastic bowing fractures of the fibula. Radiology 131:639–640.

Martin DL, Armelagos GJ, Goodman AH, and Van Gerven D. 1984. The effects of socio-economic change in prehistoric Africa: Sudanese Nubia as a case study. In: Cohen MN, Armelagos GJ, editors. Paleopathology at the origins of agriculture. New York: Academic Press. pp. 93–213.

Martin DL, Goodman AH, and Armelagos GJ. 1985. Skeletal pathologies as indicators of the quality and quantity of diet. In: Gilbert RI, Mielke JH, editors. The analysis of prehistoric diets. San Diego: Academic Press. pp. 227–279.

Martin DL, Magennis AL, and Rose JC. 1987. Cortical bone maintenance in an historic Afro-American cemetery sample from Cedar Grove, Arkansas. American Journal of Physical Anthropology 74:255–264.

Martin R, Burr D, and Sharkey N. 1998. Skeletal tissue mechanics. New York: Springer-Verlag.

Martínez-Mier EA, Soto-Rojas AE, Ureña-Cirett JL, Katz BP, Stookey GK, and Dunipace AJ. 2004. Dental fluorosis and altitude: A preliminary study. Oral Health and Preventative Dentistry 2:39–48.

Mason JB. 2002. Lessons on nutrition of displaced people. Journal of Nutrition 132:2096S–2103S.

Massey LK. 2003. Dietary animal and plant protein and human bone health: A whole foods approach. Journal of Nutrition 133:862S–865S.

Mayhew PM, Thomas CD, Clement JG, Loveridge N, Beck TJ, Bonfield W, Burgoyne CJ, and Reeve J. 2005. Relation between age, femoral neck cortical stability, and hip fracture risk. Lancet 366:129–135.

Mays S. 1992. Taphonomic factors in a human skeletal assemblage. Circaea 9:54–58.

Mays S. 1996a. Healed limb amputations in human osteoarchaeology and their causes: A case study from Ipswich, UK. International Journal of Osteoarchaeology 6:101–113.

Mays S. 1996b. Age-dependent cortical bone loss in a medieval population. International Journal of Osteoarchaeology 6:144–154.

Mays S. 1997. Carbon stable isotope ratios in medieval and later human skeletons from northern England. Journal of Archaeological Science 24:561–567.

Mays S. 1999. Osteoporosis in earlier human populations. Journal of Clinical Densitometry 2:71–78.

Mays S. 2000. Age-dependent cortical bone loss in women from 18th and early 19th century London. American Journal of Physical Anthropology 112:349–361.

Mays S. 2001. Effects of age and occupation on cortical bone in a group of 18th–19th century British men. American Journal of Physical Anthropology 116:34–44.

Mays S. 2003. The rise and fall of rickets in England. In: Murphy P, Wiltshire PEJ, editors. The environmental archaeology of industry. Oxford: Oxbow. pp. 144–153.

Mays S. 2006a. Age-related cortical bone loss in women from a 3rd–4th century AD population from England. American Journal of Physical Anthropology 129:518–528.

Mays S. 2006b. A palaeopathological study of Colles' fracture. International Journal of Osteoarchaeology 16:415–428.

Mays S. 2008. A likely case of scurvy from early bronze age Britain. International Journal of Osteoarchaeology 18:178–187.

Mays S, Lees B, and Stevenson JC. 1998. Age-dependent bone loss in the femur in a medieval population. International Journal of Osteoarchaeology 8:97–106.

Mays S, Rogers J, and Watt I. 2001. A possible case of hyperparathyroidism in a burial of 15th–17th century AD date from Wharram Percy, England. International Journal of Osteoarchaeology 11:329–335.

Mays S, Brickley M, and Ives R. 2006a. Skeletal manifestations of rickets in infants and young children in a historic population from England. American Journal of Physical Anthropology 129:362–374.

Mays S, Turner-Walker G, and Syversen U. 2006b. Osteoporosis in a population from Medieval Norway. American Journal of Physical Anthropology 131:343–351.

Mays S, Brickley M, and Ives R. 2007. Skeletal evidence for hyperparathyroidism in a 19th century child with rickets. International Journal of Osteoarchaeology 17:73–81.

Mays S, Brickley M, and Ives R. in press. Growth and vitamin D deficiency in a population from 19th century Birmingham, England. International Journal of Osteoarchaeology.

Mazess R, Barden H, Christiansen C, Harper A, and Laughlin W. 1985. Bone mineral and vitamin D in Aleutian Islanders. American Journal of Clinical Nutrition 42:143–146.

McCann P. 1962. The incidence and value of radiological signs of scurvy. British Journal of Radiology 35:683–686.

McKibbin B. 1978. The biology of fracture healing in long bones. Journal of Bone and Joint Surgery 60:150–162.

McKinley J. 2000. The analysis of cremated bone. In: Cox M, Mays S, editors. Human osteology in archaeology and forensic sciences. London: Greenwich Medical Media. pp. 403–422.

McMurray DN, Bartow RA, Mintzer CL, and Hernandez-Frontera E. 1990. Micronutrient status and immune function in tuberculosis. Annals of New York Academy of Science 587:59–69.

Meade JL, and Brissie RM. 1985. Infanticide by starvation: Calculation of caloric deficit to determine degree of deprivation. Journal of Forensic Sciences 30:1263–1268.

Means HJ. 1925. A roentgenological study of the skeletal remains of the prehistoric mound builder Indians of Ohio. American Journal of Roentgenology 13:359–367.

Mehrotra M, Gupta SK, Kumar K, Awasthi PK, Dubey M, Pandy CM, and Godbole MM. 2006. Calcium deficiency-induced secondary hyperparathyroidism and osteopenia are rapidly reversible with calcium supplementation in growing rabbit pups. British Journal of Nutrition 95:582–590.

Meiklejohn C, Schentag C, Venema A, and Key P. 1984. Socio-economic change and patterns of pathology and variation in the mesolithic and neolithic of Western Europe. Some suggestions. In: Cohen MN, Armelagos G, editors. Paleopathology at the origins of agriculture. New York: Academic Press. pp. 75–100.

Melikian M, and Waldron T. 2003. An examination of skulls from two British sites for possible evidence of scurvy. International Journal of Osteoarchaeology 13:207–212.

Melton LJ III. 2001. The prevalence of osteoporosis: Gender and racial comparison. Calcified Tissue International 69:179–181.

Melton LJ III, and Kallmes D. 2006. Epidemiology of vertebral fractures: Implications for vertebral augmentation. Academic Radiology 13:538–545.

Melton LJ III, and Riggs B. 1986. Impaired bone strength and fracture patterns at different skeletal sites. In: Uhthoff HK, editor. Current concepts of bone fragility. Berlin: Springer-Verlag. pp. 149–157.

Melton LJ III, Chao EYS, and Lane J. 1988. Biomechanical aspects of fractures. In: Riggs BL, Melton LJ III, editors. Osteoporosis. Etiology, diagnosis and management. New York: Raven Press. pp. 111–131.

Melton LJ III, Thamer M, Ray NF, Chan JK, Chesnut CH III, Einhorn TA, Johnston CC, Raisz LG, Silverman SL, and Siris ES. 1997. Fractures attributable to osteoporosis: Report from the national osteoporosis foundation. Journal of Bone and Mineral Research 12:16–23.

Melton LJ III, Crowson CS, O'Fallon WM, Wahner HW, and Riggs BL. 2003. Relative contributions of bone density, bone turnover, and clinical risk factors to long-term fracture prediction. Journal of Bone and Mineral Research 18:312–318.

Melton LJ III, Kyle RA, Achenbach SJ, Oberg AL, and Rajkumar SV. 2005. Fracture risk with multiple myeloma: A population-based study. Journal of Bone and Mineral Research 20:487–493.

Mensforth RP. 2002. Vitamin D deficiency mortality: Impaired immune response in infants and elevated cancer risk in adults. American Journal of Physical Anthropology 34 (Supplement):112.

Mensforth RP, and Latimer BM. 1989. Hamann-Todd collection aging studies: Osteoporosis fracture syndrome. American Journal of Physical Anthropology 80:461–479.

Meyer HE, Falch JA, Sogaard AJ, and Haug E. 2004. Vitamin D deficiency and secondary hyperparathyroidism and the association with bone mineral density in persons with Pakistani and Norwegian background living in Oslo, Norway. The Oslo health study. Bone 35:412–417.

Miklíková Z, and Thomas R, editors. in press. Current research in animal palaeopathology. Proceedings of the second ICAZ animal palaeopathology working group conference. Slovak University of Agriculture, Nitra, Slovakia, 23rd–24th September 2004, British Archaeological Reports International Series. Oxford: Archaeopress.

Milanesi Q. 1962. Su due ossa pathologiche del sepolcret eneolitico di Ponte di S. Pietro. Archivo per l'Antropologia è le Etnologia 93:215–233.

Miles A, Powers N, Wroe-Brown R, and Walker D. in preparation. St. Marylebone Church and burial ground: Excavations at St Marylebone School 1993 and 2004–2006. MoLAS Monograph Series.
Milgram JW. 1990. Radiologic and histologic pathology of nontumorous diseases of bone and joints. Illinois: Northbrook Publishing Company.
Milkman L. 1930. Pseudofractures hunger osteopathy, late rickets, osteomalacia. Report of a case. American Journal of Roentgenology 24:29–37.
Milkman L. 1934. Multiple spontaneous idiopathic symmetrical fractures. American Journal of Roentgenology 32:622–634.
Miller G, Jarvis J, and McBean L. 2001. The importance of meeting calcium needs with foods. Journal of the American College of Nutrition 20:168s–185s.
Miller R, Seqal J, Ashar B, Leung S, Ahmed S, Siddique S, Rice T, and Lanzkron S. 2006. High prevalence and correlates of low bone mineral density in young adults with sickle cell disease. American Journal of Hematology 81:236–241.
Mirra JM, Brien EW, and Tehranzadeh J. 1995a. Paget's disease of bone: Review with emphasis on radiologic features, part I. Skeletal Radiology 24:163–171.
Mirra JM, Brien EW, and Tehranzadeh J. 1995b. Paget's disease of bone: Review with emphasis on radiologic features, part II. Skeletal Radiology 24:173–184.
Mitchell PD. 1999. The integration of the paleopathology and medical history of the Crusades. International Journal of Osteoarchaeology 9:333–343.
Mitchell PD. 2004. Medicine in the crusades. Warfare, wounds and the medieval surgeon. Cambridge: Cambridge University Press.
Mitchell PD. 2006. Child health in the crusader period inhabitants of Tel Jezreel, Israel. Levant 38:37–44.
Mitchell PD, and Redfern RC. 2008. Diagnosistic criteria for development dislocation of the hip in human skeletal remains. International Journal of Osteoarchaeology 18:61–71.
Mitra D, and Bell N. 1997. Racial, geographic, genetic and body habitus effects on vitamin D metabolism. In: Feldman D, Glorieux F, Pike J, editors. Vitamin D. San Diego: Academic Press. pp. 521–532.
Mogle P, and Zias J. 1995. Trephanation as a possible treatment for scurvy in a middle Bronze Age (ca. 2200 BC) skeleton. International Journal of Osteoarchaeology 5:77–81.
Møller-Christensen V. 1958. Bogen om aebelholt kloster. Copenhagen: Dansk Videnskabs Forlag.
Molleson T, and Cox M. 1993. The Spitalfields project. Volume 2 – the anthropology. The middling sort. York: Council for British Archaeology Research Report 86.
Molto J. 2000. Humerus varus deformity in Roman period burials from Kellis 2, Dakhleh, Egypt. American Journal of Physical Anthropology 118:103–109.
Montiel R, Malagosa A, and Francalacci P. 2001. Authenticating ancient human mitochondrial DNA. Human Biology 73:689–713.
Moodie Rl. 1931. Roentgenologic studies of Egyptian and Peruvian mummies. Field Museum of Natural History and Anthropology Memoirs 3.
Moore TE, Kathol MH, El-Khoury GY, Walker CW, Gendall PW, and Whitten CG. 1994. Unusual radiological features in Paget's disease of bone. Skeletal Radiology 23:257–260.
Morel P, and Demetz JL. 1961. Pathologie osseuse du haut moyen-age (Contributions aux problèmes des burgondes). Paris: Masson.

Morgan JP, and Eisele PH. 1992. Radiographic changes in rhesus macaques affected by scurvy. Veterinary Radiology and Ultrasound 33:334–339.

Mosekilde L. 2005. Vitamin D and the elderly. Clinical Endocrinology 62:265–281.

Moss WJ, Ramakrishnan M, Storms D, Henderson Siegle A, Weiss WM, Lejnev I, and Muhe L. 2006. Child health in complex emergencies. Bulletin of the World Health Organization 84:58–64.

Moulton FR, editor. 1942. Fluoride and dental health. Washington, DC: American Association for Advancement of Science.

Moyad M. 2005. An introduction to dietary/supplemental omega-3 fatty acids for general health and prevention: Part II. Urologic Oncology: Seminars and Original Investigations 23:36–48.

Mulhern D, and Jones E. 2005. Test of revised method of age estimation from the auricular surface of the ilium. American Journal of Physical Anthropology 126:61–65.

Mulhern D, and Van Gerven D. 1997. Patterns of femoral bone remodelling dynamics in a Medieval Nubian population. American Journal of Physical Anthropology 104:133–146.

Mundy GR. 1999. Bone remodelling. In: Favus M, editor. Primer on the metabolic bone diseases and disorders of mineral metabolism. Fourth edition. Philadelphia: Lippincott William & Wilkins. pp. 30–38.

Murphy EM. 2000. Developmental defects and disability: The evidence from the Iron-Age semi-nomaidc peoples of Aymyrlyg, South Siberia. In: Hubert J, editor. Madness, disability and social exclusion: The archaeology and anthropology of difference. One World Archaeology Series Volume 40. London: Routledge. pp. 60–80.

Murphy P, and Wiltshore PEJ, editors. 2003. The environmental archaeology of industry. Oxford: Oxbow.

Muthukumar N. 2004. Ossification of the ligamentum flavum as a result of fluorosis causing myelopathy. Neurosurgery 56:622.

Nagant de Deuxchaisnes C, and Krane SM. 1964. Paget's disease of bone: Clinical and metabolic observations. Medicine (Baltimore) 43:233–266.

Namkung-Matthal H, Appleyard R, Jansen J, Hao Lin J, Maastricht S, Swain M, Mason RS, Murrell GAC, Diwan AD, and Diamond T. 2001. Osteoporosis influences the early period of fracture healing in a rat osteoporotic model. Bone 28:80–86.

National Osteoporosis Foundation. 2002. America's bone health: The state of osteoporosis and low bone mass in our nation. Accessed on 14th June 2007: http://www.nof.org/.

Nawaz Khan A, Chandramohan M, Turnabll I, and Macdonald S. 2003. Neuropathic arthropathy (Charcot joint). eMedicine. Accessed on 28th September 2006: http://www.emedicine.com/radio/topic476.htm.

Neiburger E. 1989. A prehistoric case of scalping. 600AD. Central Sites Archaeological Journal 36:204–208.

Nemeskéri J, and Harsányi L. 1959. Die Bedeutung paläopathologischer Untersuchungen für die historiche Anthropologie. Homo 10:203–226.

Nerlich A, Parsche F, Von Den Driesch A, and Lohrs U. 1993. Osteopathological findings in mummified baboons from ancient Egypt. International Journal of Osteoarchaeology 3:189–198.

Nerlich A, Zink A, Hagedorn H, Szeimies U, and Weyss C. 2000. Anthropological and palaeopathological analysis of the human remains from three 'Tombs of the nobles' of the necropolis of Thebes-West, Upper Egypt. Anthropology Anz 4:321–343.

Nesby-O'Dell S, Scanlon K, Cogswell M, Gillespie C, Hollis B, Looker A, Allen C, Doughertly C, Gunter E, and Bowman A. 2002. Hypovitaminosis D prevalence and determinants among African American and white women of reproductive age: Third national health and nutrition examination survey, 1988–1994. American Journal of Clinical Nutrition 76:187–192.

Nestle M. 2000. Paleolithic diets: A sceptical view. Nutrition Bulletin 25:43–47.

New S. 2001. Exercise, bone and nutrition. Proceedings of the Nutrition Society 60:265–274.

New S. 2003. Intake of fruit and vegetables: Implications for bone health. Proceedings of the Nutrition Society 62:889–899.

Nguyen TV, Center JR, and Eisman JA. 2000. Osteoporosis in elderly men and women: Effects of dietary calcium, physical activity and body mass index. Journal of Bone and Mineral Research 15:322–331.

Nielsen HA. 1911. Yderligere bidrag til stenalderfolkets anthropologi. Aarb Nord Oldk Hist. 3er series 1:81–205.

NIH Consensus Development Panel on Optimal Calcium Intake. 1994. Optimal calcium intake. Journal of the American Medical Association 272:1942–1948.

Nikander R, Sievänen H, Heinonen A, and Kannus P. 2005. Femoral neck structure in adult female athletes subjected to different loading modalities. Journal of Bone and Mineral Research 20:520–528.

Nishikimi M, and Udenfriend S. 1977. Scurvy as an inborn error of ascorbic acid biosynthesis. Trends in Biochemical Science 2:111–113.

Norris J. 1983. The 'Scurvy Disposition': Heavy exertion as an exacerbating influence on scurvy in modern times. Bulletin of History of Medicine 57:325–338.

Nystrom KC, Buikstra JE, and Braunstein EM. 2005. Radiographic evaluation of two early classic elites from Copan, Honduras. International Journal of Osteoarchaeology 15:196–207.

O'Connor NJ. 1977. A methodological study of the paleodemography of various Late Archaic populations of Eastern North America. Master's thesis. Indiana University, US.

Olsen BR. 1999. Bone morphogenesis and embryologic development. In: Favus M, editor. Primer on the metabolic bone diseases and disorders of mineral metabolism. Fourth edition. Philadelphia: Lippincott William & Wilkins. pp. 11–14.

Orden A, Oyhenart E, Cesani M, Zucchi M, Muñe M, Villanueva M, Rodriguez R, Pons E, and Pucciarelli H. 2002. Effect of moderate undernutrition the functional components of the axial skeleton. American Journal of Physical Anthropology 34 (Supplement):120.

Orenstein E, Dvonch V, and Demos T. 1985. Acute traumatic bowing of the tibia without fracture. Journal of Bone and Joint Surgery 67A:965–967.

O'Riordan JLH. 2006. Rickets in the 17th Century. Journal of Bone and Mineral Research 21:1506–1510.

Ortner D. 1984. Bone lesions in a probable case of scurvy from Metlatavic, Alaska. MASCA Journal 3:79–81.

Ortner D. 2003. Identification of pathological conditions in human skeletal remains. Second edition. San Diego: Academic Press.

Ortner DJ, and Aufderheide AC, editors. 1991. Palaeopathology: Current syntheses and future options. Washington: Smithsonian Institution Press.

Ortner D, and Ericksen M. 1997. Bone changes in the human skull probably resulting from scurvy in infancy and childhood. International Journal of Osteoarchaeology 7:212–220.

Ortner D, Garofalo E, and Frohlich B. 2006. Metabolic disease in the early Bronze Age people of Bab edh-Dhra, Jordan. 16th Paleopathology Association European Meeting, Santorini Island. pp. 100.

Ortner D, and Mays S. 1998. Dry-bone manifestations of rickets in infancy and early childhood. International Journal of Osteoarchaeology 8:45–55.

Ortner DJ, and Putschar W. 1981. Identification of pathological conditions in human skeletal remains. Washington: Smithsonian Institution Press.

Ortner D, and Utermohle CJ. 1981. Polyarticular inflammatory arthritis in a pre-columbian skeleton from Kodiak Island, Alaska, USA. American Journal of Physical Anthropology 56:23–31.

Ortner D, Kimmerle E, and Diez M. 1999. Probable evidence of scurvy in subadults from archaeological sites in Peru. American Journal of Physical Anthropology 108:321–331.

Ortner D, Butler W, Cafarella J, and Milligan L. 2001. Evidence of probable scurvy in subadults from archaeological sites in North America. American Journal of Physical Anthropology 114:343–351.

Ortner DJ, Garofalo EM, and Zuckerman MK. 2007. The EBIA burials of BAB EDH-DHRA', Jordan: Bioarchaeological evidence of metabolic bone disease. In: Faerman M, Horwitz LK, Kahana T, Zilberman U, editors. Faces from the past. Papers in honor of Patricia Smith. British Archaeological Reports International Series 1603. Oxford: Archeopress. pp. 181–194.

Orwoll E, and Klein R. 2001. Osteoporosis in men. Epidemiology, pathophysiology and clinical characterization. In: Marcus R, Feldman D, Kelsey J, editors. Osteoporosis. Volume 2. Second edition. San Diego: Academic Press. pp. 103–150.

Orwoll ES, Belknap JK, and Klein RF. 2001. Gender specificity in the genetic determinants of peak bone mass. Journal of Bone and Mineral Research 16:1962–1971.

Owsley DW, Orser CE, Mann RW, Moore-Jansen PH, and Montgomery RL. 1987. Demography and pathology of an urban slave population. American Journal of Physical Anthropology 74:185–197.

Özbek M. 2005. Skeletal pathology of a high-ranking official from Thrace (Tukey, last quarter of the 4th century BC). International Journal of Osteoarchaeology 15:216–225.

Pacifici R. 2001. Postmenopausal osteoporosis. How the hormonal changes of menopause cause bone loss. In: Marcus R, Feldman D, Kelsey J, editors. Osteoporosis. Volume 2. Second edition. San Diego: Academic Press. pp. 85–101.

Paget J. 1882. The Bradshaw lecture. Some rare and new diseases. British Medical Journal 16 (Dec):1189–1193.

Paine RR, and Brenton BP. 2006a. The paleopathology of pellagra: Investigating the impact of prehistoric and historical transitions to maize. Journal of Anthropological Sciences 84:125–135.

Paine RR, and Brenton BP. 2006b. Dietary health does affect histological age assessment: An evaluation of the Stout and Paine (1992) age estimation equation using secondary osteons from the rib. Journal of Forensic Science 51:489–492.

Pales L. 1929. Maladie de Paget prehistorique. Avec un note additionelle du Prof. R. Verneau. L'anthropologie 39:263–270.

Pálfi G, Dutour O, Deak J, and Hutas I, editors. 1999. Tuberculosis past and present. Hungary: Golden Book Publisher Ltd/Tuberculosis Foundation.

Palmer CH, and Weston JT. 1976. Several unusual cases of child abuse. Journal of Forensic Sciences 21:851–855.

Pangan AL, and Robinson D. 2001. Hemarthrosis as initial presentation of scurvy. Journal of Rheumatology 28:1923–1925.

Pankovich A, Simmons D, and Kulkarni V. 1974. Zonal osteons in cortical bone. Clinical Orthopedics and Related Research 100:356–363.

Panter-Brick C, editor. 1998. Biosocial perspectives on children. Cambridge: Cambridge University Press.

Panter-Brick C. 1998. Introduction: Biosocial research on children. In: Panter-Brick C, editor. Biosocial perspectives on children. Cambridge: Cambridge University Press. pp. 1–9.

Panuel M, Portier F, Pálfi G, Chaumoitre K, and Dutour O. 1999. Radiological differential diagnosis of skeletal tuberculosis. In: Pálfi G, Dutour O, Deak J, Hutas I, editors. Tuberculosis past and present. Hungary: Golden Book Publisher Ltd/ Tuberculosis Foundation. pp. 229–234.

Papathanasiou A. 2005. Health status of the Neolithic population of Alepotrypa Cave, Greece. American Journal of Physical Anthropology 126:377–390.

Parfitt AM. 1986a. Bone fragility in osteomalacia: Mechanisms and consequences. In: Uhthoff HK, editor. Current concepts of bone fragility. Berlin: Springer Verlag. pp. 265–270.

Parfitt AM. 1986b. Cortical porosity in postmenopausal and adolescent wrist fractures. In: Uhthoff HK, editor. Current concepts of bone fragility. Berlin: Springer Verlag. pp. 167–172.

Parfitt AM. 1988. Bone remodelling: Relationship to the amount and structure of bone, and the pathogenesis and prevention of fractures. In: Riggs BL, Melton LJ III, editors. Osteoporosis. Etiology, diagnosis and management. New York: Raven Press. pp. 45–93.

Parfitt AM. 1994. Osteonal and hemi-osteonal remodelling: The spatial and temporal framework for signal traffic in adult human bone. Journal of Cellular Biochemistry 55:273–286.

Parfitt AM. 1997. Vitamin D and the pathogenesis of rickets and osteomalacia. In: Feldman D, Glorieux F, Pike J, editors. Vitamin D. San Diego: Academic Press. pp. 645–662.

Parfitt AM. 1998. Osteomalacia and related disorders. In: Avioli L, Krane S, editors. Metabolic bone disease and clinically related disorders. Third edition. San Diego: Academic Press. pp. 327–386.

Parfitt AM. 2000. The mechanism of coupling: A role for the vasculature. Bone 26:319–323.

Parfitt AM. 2003. New concepts of bone remodelling: A unified spatial and temporal model with physiologic and pathophysiologic implications. In: Agarwal S,

Stout SD, editors. Bone loss and osteoporosis: An anthropological perspective. New York: Kluwer Academic/Plenum Publishers. pp. 3–17.

Parfitt AM. 2004. What is the normal rate of bone remodelling?. Bone 35:1–3.

Park EA. 1923. The etiology of rickets. Physiological Reviews 3:106–163.

Park EA. 1964. The imprinting of nutritional disturbances on the growing bone. Pediatrics 33:815–862.

Park EA, Guild HG, Jackson D, and Bond M. 1935. The recognition of scurvy with especial reference to the early X-ray changes. Archives of Disease in Childhood 10:265–294.

Patra RC, Dwivedi SK, Bhardwaj B, and Swarup D. 2000. Industrial fluorosis in cattle and buffalo around Udaipur, India. The Science of the Total Environment 235:145–150.

Pearson O, and Lieberman D. 2004. The aging of Wolff's 'Law': Ontogeny and responses to mechanical loading in cortical bone. Yearbook of Physical Anthropology 47:63–99.

Peccei J. 2001. Menopause: Adaptation or epiphenomenon? Evolutionary Anthropology 10:43–57.

Pechenkina EA, and Delgado M. 2006. Dimensions of health and social structure in the early intermediate cemetery at Villa El Salvador, Peru. American Journal of Physical Anthropology 131:218–235.

Peck J, and Stout SD. 2007. Intra-skeletal variability in bone mass. American Journal of Physical Anthropology 132:89–97.

Peck WA, and Woods Wl. 1988. The cells of bone. In: Riggs BL, Melton LJ III, editors. Osteoporosis. Etiology, diagnosis and management. New York: Raven Press. pp. 1–44.

Pegram R, Indar L, Eddi C, and George J. 2004. The Caribbean amblyomma program: Some ecologic factors affecting its success. Annals of the New York Academy of Science 1026:302–311.

Perrien D, Akel N, Dupont-Versteegden E, Skinner R, Siegel E, Suva L, and Gaddy D. 2007. Aging alters the skeletal response to disuse in the rat. American Journal of Physiology and Regulatory Integrative Comparative Physiology 292:R988–R996.

Perzigian AJ. 1973. Osteoporotic bone loss in two Prehistoric Indian populations. American Journal of Physical Anthropology 39:87–96.

Peters CR. 2007. Theoretical and actualistic ecobotanical perspectives on early hominin diets and paleoecology. In: Ungar PS, editor. Evolution of the human diet. Oxford: Oxford University Press. pp. 233–261.

Petit M, Beck T, Lin HM, Bentley C, Legro R, and Lloyd T. 2004. Femoral bone structural geometry adapts to mechanical loading and is influenced by sex steroids: The Penn state young women's health study. Bone 35:750–759.

Pettifor JM. 2003. Nutritional rickets. In: Glorieux FH, Pettifor JM, Jüppner H, editors. Pediatric bone. Biology and diseases. San Diego: Academic Press. pp. 541–565.

Pettifor JM. 2004. Nutritional rickets: Deficiency of Vitamin D, calcium, or both?. American Journal of Clinical Nutrition 80 (6 Suppl.):1725s–1729s.

Pettifor JM, and Daniels ED. 1997. Vitamin D deficiency and nutritional rickets in children. In: Feldman D, Glorieux F, Pike J, editors. Vitamin D. San Diego: Academic Press. pp. 663–679.

Pfeiffer S. 2000. Palaeohistology: Health and disease. In: Katzenberg MA, Saunders S, editors. Biological anthropology of the human skeleton. New York: Wiley-Liss. pp. 287–302.

Pfeiffer S, and Crowder C. 2004. An ill child among mid-holocene foragers of Southern Africa. American Journal of Physical Anthropology 123:23–29.

Pfeiffer S, and King P. 1983. Cortical bone formation and diet among protohistoric Iroquoians. American Journal of Physical Anthropology 60:23–28.

Pfeiffer S, and Lazenby RA. 1994. Low bone mass in past and present aboriginal populations. In: Draper H, editor. Advances in nutritional research. Vol. 9. New York: Plenum Press. pp. 35–51.

Pfeiffer S, Crowder C, Harrington L, and Brown M. 2006. Secondary osteon and Haversian canal dimensions as behavioral indicators. American Journal of Physical Anthropology 131:460–468.

Phan T, Xu J, and Zheng M. 2004. Interaction between osteoblast and osteoclast: Impact in bone disease. Histology and Histopathology in Cellular and Molecular Biology 19:1325–1344.

Phemister DB. 1934. Fractures of neck of femur, dislocations of hip, and obscure vascular disturbances producing aseptic necrosis of head of femur. Surgery of Gynaecology and Obstetrics 59:415–440.

Picard L. 2000. Dr. Johnson's London. Everyday life in London 1740–1770. London: Phoenix.

Picard L. 2005. Victorian London: The life of a city 1840–1870. London: Weidenfeld & Nicolson.

Piercecchi-Marti M, Louis-Borrione C, Bartoli C, Sanvoisin A, Panuel M, Pelissier-Alicot A, and Leonetti G. 2006. Malnutrition, a rare form of child abuse: Diagnostic criteria. Journal of Forensic Science 51:670–673.

Pimentel L. 2003. Scurvy: Historical review and current diagnostic approach. American Journal of Emergency Medicine 21:328–332.

Pinhasi R, Shaw P, White B, and Ogden AR. 2006. Morbidity, rickets and long-bone growth in post-Medieval Britain – a cross-populational analysis. Annals of Human Biology 33:372–389.

Pitt M. 1988. Rickets and osteomalacia. In: Resnick D, Niwayama G, editors. Diagnosis of bone and joint disorders. Volume 8. Second edition. Philadelphia: WB. Saunders. pp. 2086–2119.

Plehwe W. 2003. Vitamin D deficiency in the 21st century: An unnecessary pandemic?. Clinical Endocrinology 59:22–24.

Ponce P, Arabaolaza I, and Boylston A. 2005. Industrial accident or deliberate amputation? Three case studies from a Victorian cemetery in Wolverhampton, West Midlands. Poster presented to British Association of Biological Anthropology and Osteoarchaeology, Museum of London.

Potts JT. 1989. Primary hyperparathyroidism. In: Avioli LV, and Krane SM, editors. Metabolic bone disease and clinically related disorders. San Diego: Academic Press. pp. 411–442.

Poulsen L, Qvesel D, Brixen K, Vesterby A, and Boldsen J. 2001. Low bone mineral density in femoral neck of medieval women. Bone 28:454–458.

Pountos I, and Giannoudis PV. 2005. Biology of mesenchymal stem cells. Injury. International Journal of Care of the Injured 36S:S8–S12.

Powell M. 1988. Status and health in prehistory. A case study of the Moundville chiefdom. Washington: Smithsonion Institution Press.

Powell M. 2000. Ancient diseases, modern perspectives: Treponematosis and tuberculosis in the age of agriculture. In: Lambert P, editor. Bioarchaeological studies of life in the age of agriculture. A view from the Southeast. Tuscaloosa: University of Alabama Press. pp. 6–34.

Power C. 1994. A demographic study of human skeletal populations from historic Munster. Ulster Journal of Archaeology 57:95–118.

Power J, Loveridge N, Lyon A, Rushton N, Parker M, and Reeve J. 2003. Bone remodelling at the endocortical surface of the human femoral neck: A mechanism for regional cortical thinning in cases of hip fracture. Journal of Bone and Mineral Research 18:1775–1780.

Powers N. 2005. Cranial trauma and treatment: A case study from the medieval cemtery of St. Mary Spital, London. International Journal of Osteoarchaeolgy 15:1–14.

Pratt EL. 1984. Historical perspectives: Food, feeding and fancies. Journal of the American College of Nutrition 3:115–121.

Preece MM, Tomlinson S, and Ribot S. 1975. Studies of vitamin D deficiency in man. Quarterly Journal Medicine 44:575–589.

Prentice A. 2003. Pregnancy and lactation. In: Glorieux F, Pettifor J, Jüppner H, editors. Pediatric bone. Biology and diseases. New York: Academic Press. pp. 249–270.

Price JL. 1975. The radiology of excavated Saxon and Medieval human remains from Winchester. Clinical Radiology 26:363–370.

Prince R. 1999. The rationale for calcium supplementation in the therapeutics of age-related osteoporosis. In: Rosen C, Glowacki J, Bilezikian J, editors. The aging skeleton. San Diego: Academic Press. pp. 479–494.

Prinzo ZW, and de Benoist B. 2002. Meeting the challenges of micronutrient deficiencies in emergency-affected populations. Proceedings of the Nutrition Society 61:251–257.

Prowse T, Schwarcz H, Saunders S, Macchiarelli R, and Bondioli I. 2005. Isotopic evidence for age-related variation in diet from Isola Sacra, Italy. American Journal of Physical Anthropology 128:2–13.

Punpilai S, Sujitra T, Ouyporn T, Teraporn V, and Sombut B. 2005. Menstrual status and bone mineral density among female athletes. Nursing Health Science 7:259–265.

Pusch CM, and Czarnetzki A. 2005. Archaeology and prevalence of Paget's disease. Journal of Bone and Mineral Research 20:1484.

Putnam SE, Scutt AM, Bicknell K, Priestley CM, and Williamson EM. 2007. Natural products as alternative treatments for metabolic bone disorders and for maintenance of bone health. Physiotherapy Research 21:99–112.

Qiu S, Sudhaker Rao D, Fyhrie D, Palnitkar S, and Parfitt AM. 2005. The morphological association between microcracks and osteocyte lacunae in human cortical bone. Bone 37:10–15.

Quinlan R, Quinlan M, and Flinn M. 2005. Local resource enhancement and sex-biased breastfeeding in a Caribbean community. Current Anthropology 46:471–480.

Raisz LG. 1999. Physiology and pathophysiology of bone remodelling. Clinical Chemistry 45:1353–1358.

Raisz LG, and Seeman E. 2001. Causes of age-related bone loss and bone fragility: An alternative view. Journal of Bone and Mineral Research 16:1948–1952.

Rajakumar K. 2001. Infantile scurvy: A historical perspective. Pediatrics 108:76–78.

Rajakumar K. 2003. Pellagra. eMedicine. Accessed on 18th May 2006: http://www.emedicine.com/ped/topic1755.htm.

Rajakumar K. 2005. Scurvy. eMedicine. Accessed on 24th June 2005: http://www.emedicine.com/ped/topic2073.htm.

Ralston SH. 2005. Genetic determinants of osteoporosis. Current Opinion in Rheumatology 17:475–479.

Ramachandran M. 2006. Osteogenesis imperfecta. eMedicine. Accessed on 8th August 2006: http://www.emedicine.com/ped/topic2073.htm.

Ratanachu-Ek S, Sukswai P, Jeerathanyasakun Y, and Wongtapradit L. 2003. Scurvy in pediatric patients: A review of 28 cases. Journal of Medical Association of Thailand 86 (Supplement 3):S734–S740.

Rathbun TA. 1987. Health and disease at a South Carolina plantation: 1840–1870. American Journal of Physical Anthropology 74:239–253.

Recker R, and Barger-Lux M. 2001. Bone remodeling findings in osteoporosis. In: Marcus R, Feldman D, Kelsey J, editors. Osteoporosis. Volume 2. Second edition. San Diego: Academic Press. pp. 59–69.

Reddy NB, and Prasad KSS. 2003. Pyroclastic fluoride in ground waters in some parts of Tadpatri Taluk, Anantapur District, Andhra Pradesh. Indian Journal of Environmental Health 45:285–288.

Reed KE, and Rector AL. 2007. African pliocene paleocology: Hominin habitats, resources, and diets. In: Ungar PS, editor. Evolution of the human diet. Oxford: Oxford University Press. pp. 262–288.

Reginato A, and Coquia JA. 2003. Musculoskeletal manifestations of osteomalacia and rickets. Best Practice and Research in Clinical Rheumatology 17:1063–1080.

Regöly-Mérei G. 1962. Palaepathologia. II. Az ösemberi és késöbbi emberi maradványok rendszeres kórbonctana. Budapest: Medicina Konyukiado.

Reid I, editor. 1997. Metabolic bone disease in clinical endocrinology and metabolism. London: W.B. Saunders.

Reid IR. 1999. Menopause. In: Favus M, editor. Primer on the metabolic bone diseases and disorders of mineral metabolism. Fourth edition. Philadelphia: Lippincott William & Wilkins. pp. 55–57.

Reid IR. 2001. Vitamin D and its metabolites in the management of Osteoporosis. In: Marcus R, Feldman D, Kelsey J, editors. Osteoporosis. Volume 2. Second edition. San Diego: Academic Press. pp. 553–576.

Reinhard K, Ambler J, and Szuter C. 2007. Hunter-gatherer use of small animal food resources: Coprolite evidence. International Journal of Osteoarchaeology 17:416–428.

Renier JC, Leroy E, Audran M. 1996. The initial site of bone lesions in Paget's disease: A review of two hundred cases. Revue de Rhumatisme. 63. English edition. pp. 823–829.

Resch A, Schneider B, Bernecker P, Battmann A, Wergedal J, Willvonseder R, and Resch H. 1995. Risk of vertebral fractures in men: Relationship to mineral density of the vertebral body. American Journal of Roentgenology 164:1447–1450.

Resnick D. 1988. Paget disease of bone: Current status and a look back to 1943 and earlier. American Journal of Radiology 150:1104–1114.

Resnick D, and Niwayama G, editors. 1988. Diagnosis of bone and joint disorders. Second edition. Philadelphia: W.B. Saunders.

Resnick D, and Niwayama G, editors. 1995. Diagnosis of bone and joint disorders. Third edition. Philadelphia: W.B. Saunders.

Rewekant A. 2001. Do environmental disturbances of an individual's growth and development influence the later bone involution processes? A study of two medieval populations. International Journal of Osteoarchaeology 11: 433–443.

Reynolds WA, and Karo JJ. 1972. Radiologic diagnosis of metabolic bone disease. Orthopedic Clinics of North America 3:521–546.

Rhodes B, and Jawad ASM. 2005. Paget's disease of bone: Osteitis deformans or osteodystrophia deformans. Rheumatology 44:261–262.

Richards M, Price T, and Koch E. 2003. Mesolithic and neolithic subsistence in Denmark: New stable isotope data. Current Anthropology 44:288–295.

Richman BA, Ortner DJ, and Schulter-Ellis FP. 1979. Differences in intracortical bone remodelling in three aboriginal American populations: Possible dietary factors. Calcified Tissue International 28:209–214.

Riggs BL, and Melton LJ, editors III. 1988. Osteoporosis. Etiology, diagnosis and management. New York: Raven Press.

Riggs BL, Khosla S, and Melton LJ III. 1998. A unitary model for involutional osteoporosis: Estrogen deficiency causes both type I and type II osteoporosis in postmenopausal women and contributes to bone loss in aging men. Journal of Bone and Mineral Research 13:763–773.

Riggs BL, Khosla S, and Melton LJ. III. 2001. The type I/type II model for involutional osteoporosis. Update and modifications based on new observations. In: Marcus R, Feldman D, Kelsey J, editors. Osteoporosis. Volume 2. Second edition. San Diego: Academic Press. pp. 49–59.

Riggs B, Melton L III, Robb R, Camp J, Atkinson E, Peterson J, Rouleau P, McCollough C, Bouxsein M, and Khosla S. 2004. Population-based study of age and sex differences in bone volumetric density, size, geometry, and structure at different skeletal sites. Journal of Bone and Mineral Research 19:1945–1954.

Riggs BL, Melton LJ III, Robb RA, Camp JJ, Atkinson EJ, Oberg AL, Rouleau PA, McCollough CH, Khosla S, and Bouxsein ML. 2006. Population-based analysis of the relationship of whole bone strength indices and fall-related loads to age- and sex-specific patterns of hip and wrist fractures. Journal of Bone and Mineral Research 21:315–323.

Ritchie LD, Fung EB, and Halloran BP. 1998. A longitudinal study of calcium homeostasis during human pregnancy and lactation and after the resumption of the menses. American Journal of Clinical Nutrition 67:693–701.

Rittweger J, Frost HM, Schiessl H, Ohshima H, Alkner B, Tesch P, and Felsenberg D. 2005. Muscle atrophy and bone loss after 90 days bed rest and the effects of flywheel resistive exercise and pamidronate: Results of the LTBR study. Bone 36:1019–1029.

Ritzel H, Amling M, Pösl M, Hahn M, and Delling G. 1997. The thickness of human vertebral cortical bone and its changes in aging and osteoporosis: A histomorphometric analysis of the complete spinal column from thirty-seven autopsy specimens. Journal of Bone and Mineral Research 12:89–95.

Rizzoli R, and Bonjour JP. 2004. Dietary protein and bone health. Journal of Bone and Mineral Research 19:527–531.

Robb J, Bigazzi R, Lazzarini L, Scarsini C, and Sonego F. 2001. Social 'status' and biological 'status': A comparison of grave goods and skeletal indicators from Pontecagnano. American Journal of Physical Anthropology 115:213–222.

Roberts CA. 1987. Case reports on paleopathology. Paleopathology Association Newsletter 57:14–15.

Roberts CA. 2000. Did they take sugar? The use of skeletal evidence in the study of disability in past populations. In: Hubert J, editor. Madness, disability and social exclusion: The archaeology and anthropology of difference. One World Archaeology Series. Vol. 40. London: Routledge. pp. 46–59.

Roberts CA, and Buikstra JE. 2003. The bioarchaeology of tuberculosis: A global perspective on a reemerging disease. Gainesville: University Press of Florida.

Roberts CA, and Cox M. 2003. Health and disease in Britain. From prehistory to the present day. Stroud: Sutton Publishing Ltd.

Roberts CA, and Manchester K. 2005. The archaeology of disease. Third edition. Stroud: Sutton Publishing Ltd.

Roberts CA, and Wakely J. 1992. Microscopical findings associated with the diagnosis of osteoporosis in palaeopathology. International Journal of Osteoarchaeology 2:23–30.

Roberts CA, Lewis ME, and Manchester K, editors. 2002. The past and present of leprosy. Archaeological, historical, paleopathological and clinical approaches. Proceedings of the international congress on the evolution and paleoepidemiology of the infectious diseases 3 (ICEPID). University of Bradford 26th–31st July 1999. British Archaeological Reports International Series 1054. Oxford: Archeopress.

Robling A, and Stout SD. 2003. Histomorphology, geometry, and mechanical loading in past populations. In: Agarwal S, Stout SD, editors. Bone loss and osteoporosis. An anthropological perspective. New York: Kluwer Academic/Plenum Publishers. pp. 189–205.

Roches E, Blondiaux J, Cotton A, Chastanet P, and Flipo R-M. 2002. Microscopic evidence for Paget's disease in two osteoarchaeological samples from early Northern France. International Journal of Osteoarchaeology 12:229–234.

Rockwood CA, and Green DP. 1984. Fractures in Adults. Volume 1. Second edition. Philadelphia: Lippincott Williams & Wilkins.

Rodan G, Raisz L, and Bilezikian J. 2002. Pathophysiology of osteoporosis. In: Bilezikian J, Raisz L, Rodan G, editors. Principles of bone biology. Second edition. San Diego: Academic Press. pp. 1275–1289.

Rogers J. 1990. The human skeletal material. In: Saville A, editor. Hazelton North, Gloucestershire, 1979–1982. The excavation of a Neolithic long cairn of the Cotswold-Severn group. London: English Heritage Archaeological Report 13. Historic Buildings and Monuments Commission for England. pp. 182–198.

Rogers J. 2000. The palaeopathology of joint disease. In: Cox M, Mays S, editors. Human osteology in archaeology and forensic science. London: Greenwich Medical Media Ltd. pp. 163–182.

Rogers J, and Waldron T. 1988. Two possible cases of infantile cortical hyperostosis. Paleopathology Newsletter 63:9–12.

Rogers J, and Waldron T. 1995. A field guide to joint disease in archaeology. Chichester: Wiley.

Rogers J, Watt I, and Dieppe P. 1981. Arthritis in Saxon and medieval skeletons. British Medical Journal 283:1668–1670.

Rogers J, Jeffrey DR, and Watt I. 2002. Paget's disease in an archaeological population. Journal of Bone and Mineral Research 17:1127–1134.

Rokhlin DG. 1965. Diseases of ancient men. Moscow: Nauka.

Rome E, and Ammerman S. 2003. Medical complications of eating disorders: An update. Journal of Adolescent Health 33:418–426.

Roodman GD. 1994. Biology of the osteoclast in Paget's disease. Seminars in Arthritis and Rheumatism 23:235–236.

Roodman GD. 1999. Mechanisms of abnormal bone turnover in Paget's disease. Bone 24:39S–40S.

Roodman DG, and Windle JJ. 2005. Paget disease of bone. Journal of Clinical Investigation 115:200–208.

Roosens H, Beex G, and Bonenfant P, editors. 1963. Een urnenveld Grote-Brogel. Limberg 42. Archaeologia Belgica 67.

Rose CJ, editor. 1985. Gone to a better land. Little Rock: Arkansas Archaeological Survey Research Series 2.

Rosen CJ, and Kiel DP. 1999. The aging skeleton. In: Favus M, editor. Primer on the metabolic bone diseases and disorders of mineral metabolism. Fourth edition. Philadelphia: Lippincott William & Wilkins. pp. 57–59.

Rosen CJ, Glowacki J, and Bilezikian JP, editors. 1999. The aging skeleton. San Diego: Academic Press.

Rosen CJ, Donahue LR, and Beamer W. 2002. Defining the genetics of osteoporosis – using the mouse to understand man. In: Bilezikian J, Raisz L, Rodan G, editors. Principles of bone biology. Second edition. San Diego: Academic Press. pp. 1657–1665.

Rosen RS, Armbrustmacher V, and Sampson BA. 2003. Spontaneous cerebellar hemorrhage in children. Journal of Forensic Sciences 48:1–3.

Ross JA. 1982. European distribution of Paget's disease of bone. British Medical Journal 285:1427–1428.

Ross DS. 1998. Bone disease in hyperparathyroidism. In: Avioli LV, Krane SM, editors. Metabolic bone disease and clinically related disorders. Third edition. San Diego: Academic Press. pp. 531–544.

Rothschild B, Hershkovitz I, Dutour O, Latimer B, Rothschild C, and Jellema L. 1997. Recognition of leukemia in skeletal remains: Report and comparison of two cases. American Journal of Physical Anthropology 102:481–496.

Rousière M, Michou L, Cornélis F, and Orcel P. 2003. Paget's disease of bone. Best Practice and Research Clinical Rheumatology 17:1019–1041.

Rowling J. 1967. Paraplegia. In: Brothwell D, Sandison A, editors. Diseases in antiquity: A survey of the diseases, injuries and surgery of early populations. Springfield: Charles C Thomas. pp. 272–278.

Roy DK, Berry JL, Pye SR, Adams JE, Swarbrick CM, King Y, Silman A, and O'Neil TW. 2007. Vitamin D status and bone mass in UK South Asian women. Bone 40:200–204.

Ruff C. 2000. Biomechanical analyses of archaeological human skeletons. In: Katzenberg M, Saunders S, editors. Biological anthropology of the human skeleton. New York: Wiley-Liss. pp. 71–102.

Ruff CB, Scott WW, and Liu AYC. 1991. Articular and diaphyseal remodelling of the proximal femur with changes in body mass in adults. American Journal of Physical Anthropology 86:397–413.

Ruff C, Walker AL, and Trinkaus E. 1994. Postcranial robusticity in *Homo*. III: Ontogeny. American Journal of Physical Anthropology 93:35–54.

Ruff C, Holt B, and Trinkaus R. 2006. Who's afraid of the big bad Wolff?: 'Wolff's law' and bone functional adaptation. American Journal of Physical Anthropology 129:484–498.

Ruffer MA. 1914. On the disease of the Sudan and Nubia in ancient times. Mitteilungen der Gesellschaft der Medizin und der Natvrewissenschaften und der Technik 13:453–560.

Rühli F, Hotz G, and Böni T. 2003. Brief communication: The Galler collection – a little-known historic swiss bone pathology reference series. American Journal of Physical Anthropology 121:15–18.

Rühli F, Muntener M, and Henneberg M. 2005. Age-dependent changes of the normal human spine during adulthood. American Journal of Human Biology 17:460–469.

Rundle 1886. Crush of legs by tramcar: Double amputation below the knees: Recovery. Lancet 128:1222.

Sachan A, Gupta R, Das V, Agarwal A, Awasthi P, and Bhatia V. 2005. High prevalence of vitamin D deficiency among pregnant women and their newborns in Northern India. American Journal of Clinical Nutrition 81:1060–1064.

Saifuddin A, and Hassan A. 2003. Paget's disease of the spine: Unusual features and complications. Clinical Radiology 58:102–111.

Sakamoto Y, and Takano Y. 2002. Morphological influence of ascorbic acid deficiency on endochondral ossification in osteogenic disorder Shionogi rat. Anatomical Record 268:93–104.

Salter RB. 1999. Textbook of disorders and injuries of the musculoskeletal system. Third edition. Baltimore: Lippincott Williams & Wilkins.

Sambrook PN, Browne CD, Eisman JA, and Bourke SJ. 1987. A case of crush fracture osteoporosis from late roman Pella in Jordan. OSSA 13:193–197.

Sandberg A. 2002. Bioavailability of minerals in legumes. British Journal of Nutrition 88 (Supplement 3):S281–S285.

Sato K, and Byers P. 1981. Quantitative study of tunnelling and hook resorption in metabolic bone disease. Calcified Tissue International 33:459–466.

Saul F. 1972. Human skeletal remains of altar de sacrificios. Papers of the Peabody Museum of Archaeology and Ethnology 63:1–123.

Saunders SR, and Herring A, editors. 1995. Grave reflections: Portraying the past through cemetery studies. Toronto: Canadian Scholars Press.

Saunders SR, Herring A, Sawchuk L, Boyce G, Hoppa R, and Klepp S. 2002. The health of the middle class. The St. Thomas' Anglican Church cemetery project. In: Steckel RH, Rose JC, editors. The backbone of history: Health and nutrition in the Western Hemisphere. Cambridge: Cambridge University Press. pp. 130–161.

Savas S, Cetin M, and Akdogan M. 2001. Endemic fluorosis in Turkish patients: Relationship with knee osteoarthritis. Rheumatology International 21:30–35.

Saville A, editor. 1990. Hazelton North, Gloucestershire, 1979–1982. The excavation of a neolithic long cairn of the cotswold-severn group. London: English Heritage

Archaeological Report 13. Historic Buildings and Monuments Commission for England.

Schaffler MB, Choi K, and Milgrom C. 1995. Aging and matrix microdamage accumulation in human compact bone. Bone 17:521–525.

Schamall D, Kneissel M, Wiltschke-Schrotta K, and Teschler-Nicola M. 2003a. Bone structure and mineralization in a late antique skeleton with osteomalacia. American Journal of Physical Anthropology 120 (Supplement 36):184.

Schamall D, Teschler-Nicola M, Kainberger F, Tangl St, Brandstätter F, Patzak B, Muhsil J, and Plenk H Jr.. 2003b. Changes in trabecular bone structure in rickets and osteomalacia: The potential of a medico-historical collection. International Journal of Osteoarchaeology 13:283–288.

Scheuer L, and Black S. 2000a. Developmental juvenile osteology. San Diego: Elsevier.

Scheuer L, and Black S. 2000b. Development and ageing of the juvenile skeleton. In: Cox M, Mays S, editors. Human osteology in archaeology and forensic science. London: Greenwich Medical Media Ltd. pp. 9–21.

Schmidt-Schultz TH, and Schultz M. 2004. Bone protects proteins over thousands of years: Extraction, analysis and interpretation of extracellular matrix proteins in archaeological skeletal remains. American Journal of Physical Anthropology 123:30–39.

Schultz M. 2001. Paleohistopathology of bone: A new approach to the study of ancient diseases. Yearbook of Physical Anthropology 44:106–147.

Schultz M. 2003. Differential diagnosis of intravitam and postmortem bone loss at the micro-level. In: Agarwal S, Stout SD, editors. Bone loss and osteoporosis. An anthropological perspective. New York: Kluwer Academic/Plenum Publishers. pp. 173–187.

Schutkowski H, Schultz M, and Holzgraefe M. 1996. Fatal wounds in a late Neolithic double inhumation – a probable case of meningitis following trauma. International Journal of Osteoarchaeology 6:179–184.

Schwartz JH. 1995. Skeleton keys. An introduction to human skeletal morphology, development and analysis. New York: Oxford University Press.

Scott HS, and Helton HJ. 1953. Osseous manifestations of systemic diseases ininfancy and children. Journal of the American Osteopathic Association 52:636–637.

Sealy J. 2006. Diet, mobility and settlement pattern among Holocene hunter-gatherers in Southernmost Africa. Current Anthropology 47:569–595.

Sebastian A, Frassetto L, Sellmeyer D, Merriam R, and Morris R. 2002. Estimation of the net acid load of the diet of ancestral pre-agricultural *Homo sapiens* and their hominid ancestors. American Journal of Clinical Nutrition 76:1308–1316.

Seeman E. 1999a. Genetic determinants of the population variance in bone mineral density. In: Rosen C, Glowacki J, Bilezikian J, editors. The aging skeleton. San Diego: Academic Press. pp. 77–94.

Seeman E. 1999b. The structural basis of bone fragility in men. Bone 25:143–147.

Seeman E. 2001. Sexual dimorphism in skeletal size, density and strength. Journal of Clinical Endocrinology and Metabolism 86:4576–4584.

Seeman E, Duan Y, Fong C, and Edmonds J. 2001. Fracture site-specific deficits in bone size and volumetric density in men with spine or hip fractures. Journal of Bone and Mineral Research 16:120–127.

Selinus O, Alloway B, Centeno JA, Finkelman RB, Fuge R, Lindh U, and Smedley P, editors. 2005. Essentials of medical geology. Impacts of the natural environment on public health. San Diego: Elsevier.

Serenius F, Eldrissy A, and Dandona P. 1984. Vitamin D nutrition in pregnant women at term and in newly born babies in Saudi Arabia. Journal of Clinical Pathology 37:444–447.

Severs D, Williams T, and Davies JW. 1961. Infantile scurvy – a public health problem. Canadian Journal of Public Health 52:214–220.

Sevitt S. 1981. Bone repair and fracture healing in man. Edinburgh: Churchill Livingstone.

Shackelford LL, and Trinkaus E. 2002. Late pleistocene human femoral diaphyseal curvature. American Journal of Physical Anthropology 118:359–370.

Shanley D, and Kirkwood T. 2001. Evolution of the human menopause. BioEssays 23:282–287.

Sheldrick P. 2007. Hip fracture in burials from Kellis 2 Dakhleh Oasis. Paper presented at the 34th Annual meeting of the Paleopathology Association, Philadelphia, 27th–28th March 2007.

Shields LBE, Hunsaker D, and Hunsaker JC III. 2004. Abuse and neglect: A ten-year review of mortality and morbidity in our elders in a large metropolitan area. Journal of Forensic Sciences 49:1–6.

Shires R, Avioli LV, Bergfeld MA, Fallon MD, Slatopolsky E, and Teitelbaum SL. 1980. Effects of semi-starvation on skeletal homeostasis. Endocrinology 107:1530–1535.

Shomar B, Müller G, Yahya A, Askar S, and Sansur R. 2004. Fluorides in groundwater, soil and infused black tea and the occurrence of dental fluorosis among school children of the Gaza Strip. Journal of Water and Health 2:23–25.

Shorbe HB. 1953. Infantile scurvy. Clinical Orthopedics 1:49–55.

Silver J, and Naveh-Many T. 1997. Vitamin D and the parathyroid glands. In: Feldman D, Glorieux F, Pike J, editors. Vitamin D. San Diego: Academic Press. pp. 353–367.

Silverman FN, editor. 1985. Caffey's pediatric X-ray diagnosis. An integrated imaging approach. Volume 1. Eighth edition. Chicago: Year Book Medical Publishers.

Simopoulos A. 2006. Evolutionary aspects of diet, the omega-6/omega-3 ratio and genetic variation: Nutritional implications for chronic diseases. Biomedicine and Pharmacotherapy 60:502–507.

Singh M, Nagrath AR, Maini PS, and Haryana R. 1970. Changes in the trabecular pattern of the upper end of the femur as an index of osteoporosis. Journal of Bone and Joint Surgery 52A:457–467.

Sissons HA. 1952. Osteoporosis and epiphyseal arrest in joint tuberculosis; an account of the histological changes in involved tissues. Journal of Bone and Joint Surgery 34B:275–290.

Sledzik PS, and Sandberg LG. 2002. The effects of nineteenth century military service on health. In: Steckel RH, Rose JC, editors. The backbone of history. Health and nutrition in the Western Hemisphere. New York: Cambridge University Press. pp. 185–207.

Slemenda CW, Turner C, Peacock M, Christian J, Sorbel J, Hui S, and Johnston C. 1996. The genetics of proximal femur geometry, distribution of bone mass and bone mineral density. Osteoporosis International 6:178–182.

Slemenda CW, Peacock M, Hui S, Zhou L, and Johnston CC. 1997. Reduced rates of skeletal remodelling are associated with increased bone mineral density during the

development of peak skeletal bone mass. Journal of Bone and Mineral Research 12:676–682.
Sloan B, Kulwin DR, and Kersten RC. 1999. Scurvy causing bilateral orbital hemorrhage. Archives of Ophthalmology 117:842–843.
Solomon L. 1986. Hip fracture and cortical bone density in aging African and Caucasian populations. In: Uhthoff HK, editor. Current concepts of bone fragility. Berlin: Springer-Verlag. pp. 379–384.
Sommerfeldt D, and Rubin C. 2001. Biology of bone and how it orchestrates the form and function of the skeleton. European Spine Journal 10:S86–S95.
Sowers M, Crutchfield M, Jannausch ML, Updike S, and Corton G. 1991. Bone mineral change in pregnancy. Obstetrics and Gynaecology 77:841–845.
Spencer H, O'Sullivan V, and Sontag SJ. 1992. Does lead play a role in Paget's disease of bone? A hypothesis. Journal of Laboratory and Clinical Medicine 120:798–800.
St-Arnaud R, and Glorieux FH. 1997. Vitamin D and bone development. In: Feldman D, Glorieux F, Pike J, editors. Vitamin D. San Diego: Academic Press. pp. 293–303.
Stanbury M, Reilly MJ, and Rosenman KD. 2003. Work-related amputations in Michigan, 1997. American Journal of Industrial Medicine 44:359–367.
Start H, and Kirk L. 1998. 'The bodies of Friends' – the osteological analysis of a quaker burial ground. In: Cox M, editor. Grave concerns. Death and burial in England 1700–1850. York: Council of British Archaeology. pp. 167–177.
Steckel RH, and Rose JC, editors. 2002. The backbone of history. Health and nutrition in the Western Hemisphere. New York: Cambridge University Press.
Steinbach H, and Noetzli M. 1964. Roentgen appearance of the skeleton in osteomalacia and rickets. American Journal of Roentgenology 91:955–972.
Steinbach H, Kolb F, and Gilfillan R. 1954. A mechanism of the production of pseudofractures in osteomalacia (Milkman's syndrome). Radiology 62:388–395.
Steinbach H, Kolb F, and Crane J. 1959. Unusual roentgen manifestations of osteomalacia. Radiology 82:875–886.
Steinbock RT. 1993. Rickets and osteomalacia. In: Kiple KF, editor. The Cambridge world history of human disease. Cambridge: Cambridge University Press. pp. 978–981.
Stevens D. 2001. Outbreak of micronutrient deficiency disease: Did we respond appropriately? Field Exchange 12:1–7 online edition. Accessed on 23rd May 2005: http://www.ennonline.net.
Stewart T, and Ralston S. 2000. Role of genetic factors in the pathogenesis of osteoporosis. Journal of Endocrinology 166:235–245.
Stini WA. 1990. 'Osteoporosis': Etiologies, prevention, and treatment. Yearbook of Physical Anthropology 33:151–194.
Stini WA. 1995. Osteoporosis in biocultural perspective. Annual Review of Anthropology 24:397–421.
Stinson S. 1985. Sex differences in environmental sensitivity during growth and development. Yearbook of Physical Anthropology 28:123–147.
Stirland A. 1985. The human bones. East Anglian Archaeology 28:49–57.
Stirland A. 1991. Paget's disease (osteitis deformans): A classic case? International Journal of Osteoarchaeology 1:173–177.
Stirland A. 1997. Care in the medieval community. International Journal of Osteoarchaeology 7:587–590.
Stirland A. 2000. Raising the dead: The skeleton crew of King Henry VIII's great ship the Mary Rose. Chichester: Wiley.

Stojanowski CM, Seidemann RM, and Doran GH. 2002. Differential skeletal preservation at windover pond: Causes and consequences. American Journal of Physical Anthropology 119:15–26.

Stone I. 1965. Studies of a mammalian enzyme system for producing evolutionary evidence in man. American Journal of Physical Anthropology 23:83–85.

Stone M, Briody J, Kohn M, Clarke S, Madden S, and Cowell C. 2006. Bone changes in adolescent girls with anorexia nervosa. Journal of Adolescent Health 39:835–841.

Storey R. 2007. An elusive paleodemography? A comparison of two methods for estimating the adult age distribution of deaths at late classic Copan, Honduras. American Journal of Physical Anthropology 132:40–47.

Stout SD. 1978. Histological structure and its preservation in ancient bone. Current Anthropology 19:601–604.

Stout SD, and Teitelbaum SL. 1976a. Histological analysis of undecalcified thin sections of archeological bone. American Journal of Physical Anthropology 44:263–270.

Stout SD, and Teitelbaum SL. 1976b. Histomorphometric determination of formation rates of archaeological bone. Calcified Tissue Research 21:163–169.

Strachan D, Powell K, Thaker A, Millard F, and Maxwell J. 1995. Vegetarian diet as a risk factor for tuberculosis in immigrant South London Asians. Thorax 50:175–180.

Strouhal E, Němečková A, and Kouba M. 2003. Paleopathology of Iufaa and other persons found beside his shaft tomb at Abusir (Egypt). International Journal of Osteoarchaeology 13:331–338.

Stuart-Macadam PL. 1989. Nutritional deficiency diseases: A survey of scurvy, rickets and iron-deficiency anemia. In: İşcan MY, Kennedy KAR, editors. Reconstruction of life from the skeleton. New York: Wiley-Liss. pp. 201–222.

Stuart-Macadam P, Glencross B, and Kricun M. 1998. Traumatic bowing deformities in tubular bones. International Journal of Osteoarchaeology 8:252–262.

Sugiyama L. 2004. Illness, injury and disability among the Shiwiar forager-horticulturalists: Implications of health-risk buffering for the evolution of human life history. American Journal of Physical Anthropology 123:371–389.

Sumner DR, Morbeck ME, and Lobick JJ. 1989. Apparent age-related bone loss among adult female gombe chimpanzees. American Journal of Physical Anthropology 79:225–234.

Swann GF. 1954. Pathogenesis of bone lesions in neurofibromatosis. British Journal of Radiology 27:623–629.

Sy MH, Toure-Fall A, Diop-Sall N, Dangou J, and Seye SI. 2000. Concomitant sickle cell disease and skeletal fluorosis. Joint Bone Spine 67:478–480.

Szulc P, Beck TJ, Marchand F, and Delmas PD. 2005. Low skeletal muscle mass is associated with poor structural parameters of bone and impaired balance in elderly men – the MINOS study. Journal of Bone and Mineral Research 20:721–729.

Tamura Y, Welch DC, Zic AJ, Cooper WO, Stein SM, and Hummell DS. 2000. Scurvy presenting as a painful gait with bruising in a young boy. Archives of Pediatric and Adolescent Medicine 154:732–735.

Taneja K, Goel DP, and Taneja A. 1990. Childhood Paget's disease of bone. Indian Pediatrics 27:866–869.

Tangpricha V, Pearce E, Chen T, and Holick M. 2002. Vitamin D insufficiency among free-living healthy young adults. American Journal of Medicine 112:659–662.

Tayles N. 1996. Anemia, genetic diseases, and malaria in prehistoric mainland Southeast Asia. American Journal of Physical Anthropology 101:11–27.

Termine JD, and Robney PG. 1996. Bone matrix proteins and the mineralization process. In: Favus MJ, editor. Primer on the metabolic bone diseases and disorders of mineral metabolism. Third edition. Philadelphia: Lippincott-Raven Press. pp. 24–28.

Thacher TD, Fischer PR, Strand MA, and Pettifor JM. 2006. Nutritional rickets around the world: Causes and future directions. Annals of Tropical Pediatrics 26:1–16.

Thane P. 2000. Old age in English history. Past experience, present issues. Oxford: Oxford University Press.

Thane P. 2005. The age of old age. In: Thane P, editor. The long history of old age. London: Thames & Hudson Ltd. pp. 9–29.

Thane P, editor. 2005. The long history of old age. London: Thames & Hudson Ltd.

Thompson DD, and Gunness-Hey M. 1981. Bone mineral-osteon analysis of Yupik-Inupiaq skeletons. American Journal of Physical Anthropology 55:1–7.

Thould AK, and Thould BT. 1983. Arthritis in Roman Britain. British Medical Journal 287:1909–1911.

Tickner FJ, and Medvei VC. 1958. Scurvy and the health of European crews in the Indian Ocean in the seventeenth century. Medical History 2:36–46.

Toole MJ. 1992. Micronutrient deficiencies in refugees. Lancet 339:1214–1216.

Torwalt CR, Balachandra AT, Youngson C, and de Nanassy J. 2002. Spontaneous fractures in the differential diagnosis of fractures in children. Journal of Forensic Sciences 47:1–5.

Treves F. 1887a. A sixteenth century amputation. Lancet 129:63–64.

Treves F. 1887b. London hospital: A rare case of general tuberculosis; necropsy remarks. Lancet 129:18–19.

Troyansky D. 2005. The eighteenth century. In: Thane P, editor. The long history of old age. London: Thames & Hudson Ltd. pp. 175–210.

Turner C. 1998. Three rules for bone adaptation to mechanical stimuli. Bone 23:399–407.

Turner-Walker G, Syversen U, and Mays S. 2001. The archaeology of osteoporosis. European Journal of Archaeology 4:263–269.

Tyler P, Madani G, Chaudhuri R, Wilson L, and Dick E. 2006. The radiological appearances of thalassaemia. Clinical Radiology 61:40–52.

Ubelaker DH, and Newson LA. 2002. Patterns of health and nutrition in prehistoric and historic Ecuador. In: Steckel RH, Rose JC, editors. The backbone of history. Health and nutrition in the Western Hemisphere. New York: Cambridge University Press. pp. 343–375.

Ueland T. 2004. Bone metabolism in relation to alterations in systemic growth hormone. Growth Hormone and IGF Research 14:404–417.

Uetani M, Hashmi R, and Hayashi K. 2004. Malignant and benign compression fractures: Differentiation and diagnostic pitfalls on MRI. Clinical Radiology 59:124–131.

Uhthoff HK, editor. 1986. Current concepts of bone fragility. Berlin: Springer-Verlag.

Ulijaszek SJ. 1991. Human dietary change. Philosophical Transactions of the Royal Society of London 334:271–279.

Ulijaszek SJ. 1998. Long term consequences of early environmental influences. In: Ulijaszek SJ, Johnston FE, Preece MA, editors. The Cambridge encyclopaedia of human growth and development. Cambridge: Cambridge University Press. pp. 417–418.

Ulijaszek SJ, and Strickland SS. 1993. Nutritional anthropology: Prospects and perspectives. London: Smith-Gordon Publishing.

Ulijaszek SJ, Johnston FE, and Preece MA, editors. 1998. The Cambridge encyclopaedia of human growth and development. Cambridge: Cambridge University Press.

Ungar PS, editor. 2007. Evolution of the human diet. Oxford: Oxford University Press.

Ustianowski A, Shaffer R, Collins S, Wilkinson RJ, and Davidson RN. 2005. Prevalence and associations of vitamin D deficiency in foreign-born persons with tuberculosis in London. Journal of Infection 50:432–437.

USDA National nutrient database for standard reference (Release 19) 2006. Accessed on 12th April 2007: http://www.nal.usda.gov/fnic/foodcomp/search/.

Vahter M, Berglund M, and Åkesson A. 2004. Toxic metals and the menopause. Journal of the British Menopause Society 10:60–64.

Van Der Meulen M, and Carter D. 1999. Mechanical determinants of peak bone mass. In: Rosen C, Glowacki J, Bilezikian J, editors. The aging skeleton. San Diego: Academic Press. pp. 105–115.

Van der Merwe A, and Steyn M. 2006. Human skeletal remains from Kimberley, South Africa: An assessment of health in a 19th century mining community. Paper presented at the 16th Paleopathology Association European Meeting, Santorini Island, August 28–September 1, 2006.

Van der Sluis IM, and de Muinck Keizer-Schrama SMPF. 2001. Osteoporosis in childhood: Bone density of children in health and disease. Journal of Pediatric Endocrinology and Metabolism 14:817–832.

Van der Sluis IM, van der Heuvel-Elbrink M, Hahlen K, Jrenning E, and de Melnck Keizer-Schrama S. 2002. Altered bone mineral density and body composition and increased fracture risk in childhood acute lytic lymphoblastic leukemia. Journal of Pediatrics 141:204–210.

Van der Voort DJM, Geusens PP, and Dinant GJ. 2001. Risk factors for osteoporosis related to their outcome: Fractures. Osteoporosis International 12:630–638.

Vaughan J. 1975. The physiology of bone. Oxford: Clarendon Press.

Verano JW, Anderson LS, and Franco R. 2000. Foot amputation by the Moche of ancient Peru: Osteological evidence and archaeological context. International Journal of Osteoarchaeology 10:177–188.

Vercruysse J, and Claerebout E. 2001. Treatment vs. non-treatment of helminth infections in cattle: Defining the threshold. Veterinary Parasitology 98:195–214.

Vico L, Collet P, Guignandon A, Lafage-Proust M, Thomas T, Rehaillia M, and Alexandre C. 2000. Effects of long-term microgravity exposure on cancellous and cortical weight-bearing bones of cosmonauts. Lancet 355:1607–1611.

Volkert D, Kreuel K, Heseker H, and Stehle P. 2004. Energy and nutrient intake of young-old, old-old and very old elderly in Germany. European Journal of Clinical Nutrition 58:1190–1200.

Voskaridou E, Kyrtsonis MC, Terpos E, Skordili M, Theodoropoulos I, Bergele A, Diamanti E, Kalovidouris A, Loutradi A, and Loukopoulos D. 2001. Bone resorption is increased in young adults with Thalassaemia major. British Journal of Haematology 112:36–41.

Vrzgula L. 1991. Metabolic disorders and their prevention in farm animals (developments in animal veterinary science). Amsterdam: Elsevier.

Wakely J. 1996. Limits to interpretation of skeletal trauma – two case studies from medieval Abingdon, England. International Journal of Osteoarchaeology 6:76–83.

Wakely J, and Carter R. 1996. Skeletal analysis and dental report. In: Cooper L, editor. A Roman cemetery in Newarke Street, Leicester. Leicestershire Archaeology 70:33–49.

Wakely J, Manchester K, and Roberts C. 1989. Scanning electron microscope study of normal vertebrae and ribs from Early Medieval human skeletons. Journal of Archaeological Science 16:627–642.

Waldron T. 1987. The relative survival of the human skeleton: Implications for palaeopathology. In: Boddington A, Garland AN, Janaway RC, editors. Death decay and reconstruction. Manchester: Manchester University Press. pp. 55–64.

Waldron T. 1994. Counting the dead: The epidemiology of skeletal populations. Chicester: Wiley.

Waldron HA. 2004. Recalculation of secular trends in Paget's disease. Journal of Bone and Mineral Research 19:523.

Waldron HA. 2005. Reply to archaeology and prevalence of Paget's disease. Journal of Bone and Mineral Research 20:1485.

Waldron T, Rogers J, and Watt I. 1994. Rheumatoid arthritis in an English post-medieval skeleton. International Journal of Osteoarchaeology 4:165–167.

Walker M. 1995a. Menopause in female rhesus monkeys. American Journal of Primatology 35:59–71.

Walker PL. 1995b. Problems of preservation and sexism in sexing: Some lessons from historical collections for palaeodemographers. In: Saunders SR, Herring A, editors. Grave reflections: Portraying the past through cemetery studies. Toronto: Canadian Scholars Press. pp. 31–47.

Walker PL. 2005. Greater sciatic notch morphology: Sex, age and population differences. American Journal of Physical Anthropology 127:385–391.

Walker PL, and Thornton R. 2002. Health, nutrition and demographic change in native California. In: Steckel RH, Rose JC, editors. The backbone of history. Health and nutrition in the Western hemisphere. New York: Cambridge University Press. pp. 506–523.

Walker PL, Collins Cook D, and Lambert PM. 1997. Skeletal evidence for child abuse: A physical anthropological perspective. Journal of Forensic Sciences 42:196–207.

Walrath D, Turner P, and Bruzek J. 2004. Reliability test of the visual assessment of cranial traits for sex determination. American Journal of Physical Anthropology 125:132–137.

Walsh JP. 2004. Paget's disease of bone. Medical Journal of Australia 181:262–265.

Wasserman RH. 1997. Vitamin D and the intestinal absorption of calcium and phosphorus. In: Feldman D, Glorieux F, Pike J, editors. Vitamin D. San Diego: Academic Press. pp. 259–273.

Watts N. 2001. Osteoporotic vertebral fractures. Neurosurgical Focus 10:12–16.

Weaver DS, Perry GH, Macchiarelli R, and Bondioli L. 2000. A surgical amputation in second century Rome. Lancet 356:686.

Webb S. 1995. Palaeopathology of aboriginal Australians. Cambridge: Cambridge Univerisity Press.

Weinstein M, Babyn P, and Zlotkin S. 2001. An orange a day keeps the doctor away. Pediatrics 108:e55.

Wells C. 1964. Bones, bodies and disease: Evidence of disease and abnormality in early man. London: Thames and Hudson.

Wells C. 1965. Osteogenesis imperfecta from an Anglo-Saxon burial ground at Burgh Castle, Suffolk. Medical History 9:88–89.

Wells C. 1967. Weaver, tailor or shoemaker? An osteological detective story. Medical Biological Illustrations 17:39–47.

Wells C. 1975. Prehistoric and historical changes in nutritional diseases and associated conditions. Progress in Food and Nutrition Science 1:729–779.

Wells C, and Woodhouse N. 1975. Paget's disease in an Anglo-Saxon. Medical History 19:396–400.

Wengreen HJ, Munger RG, West NA, Cutler DR, Corcoran CD, Zhang J, and Sassano NE. 2004. Dietary protein intake and risk of osteoporotic hip fracture in elderly residents of Utah. Journal of Bone and Mineral Research 19:537–545.

Wescott DJ. 2006. Effect of mobility on femur midshaft external shape and robusticity. American Journal of Physical Anthropology 130:201–213.

White W. 1985. The human skeletal remains from the Broadgate site LSS85. London: Museum of London Archaeological Service, Archive Report, Unpublished Site Report HUM/REP/01/87, Unpublished.

White TD. 2000. Human osteology. Second edition. San Diego: Academic Press.

White CD, and Armelagos GJ. 1997. Osteopenia and stable isotope ratios in bone collagen of Nubian female mummies. American Journal of Physical Anthropology 103:185–199.

Whyte MP. 1998. Skeletal disorders characterised by osteosclerosis or hyperostosis. In: Avioli LV, Krane SM, editors. Metabolic bone disease and clinically related disorders. Third edition. San Diego: Academic. pp. 697–738.

Whyte MP. 1999. Osteogenesis imperfecta. In: Favus MJ, editor. Primer on the metabolic bone diseases and disorders of mineral metabolism. Fourth edition. Philadelphia: Lippincott Williams and Williams. pp. 386–389.

Whyte MP, Obrecht SE, Finnegan PM, Jones JL, Podgornik MN, McAlister WH, and Mumm S. 2002. Osteoprotegerin deficiency and juvenile Paget's disease. New England Journal of Medicine 347:175–184.

Widdowson EM. 1991. Contemporary human diets and their relation to health and growth: Overview and conclusions. Philosophical Transactions of the Royal Society of London 334:289–295.

Will R, Palmer R, Bhallo AK, Ring F, and Calin A. 1989. Osteoporosis in early ankylosing spondylitis: A primary pathological event?. Lancet 2:1483–1485.

Wingate Todd T, and Barber CG. 1934. The extent of skeletal change after amputation. Journal of Bone and Joint Surgery 16:53–64.

Wittenberg K. 2001. The blade of grass sign. Radiology 221:199–200.

Wolbach SB, and Howe PR. 1926. Intercellular substances in experimental scorbutus. Archives of Pathology and Laboratory Medicine 1:1–24.

Wondwossen F, Åstrom A, Odent Bårdsen A, Zenebe M, Redda TH, and Kjell B. 2003. Perceptions of dental fluorsis among adolescents living in urban areas of Ethiopia. Ethiopian Medical Journal 41:35–44.

Wood JW, Milner GR, Harpending HC, and Weiss KM. 1992. The osteological paradox: Problems in inferring prehistoric health from skeletal samples. Current Anthropology 33:343–358.

World Health Organization. 1999a. Report of a WHO consultation: Definition of metabolic syndrome in definition, diagnosis, and classification of diabetes mellitus and its complications. I. Diagnosis and classification of diabetes mellitus. Geneva: World Health Organization, Department of Noncommunicable Disease Surveillance.

World Health Organization. 1999b. Scurvy and its prevention and control in major emergencies. Geneva: World Health Organization.

World Health Organization. 2002. Meeting of the advisory group on maternal nutrition and low birthweight. Geneva: World Health Organization.

World Health Organization. 2006. Health action in crisis. Annual report 2005. Geneva: World Health Organization.

Wrangham R, Holland Jones R, Laden G, Pilbeam D, and Conklin-Brittain N. 1999. The raw and the stolen. Cooking and the ecology of human origins. Current anthropology 40:567–594.

Yam AA, Kane AW, Cisse D, Gueye MM, Diop L, Agboton P, and Faye M. 1999. Traditional tea drinking in Senegal a real source of fluoride intake for the population. Odontostomatologie Tropicale 22:25–28.

Yamada K, Miyahara K, Nakagawa M, Kobayashi Y, Furuoka H, Matsui T, Shigeno S, Hirose T, and Sato M. 1999. A case of a dog with thickened calvaria with neurologic symptoms: Magnetic resonance imaging (MRI) findings. Journal of Veterinary Medical Society 61:1055–1057.

Yanoff LB, Parikh SJ, Spitalnik A, Denkingert B, Sebring NG, Slaughter P, McHugh T, Remaley AT, and Yanovski JA. 2006. The prevalence of hypovitaminosis D and secondary hyperparathyroidism in obese Black Americans. Clinical Endocrinology 64:523–529.

Yildiz M, and Oral B. 2003. Effects of menopause on bone mineral density in women with endemic fluorosis. Clinical Nuclear Medicine 28:308–311.

Y'Edynak G. 1987. Mesolithic dental reduction as a possible result of sex differences in dental disease and diet. American Journal of Physical Anthropology 72:196.

Zak M, Hassager C, Lovell D, Nielsen S, Henderson C, and Pedersen F. 1999. Assessment of bone mineral density in adults with a history of juvenile chronic arthritis. A cross-sectional long-term follow-up study. Arthritis and Rheumatism 42:790–798.

Zakrzewski SR, and Clegg M, editors. 2005. Proceedings of the fifth annual conference of the British Association for Biological Anthropology and Osteoarchaeology. British Archaeological Reports International Series 1383. Oxford: Archeopress.

Zalloua PA, Hsu Y-H, Terwedow H, Zang T, Wu Di, Tang G, Li Z, Hong X, Azar S, Wang B, Bouxsein ML, Brain J, Cummings SR, Rosen CJ, and Xu X. 2007. Impact of seafood and fruit consumption on bone mineral density. Maturitas 56:1–11.

Zeder MA. 2006. Central questions in the domestication of plants and animals. Evolutionary Anthropology 15:79–118.

Zilva SS, and Still GF. 1920. Orbital haemorrhage with proptosis in experimental scurvy. Lancet 195:1008.

Zink AR, Reischl U, Wolf H, and Nerlich AG. 2002. Molecular analysis of ancient microbial infections. FEMS Microbiology Letters 213:141–147.

Zink AR, Panzer S, Fesq-Martin M, Burger-Heinrich E, Wahl J, and Nerlich AG. 2005. Evidence for a 7000-year-old case of primary hyperparathyroidism. Journal of the American Medical Association 293:40–42.

Zivanovic S. 1982. Ancient diseases. New York: Pica.

Zmuda JM, Cauley JA, and Ferrell RE. 1999. Recent progress in understanding the genetic susceptibility to osteoporosis. Genetic Epidemiology 16:356–367.

Zongping L. 2005. Studies on rickets and osteomalacia in bactrian camels (*Camelus bactrianus*). Veterinary Journal 169:444–453.

Index

C

G

H

I

J

K

L

N

O

R

S